Applied Statistical Methods

This book is designed to provide students, teachers, and researchers with a text that includes a full range of statistical methods available to address commonly encountered research problems. Many textbooks for introductory, intermediate, and advanced statistics courses focus heavily on parametric methods. However, in practice, the assumptions underlying these methods are frequently not met, therefore calling into question their use. This book addresses this issue by presenting parametric, nonparametric, robust, and Bayesian techniques that are appropriate for research scenarios often encountered in practice and typically found in statistics courses. For each of these major topics, the standard parametric approach is presented, along with the assumptions underlying it and the methods used to assess the viability of these assumptions. Next, a set of alternative techniques for the research scenario is presented and applied to the motivating example that begins each chapter. Each chapter concludes with a summary focused on how researchers should select which method to use when and a summary of the material covered in the chapter. The chapters have motivating examples that serve as an anchor for discussion of the featured methods. The focus of the chapters is intended to be conceptual (as opposed to highly technical) to make the text useful to individuals with a wide array of statistical backgrounds. More technical material is included in each chapter for interested readers and instructors who would like to focus more attention on it. Instructors will be able to use this book as a main text in introductory, intermediate, and some specialized statistics courses such as nonparametric and robust methods. In addition, researchers and data analysts from a wide array of disciplines will be able to use this book as a primary resource in their work.

Key features of this book are as follows:

- It presents a full range of statistical options available to researchers for major problems faced in the social and life sciences, health care, and business.
- It explains how to evaluate a dataset to determine which statistical approach (e.g., standard parametric, nonparametric, robust, Bayesian) may be optimal.
- It features a website containing datasets and computer code.

Applied Statistical Methods

Including Nonparametric and Bayesian Approaches

W. Holmes Finch

CRC Press
Taylor & Francis Group
Boca Raton London New York

CRC Press is an imprint of the
Taylor & Francis Group, an **informa** business

A CHAPMAN & HALL BOOK

First edition published 2025
by CRC Press
2385 Executive Center Drive, Suite 320, Boca Raton, FL 33431

and by CRC Press
4 Park Square, Milton Park, Abingdon, Oxon, OX14 4RN

CRC Press is an imprint of Taylor & Francis Group, LLC

Library of Congress Cataloging-in-Publication Data
Names: Finch, W. Holmes (William Holmes) author
Title: Statistical analysis for application / W. Holmes Finch.
Description: First edition. | Boca Raton, FL : CRC Press, 2025. |
Includes bibliographical references and index
Identifiers: LCCN 2024061306 (print) | LCCN 2024061307 (ebook) |
ISBN 9781032457574 hardback | ISBN 9781032459349 paperback |
ISBN 9781003379324 ebook
Subjects: LCSH: Mathematical statistics
Classification: LCC QA276 .F468 2025 (print) | LCC QA276 (ebook) |
DDC 001.4/22–dc23/eng/20250330
LC record available at https://lccn.loc.gov/2024061306
LC ebook record available at https://lccn.loc.gov/2024061307

ISBN: 9781032457574 (hbk)
ISBN: 9781032459349 (pbk)
ISBN: 9781003379324 (ebk)

DOI: 10.1201/9781003379324

Typeset in Palatino
by codeMantra

Contents

Preface

The idea for this book was born from my 30-plus-year career as a statistician. In my experience of teaching, consulting, and research, I repeatedly found a gulf between advances in statistical methods and their implementation by researchers in other disciplines. Frequently, I found that colleagues in other fields were constrained to use methods that relied on stringent assumptions that their data frequently did not meet. At the same time, statisticians were devising new and exciting approaches to knotty (and often common) data analysis problems, but these advancements would not make their way into the broader research universe. Thus, my main goal in writing this book was to describe techniques that non-statisticians might find useful when faced with data that do not conform to the standard assumptions underlying common statistical techniques. I do not pretend that the methods contained in this book represent an exhaustive set of alternatives. Nor do I think that these approaches are the best alternatives for every situation. However, I do hope that this book presents the reader with a set of useful tools that can be applied to a wide array of research problems and that address situations where the standard parametric statistical methods are not appropriate. I also hope that upon learning about the methods described here, readers will be excited to learn more about them (and other techniques) and further explore the literature. If that turns out to be the case for you, then I've achieved my goal in writing this text.

About the Author

W. Holmes Finch is the George and Frances Ball Distinguished Professor of Educational Psychology and a Professor in the Department of Educational Psychology at Ball State University, where he has been since 2003. He received his Ph.D. from the University of South Carolina in 2002. He teaches courses in factor analysis, structural equation modeling, categorical data analysis, regression, multivariate statistics, and measurement to graduate students in psychology and education. His research interests are in the areas of multilevel models, latent variable modeling, methods of prediction and classification, and nonparametric multivariate statistics. He is also an Accredited Professional Statistician (PStat®).

1

Introduction

The primary purpose of this chapter is to introduce you to some basic statistical concepts that we will use throughout the rest of this book. For some readers, these topics will be quite familiar and thus will serve as a basic review. For others, this chapter will introduce a new material that they haven't previously encountered; in that case, it will be beneficial to consider these topics carefully and work through them. These topics reflect some significant ideas that underpin all of statistics, including populations and samples, hypothesis testing, and levels of measurement. In addition, we will devote some attention to some basic statistical calculations that will appear throughout this book. After reading this chapter, you should have a solid foundation that will allow you to move forward with the remaining material in the text. While some of the topics presented here will be familiar to many readers, others will be quite new to most. In either case, you are encouraged to work with the concepts here as needed so that you will be prepared for the more challenging material in the remaining chapters. With that in mind, let's get started.

Levels of Measurement

When we discuss levels of measurement in this book, we are really talking about the types of values that a variable can take. This is an important point because the type of data will play a large role in determining the type of statistical analysis we use. It's also important to keep in mind that different authors will use different heuristics for defining the level of measurement. There is nothing inherently superior about the approach selected for this book. Rather, it simply reflects the organization that I believe best aligns with how we will discuss statistical analyses in later chapters. However, do keep in mind that others may present this material slightly differently.

With that in mind, we will describe levels of measurement as falling into four types: nominal, ordinal, interval, and ratio. Nominal variables are measured in discrete categories for which there is no inherent order. For example, a survey question asking respondents to indicate whether they prefer dogs, cats, birds, or snakes as pets would yield a nominal variable. There are four

DOI: 10.1201/9781003379324-1

discrete categories and no inherent order among them. Ordinal variables are also categorical, but the categories can be ordered from small to large, more to less, etc. Returning to our pet survey, consider the following item:

Rate your agreement with the following statement.

Dogs make good pets.

Disagree	Neutral	Agree

We could code Disagree=–1, Neutral=0, and Agree=1. This item produces a categorical score for each respondent where disagreement takes a lower value than neutral which takes a lower value than agreement. Also, there are no values possible between these three categories. Survey items are quite often ordinal in nature.

Both nominal and ordinal variables are categorical in nature, meaning that no values are possible between the individual categories. In contrast, we may also encounter continuous variables, which can take (theoretically) an infinite number of values. In reality, it will be rare for us to work with truly continuous variables for which the number of possible values is infinite. More frequently, we will deal with variables that can take a large number of values such that we can assume they are continuous. Test scores are such an example. In reality, test scores may only take discrete values between 0 and 100, but there are a sufficient number of such values that we can treat the scores as a continuity.

Within this broad family of continuous variables, there exist two types: interval and ratio. Interval variables are ordered, like ordinal variables, but the distance between any two values is the same regardless of where they sit on the scale. Temperature as measured in Celsius is an example of an interval variable. There are theoretically an infinite number of possible values (e.g., 5.0, 5.1, 5.11, 5.111) and the difference in temperature between 5 and 6 degrees is the same as between 20 and 21 degrees. Ratio variables take on the same traits as intervals in that they are continuous and the rule of equidistance applies. In addition, ratio variables have a true 0 value, which is not true for interval variables. An example of a ratio variable would be a bank account balance. When the account contains no money, we have a true 0.

Populations and Samples/Parameters and Statistics

In addition to levels of measurement, a second fundamental concept in statistics involves populations and samples. Indeed, it is not hyperbole to say that much of statistics focuses on learning about populations through samples. Thus, it is of key importance for us to understand these concepts. As an example, a researcher is interested in understanding the relationship between anxiety and executive functioning, which can be thought of as the ability to plan and organize one's cognitive activities. In this case, the researcher wants to gain insights into how anxiety and executive functioning are related to one another for children with elevated levels of anxiety. In an ideal (and totally unrealistic) world, the researcher would collect anxiety scale and executive functioning assessment scores from every child with an elevated level of anxiety. However, resource limitations (e.g., time, money, proper identification of individuals in the target group) will keep our researcher from gathering data from the entire target population (i.e., children with elevated levels of anxiety).

Given these limitations in obtaining data from the entire population, how can the researcher assess the relationship between anxiety and executive functioning among the entire population of children with elevated anxiety? The answer to this question lies in drawing a random sample from the target population. Specifically, the researcher can administer the anxiety and executive functioning measures to a group of children identified with elevated anxiety who have been randomly selected. They can then use these scores from the sample to make inferences about the larger population. This idea of using samples to represent populations lies at the heart of much of statistical theory and practice. Therefore, it is of key importance that the sample be representative of the target population. When this is not the case, the results of statistical analyses using the sample cannot provide us with useful information about the broader population in which we have interest. Obtaining a sample that is representative of the population is most easily accomplished through random selection of individuals from the target population.

The values of interest to us from the population are called parameters. These parameters represent, in a sense, the holy grail for researchers. They are the characteristics of the population that we want to know about. In our anxiety and executive functioning example, the parameter of interest would be the relationship between the anxiety assessment score and the executive functioning score. As we will learn in Chapter 8, this relationship can be expressed numerically in the form of a correlation coefficient. However, as we discussed in the previous paragraph, it is not possible in practice to obtain the population of interest. Therefore, we will need to draw a sample from the target population, administer the anxiety and executive functioning scales to members of the sample, and calculate the correlation between the two scores. This sample correlation coefficient is a statistic that estimates the population parameter of interest. More generally, much of what we do throughout the rest of this book involves using a sample to calculate one or more statistics of interest that will serve as estimates for the population parameters that we cannot directly observe.

Descriptive and Inferential Statistics

Statistical methods can be broadly thought of as belonging to one of two families: descriptive and inferential. Descriptive statistics refer to numbers and graphs that tell us something about one specific sample. Examples of such statistics would be the sample mean, the sample correlation, or a histogram of test scores for a single sample. We will use descriptive statistics throughout this book to gain insights into the characteristics of our sample. While quite useful, descriptive statistics don't allow us to say anything directly about the population as a whole. Indeed, we must be careful when interpreting results for a single sample that we do not make statements about the broader population. How then can we use information from our sample to learn about the population? This is where inferential statistics come in. As we will learn, starting in Chapter 3, we can use information drawn from our sample in conjunction with statistical theory to make inferences about the entire population. These inferences will typically include probabilities that provide us with a level of confidence regarding how much we can say about the population given our sample data. It is this inference that make statistics such a powerful tool in so many areas of research.

Graphing

Graphs will be an extremely important tool that we will make use of throughout this book. Indeed, we recommend that virtually all data analysis efforts start with graphing of the data, as well as the calculation of key descriptive statistics, such as the mean and standard deviation (both of which are discussed in detail below). Graphs such as histograms and density plots will provide insights into the distribution of values for variables of interest. Boxplots can be used for this purpose also, and they can also yield insights into differences in continuous variable values for two or more groups in the sample. Scatterplots allow us to examine relationships between pairs of continuous variables. We will make use of each of these graphical tools, as well as others in the coming chapters. They should be a standard part of any data analysis effort.

Outliers

In statistics, outliers are simply data points that are abnormally far from the other values in the sample. Of course, the first question that arises from this definition is: How do we define abnormally far? To this question, there is not a single, generally agreed-upon answer. We will devote considerable attention to the issue of outliers and outlier detection throughout this book. As we will see, there are many ways to identify outliers and many ways to deal with them once they have been identified. We will, therefore, not discuss these methods here but rather focus our attention on outliers more generally.

In addition to defining outliers, it is also worth considering how outliers come to be. Sometimes outliers arise in a sample because it has been contaminated by observations, not from the target population. For example, consider a researcher who is interested in estimating the average running speed of Olympic sprinters. Accidentally, data from two recreational runners are entered into the sample. Their very slow speeds (when compared to the Olympians) would be outliers. Now imagine another study in which a researcher wants to estimate the average height of male college students. They randomly sample 20 students for this purpose, two of whom play on the school's basketball team, one as a center and the other as a forward. Again, we have two individuals who are clearly outliers because their variable values will fall far from those of the remainder of the sample.

After identifying them, we need to consider what to do with the outlying data values. As noted above, much of the remainder of this book will be devoted to doing just that. However, before we even conduct our analyses, we need to decide whether the outlying observations should even be retained in the sample. This book is going to recommend the decision to retain outliers in the sample or not be made primarily based on the mechanism by which the data points became outliers. More specifically, if the outliers do not belong to the target population, as in the Olympic sprinters example, then they should be removed from the sample. The rationale behind this recommendation is that the outliers do not represent the population that we are interested in learning about. It does not make sense for us to include race times of non-Olympians if the goal of our study is to estimate the pace of Olympic sprinters. In contrast, if outliers are part of the target population but simply unusual members of it, as in the college student height example above, then they should be kept in the sample. Although their variable values are unusual, they are part of the target population. Indeed, it is for situations just such as this that many of the statistical methods discussed in this book were developed.

Measures of Central Tendency

When describing samples and populations, it is usually of interest to do so in terms of what is typical in terms of the measured variable(s). For example, school districts reporting on student performance will usually refer to the average test score or the percent of students who met an achievement standard on the test. Researchers will generally discuss the mean scores on variables that were used in the analyses. In this way, they are able to position their sample in the context of the broader population, as well as other studies done in the same area of research. In short, describing the central tendency for a sample (or a population) is usually of key importance.

Perhaps the most common measure of central tendency is the sample mean, $\bar{x}$, which is an estimate of the population mean, μ. Returning to our example of the researcher who is interested in understanding the relationship between anxiety and executive functioning, we have a random sample of 74 children who have been identified with elevated anxiety and that was drawn from the population of children with elevated anxiety. We have administered an anxiety scale to each of these individuals. The sample mean is then calculated as

$$\bar{x} = \frac{\sum_{i=1}^{N} x_i}{N} \tag{1.1}$$

where
   x_i = score of interest for person i
   N = sample size.

In other words, we calculate the sample mean by adding up the scores for our sample of individuals and then dividing this value by the total sample size.

In some situations, the variable of interest is dichotomous and coded as 1 or 0. For example, we can code whether a child has been identified as having elevated anxiety (1) or not (0). The mean for this variable is equivalent to the proportion of individuals who have been identified as having elevated anxiety. This value can also be interpreted as the probability that a randomly selected child within the sample will be identified with an elevated anxiety level. We will see proportions such as these throughout this book.

The mean is only one of three measures of central tendency that we will consider in this chapter. Indeed, the mean is not always the optimal measure to describe a typical score for a sample. Another common statistic for this purpose is the sample median, $\tilde{x}$, which is simply the middle value in an ordered set of numbers. Therefore, if we want to calculate the median for a sample with an odd number of elements, we simply order the values from smallest to largest, and the median would be the middle value of this ordered set. If the sample contains an even number of elements, $\tilde{x}$ is the mean of the two middle elements.

Another useful estimate of central tendency for a sample is the most frequently occurring value for the variable, which is known as the mode. The mode is particularly important when we work with nominal or ordinal variables, for which there are usually a very small number of discrete values. On the other hand, when there are a large number of possible values for a variable, such as with the interval and ratio scales, the mode is not a particularly good estimate of central tendency. In these situations, we would be better off using the mean or the median to characterize the central tendency for our sample. Finally, it should be noted that in addition to the mean, the median, and the mode, there are a number of other measures of central tendency available to the researcher, which we will discuss in Chapter 3, including Winsorized and Trimmed means, M-measures of location, and R-measures of location, among others. As we will see, these alternatives are particularly useful when the sample has a large number of outliers present in the sample.

Measures of Variation

In addition to measures of central tendency, we will also usually want to understand the amount of variability present for variables in our sample. Information about variability is important because it reflects the extent to which individuals cluster around the center of the sample. Larger values for these measures of variation are associated with greater dispersion of scores in the sample. Collectively, measures of central tendency and variability are used in conjunction with one another to characterize scores in a sample. As with central tendency, there are multiple statistics for estimating variation in a sample, with the most popular being the variance and standard deviation. The population variance, σ^2 is a measure of total variation in the population, and the population standard deviation, $\sigma = \sqrt{\sigma^2}$. For a sample, the variance is calculated as

$$S^2 = \sum_{i=1}^{N} (x_i - \bar{x})^2 / (N-1) \qquad (1.2)$$

where
x_i = Score for individual i
$\bar{x}$ = Sample mean
N = Sample size.

The denominator of equation (1.2), $N-1$, is known as the degrees of freedom and will reappear in other contexts throughout the book. It reflects the fact that for a given sample with a given mean, there is a limit to the number of freely varying values that a variable can take. Specifically, notice that the differences in the numerator will sum to zero before they are squared. So, if we have a sample of 30 children whom we have assessed for anxiety, once we know 29 of the 30 differences, the 30th difference can only take one particular value (i.e., it is not free to vary). Thus, the degree of freedom for the scores to vary is $N-1$, or $30 - 1 = 29$ for our example. Just as in the population, the standard deviation in the sample is the square root of the variance or

$$S = \sqrt{S^2} \tag{1.3}$$

Skewness and Kurtosis

Two other statistics that are frequently used to characterize samples are skewness and kurtosis. Skewness measures how symmetric (or not) a set of values is. When the data are perfectly symmetric, skewness=0. Skewness is calculated as

$$\tilde{u}_3 = \sum_{i=1}^{N} (x_i - \bar{x})^3 / (N-1)S \tag{1.4}$$

Figure 1.1 shows a symmetrically distributed sample with skewness=0.

On the other hand, variables with a positive skewness are unbalanced with more large values than small ones, as in Figure 1.2.

Negatively skewed variables have longer tails to the left (more small values than large ones). A negatively skewed variable appears in Figure 1.3.

In addition to skewness, data distributions can also be characterized by kurtosis, which is a measure of how large or small the tails of the distribution are. Kurtosis is calculated as

$$K = \frac{\left(\dfrac{\sum_{i=1}^{N} (x_i - \bar{x})^4}{N} \right)}{\left(\dfrac{\sum_{i=1}^{N} (x_i - \bar{x})^2}{N} \right)} \tag{1.5}$$

In this context, tails refer to the extreme values at either end of the distribution. When the tails are heavy, we see a relatively large number of values at each end of the data distribution. On the other hand, light-tailed distributions are characterized by having the bulk of the data near the center and fewer values at the ends. A light-tailed distribution is characterized by a larger kurtosis value, whereas a heavy-tailed distribution has a smaller kurtosis. Two such distributions are plotted in Figures 1.4 and 1.5.

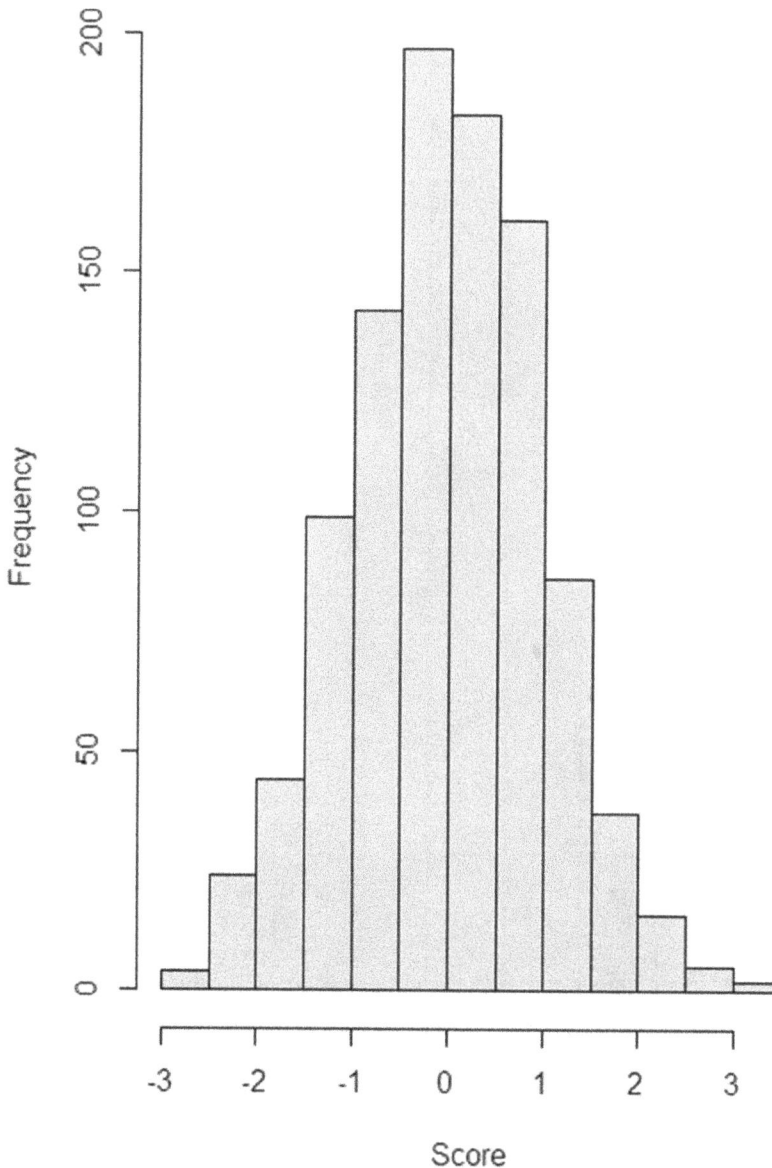

FIGURE 1.1
Symmetrically distributed variable with skewness=0.

Skewness and kurtosis will be central players in this book, as they have a direct impact on many of the standard statistical methods that researchers rely upon. We will see how they influence the performance of these methods and learn about alternative techniques that are more resistant to (robust) the presence of highly skewed and kurtotic data.

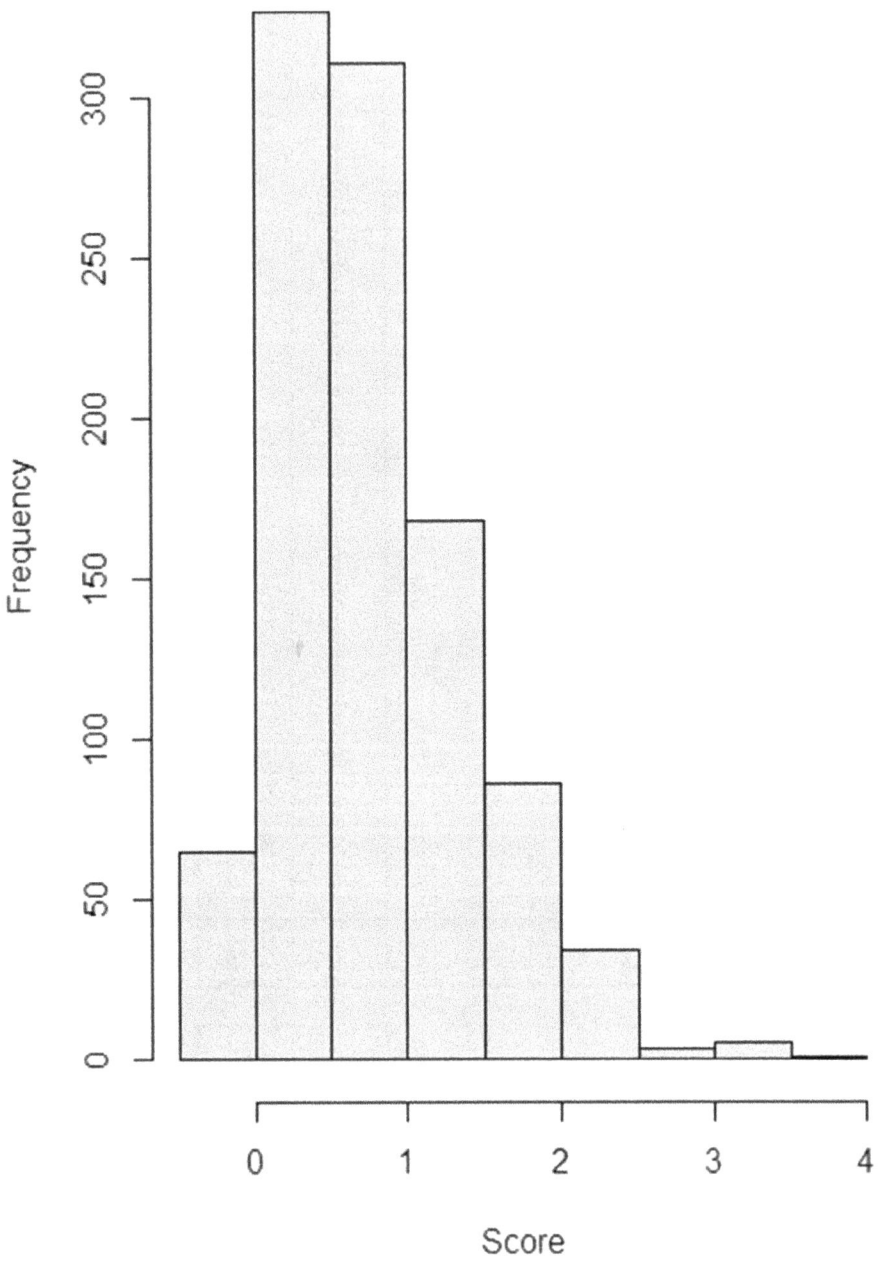

FIGURE 1.2
Positively skewed variable with skewness=4.5.

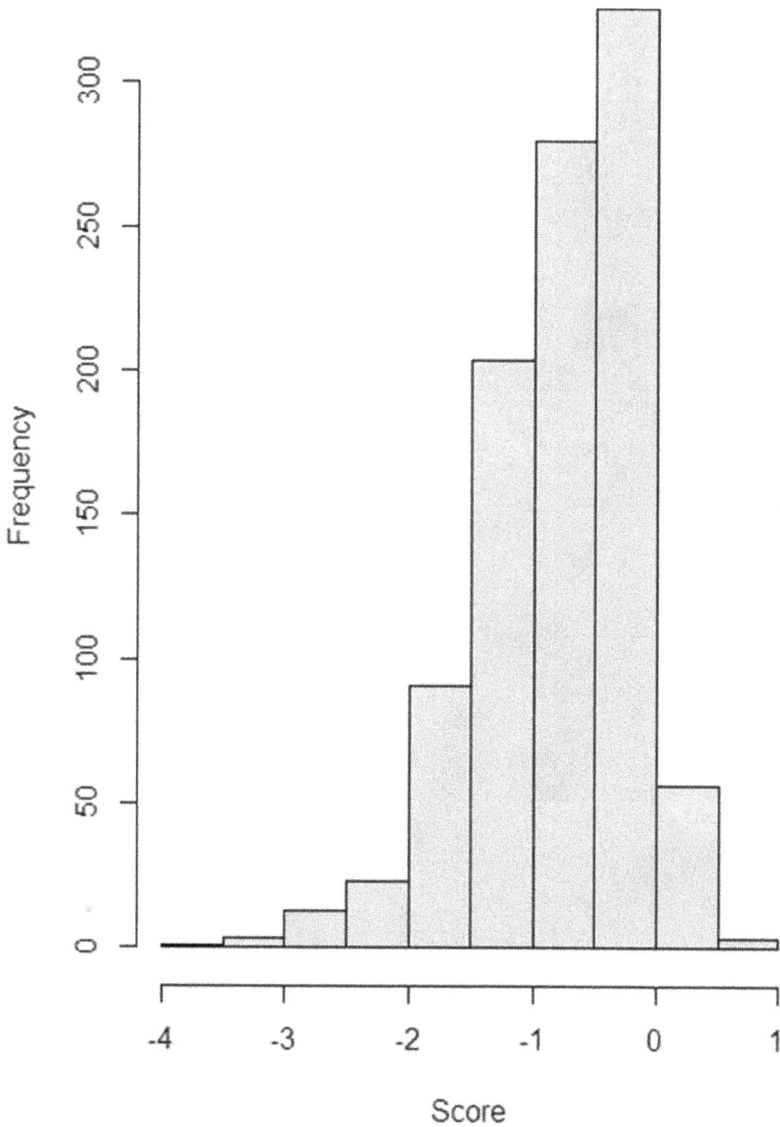

FIGURE 1.3
Negatively skewed variable with skewness=−4.5.

The Normal Distribution

A core concept in statistics is that of sampling distributions. Sampling distributions are simply functional representations of variables, reflecting the likelihood of obtaining a specific value within the range of all possible values.

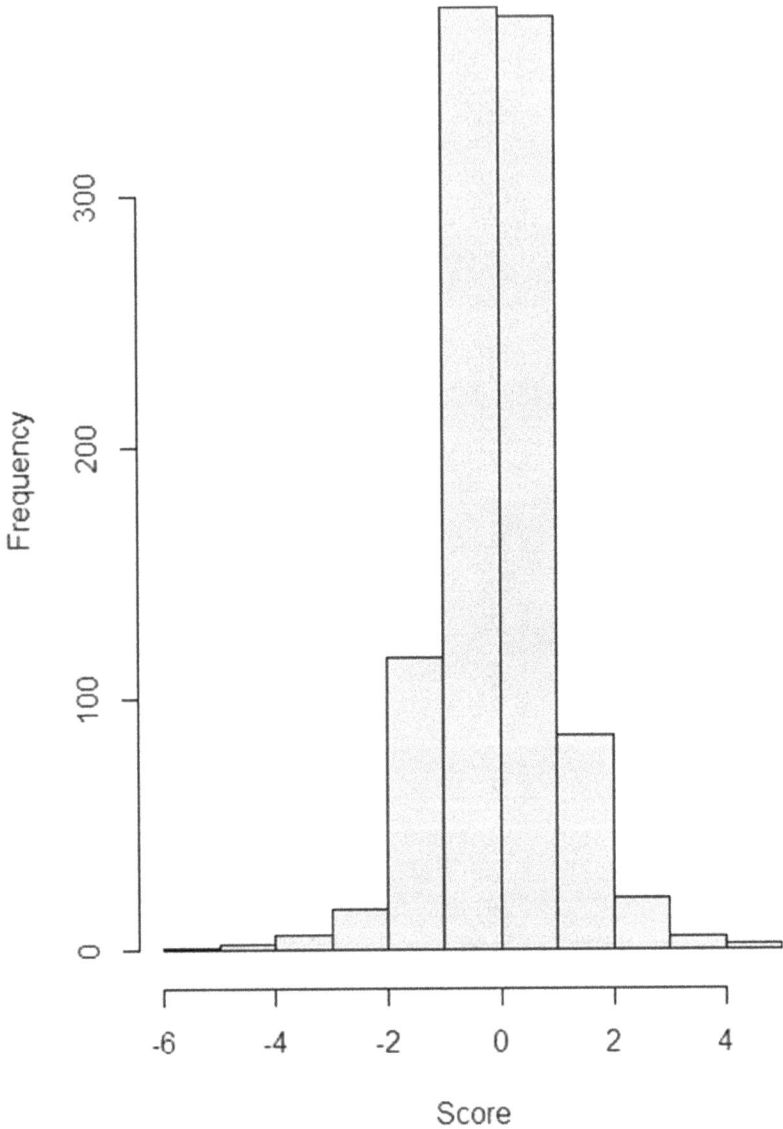

FIGURE 1.4
Symmetric distribution with Kurtosis=4.

For example, if we have access to all data points in a population, we could construct a table with the frequency of each. Using this table, we could calculate the proportion of each possible value, which would simply correspond to the probability that a randomly selected individual from the population would have a particular data value. Of course, as we have already discussed, in the real world, we do not have access to populations and thus cannot construct

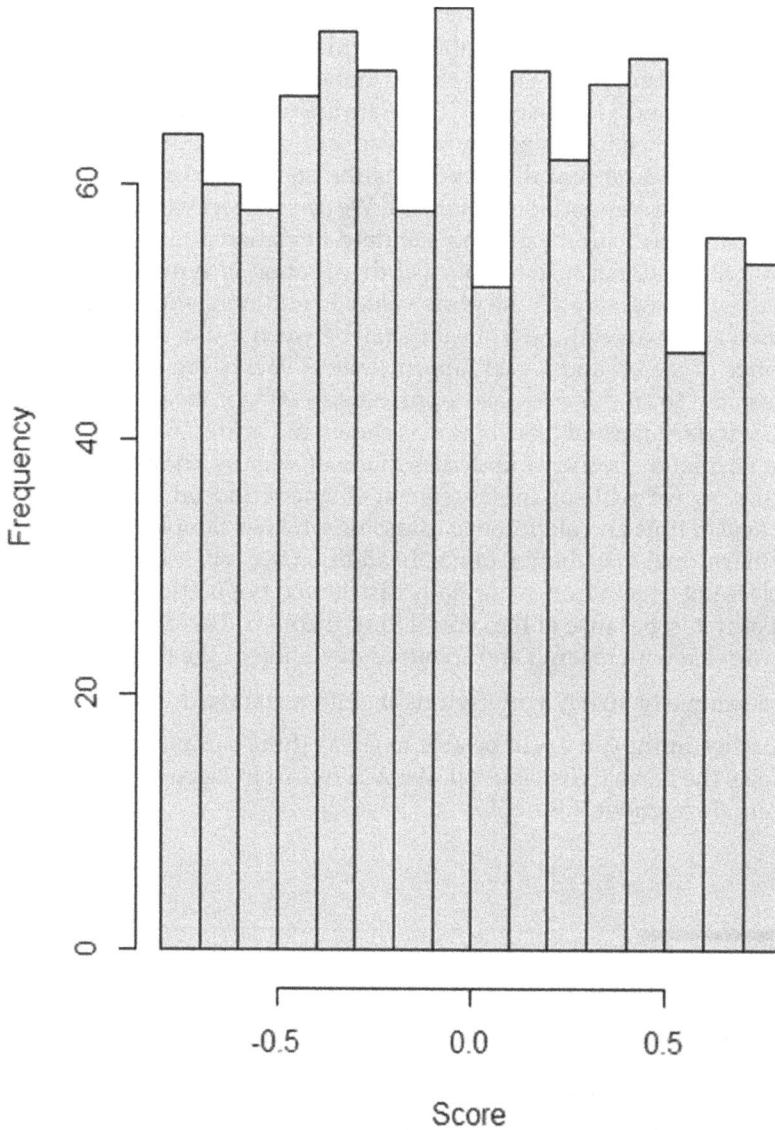

FIGURE 1.5
Symmetric distribution with Kurtosis=−4.

these sampling distributions. However, there exist many statistical sampling distributions with known qualities. These sampling distributions, which we will discuss in more detail in Chapter 2, can provide us with the probability of any particular value being randomly selected. As we will see in future chapters, these statistical sampling distributions are extremely important in the conduct of nearly all statistical analyses.

One sampling distribution that has proven to be especially central in statistical practice is the normal distribution, which is commonly associated with the bell-shaped curve. The normal distribution is perfectly symmetric with a skewness of 0 and a kurtosis of 3. There are an infinite number of normal distributions, each characterized by its mean and standard deviation. One very commonly used normal distribution is the standard normal, which has a mean of 0 and a standard deviation of 1. We can convert values from any normal distribution with mean μ and standard deviation σ to the standard normal distribution. Because the normal distribution is symmetric, the mean, median, and mode are all the same value. In addition, with the normal distribution approximately 68% of values fall between $\mu - 1\sigma$ and $\mu + 1\sigma$. Within the range of $\mu - 2\sigma$ and $\mu + 2\sigma$, approximately 95% of the data will fall, and between $\mu - 3\sigma$ and $\mu + 3\sigma$ lies approximately 99% of the data. A histogram of the standard normal distribution is shown in Figure 1.6.

The normal distribution underlies much of what is known as parametric statistics, as we will see in subsequent chapters. Indeed, we will devote a great deal of time and attention to assessing whether sample data likely come from the normal distribution or not. In addition, we will see that even when a sample is not drawn from a normally distributed population, we may be able to act as if it is, because of the central limit theorem. This theorem states that for a variable with mean μ and standard deviation σ, the mean of the means with a sample of size N will itself be μ, with a standard error of $\dfrac{\sigma}{\sqrt{N}}$. As N approaches infinity, we will be able to act as if the sampling distribution of μ follows the normal distribution. We will return to discuss the central limit theorem throughout this book.

Standardized Scores

Researchers working with data sometimes want to express the values of a particular variable for their sample in terms of the number of standard deviations it lies from the mean. Standardizing the scores in this way allows for comparisons of the relative values of variables on different scales (e.g., GRE scores versus college grade point average). Converting variables to these standard scores can also make interpretation of the results from various statistical techniques easier, such as with regression (Chapter 9) or multilevel models (Chapter 13).

Such standardized values are known as z score. They are calculated using the sample mean and standard deviation as

$$z_i = \frac{x_i - \bar{x}}{s} \tag{1.6}$$

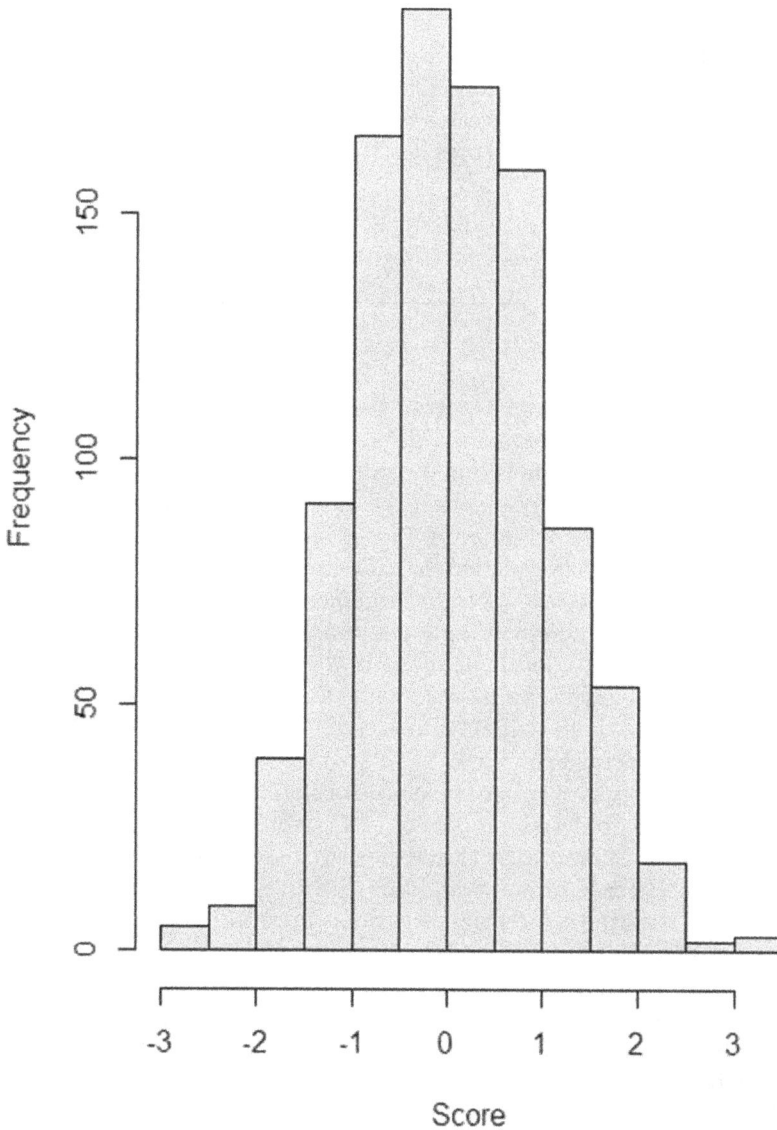

FIGURE 1.6
Histogram of a standard normal distribution.

To demonstrate this calculation, let's use a variable with a sample mean of 106 and a standard deviation of 10.2. For an individual who has a score on the variable of 108, the z value would be

$$z = \frac{108 - 106}{10.2} = \frac{2}{10.2} = 0.196$$

Remember that the z score represents the number of standard deviations away from the mean an individual's score lies. In this example, a score of 108 is approximately 0.2 standard deviations (or 1/5th) above the mean for the sample. In general, z scores greater than 0 reflect raw scores that are above the sample mean, whereas z scores less than 0 correspond to raw scores that are below the mean.

Statistical Inference

Statistical inference is a core aspect of statistical practice and will be central to much of what we do together in this book. Statistical inference refers to the process of using information from the sample to make some inferences about what might be true in the population. As we discussed earlier in this chapter, describing our sample, while important and useful, cannot by itself tell us what is true about the population. In order to do that, we will need to make use of a variety of statistical tools that together can be referred to as statistical inference. At this point, we will not get into the details of how this is done. Those discussions will dominate much of the remainder of this book, starting in Chapter 3. However, it is worthwhile for us to introduce the concept of statistical inference here so that we are ready to dive into the specifics of how it is done in subsequent chapters.

Rather than discuss every possible type of statistical inference (which even one book cannot do, let alone part of one chapter), let's focus on one very common type of inference. Imagine a research scenario in which we would like to know whether a new method of teaching math results in better test performance than the standard instructional approach. We randomly assign 100 students to the new treatment and another 100 students to the standard method that we call the control group. After 4 weeks of instruction, we give all of the students a math test. Ideally, we would like to know whether the new treatment results in better math test scores than does the control for all students in the population. In other words, our main interest is in comparing the population treatment mean on the math test (μ_T) with the population control math test mean (μ_C).

A very common way to frame our research goal regarding the comparison of population means for the two groups is in the form of hypotheses. For example, if it turns out that the new teaching method doesn't result in better math test performance than that of the control group, then we would see the following:

$$\mu_T = \mu_C$$

We will see in Chapter 4 that this result of no difference is called the null hypothesis and is commonly written as

$$H_0 : \mu_T = \mu_C$$

A core element of statistics, and one to which we will devote a great deal of attention in the following chapters, involves testing whether or not to reject or retain H_0. As we will learn, although we do not have direct access to the population, we can make inferences about it using our sample data and statistical tools known as hypothesis tests and confidence intervals. When the weight of statistical evidence suggests that the null hypothesis is unlikely to be true in light of evidence from our sample, we will reject H_0 and conclude, for this example, that the population means are likely to differ for the two treatments. We will also be able to construct a variety of null and alternative hypotheses that will directly address our research goals and questions.

This basic paradigm for statistical inference can be applied to a much wider set of problems than simply mean comparisons, as will be discussed in subsequent chapters. For example, we can use it to make inferences about whether two or more variables are related to one another in the population and whether such relationships are present after we account for their relationships with other variables. In short, statistical inference can take many forms and can be applied in a wide variety of contexts, as we will see throughout the remainder of this book. The goal of this discussion is simply to provide you with a sense of where we are headed.

Type I Error, Type II Error, and Statistical Power

We will conclude this chapter by outlining a final set of core ideas associated with statistical inference. In describing statistical inference, we alluded to the assessment of a null hypothesis in light of evidence arising from our sample and the statistical machinery that we apply to it. If the evidence is sufficiently convincing, we will reject the null hypothesis; otherwise, we will retain it. Retaining the null does not mean that we accept it as representing the truth about the population. Rather, it means that we do not have enough evidence, based on the current sample, to conclude that the null is false. This phrasing may appear a bit pedantic, but it actually reflects an important aspect of statistical thought. Namely, decisions based on statistical inference are probabilistic in nature and not definitive representations of absolute truth. For example, when we reject the null hypothesis, we are doing so based on a balance of probabilities for whether it is true or not given the evidence at hand.

In the same way, when we fail to reject the null, we cannot say that it must be true, because our decision is derived from an assessment of probabilities.

What are these probabilities and how are they used? As we will see in Chapters 2 and 3, the inferential statistical analysis that we conduct to assess the null hypothesis yields a probability that is commonly referred to as the p-value. Colloquially, we can think of the p-value as the probability of obtaining the sample that we have (along with the associated statistical results) if the null hypothesis were true. Thus, smaller p-values indicate that it is less likely we would have this statistical result when the null is true. A primary question in hypothesis testing revolves around how small this probability needs to be for us to conclude that the null hypothesis is likely to be false. This threshold is known as α. In the context of statistical hypothesis testing, when the p-value is less than α, we reject the null. An extremely common value for α is 0.05. So, when we conduct a statistical analysis (e.g., compare to means) and obtain a p-value less than 0.05, we would reject the null and conclude that the means are likely to differ in the population. In later chapters, we will discuss some of the problems with hypothesis testing and present alternative techniques such as confidence intervals that can provide similar information, as well as more depth and nuance to our findings. We will also learn about effect sizes, which can amplify and contextualize the results of hypothesis testing. All of that discussion is in our future, however. First, we need to finish working through the logic of hypothesis testing.

In general, the larger the value of α, the more likely we are to reject the null hypothesis. However, larger α values are also associated with a higher probability of making an error when we reject the null. Such errors are called Type I errors. They occur when we reject the null hypothesis based on the statistical results from our sample, but in fact, the null hypothesis is true in the population. Thinking back to our comparison of math test means from earlier in the chapter, a Type I error would occur if we reject the null hypothesis that the two groups had different means on the test, but in the population, the means were actually the same. Our value of α is the probability of making a Type I error. We could also make an error by not rejecting the null hypothesis when it is in fact false in the population. Such a Type II error would occur in the math test example if we didn't reject the null hypothesis and concluded that the groups had equal mean scores in the population when in truth the means differed. The probability of correctly rejecting the null hypothesis when it is false is known as statistical power. Power is simply a 1-Type II error rate.

When conducting statistical analyses, we are always trying to minimize the Type I and Type II error rates and maximize power. Of course, as with most things in life, there are necessary trade-offs among these competing forces. If we increase α, we are more likely to make a Type I error but are less likely to make a Type II error. In other words, we've made it easier to reject the null hypothesis in general. In contrast, if we make α very small (e.g., 0.01), we are highly unlikely to make a Type I error. However, we've also made it difficult to reject the null hypothesis regardless of whether it's true or not and have therefore reduced our power. One final note to make here is that power is positively related to sample size. The larger our sample, the higher our power and the lower our Type II error rate. In addition, increasing our sample size does not impact the Type I error rate at all. Thus, a powerful tool in the data analysis toolbox is a large sample. Of course, obtaining a large sample is frequently easier said than done!

Summary

The purpose of this chapter is to prepare you for the material in the rest of this book. It is crucial that we have a solid understanding of several core statistical concepts before we discuss the methods presented in the remainder of this book. The topics covered in this chapter will give us the core competencies that we need to move through the rest of the topics covered in Chapters 2–13. We will build on the ideas presented here in order to obtain a full set of tools that can be used to address a wide array of research questions using statistics. But regardless of how far into statistics we dive, the ideas contained in this first chapter will be with us. Thus, it is worth being sure that you are comfortable with these concepts before you move on.

In this chapter, we became acquainted with a number of descriptive statistics for measuring the central tendency, variability, skewness, and kurtosis of a sample. We also saw how graphs can be used to describe our sample, providing us with insights into all four of these aspects of our sample distribution. We then laid the basis for inferential statistics with our discussion of the normal distribution and the idea of sampling distributions. Finally, we expanded on these ideas around inference with our discussion of hypothesis testing, statistical error, and power. We will move much more deeply into these concepts in subsequent chapters.

2

Theoretical Foundation

The goal of this chapter is to present the foundational concepts that will inform our discussions throughout the rest of this book. These ideas are key to understanding the various inferential statistical methods that we will be applying together. The following presentation is intended to be conceptual in nature and not to delve deeply into the mathematical underpinnings of the various topics. Equations and more technical material are included where necessary. However, the overarching purpose of this chapter is to provide readers, regardless of their technical background, with an understanding of the core ideas underlying the statistical methods that we will be discussing throughout the following chapters so that you can apply them in your own work. We will first discuss sampling distributions and how they can be used as proxies for the population level data about which we would like to make inferences. Next, we will examine how assumptions need to be made about the population of interest if we are to use such parametric methods based on sampling distributions. We will then turn our attention to a variety of approaches that have been suggested for cases when key assumptions underlying the parametric approach do not hold and when outliers are present in the data. These alternatives include approaches based on ranks of the original data, trimmed and Winsorized samples, and alternative estimators such as M and S. In addition, this chapter will also feature alternative approaches to inference based on resampling strategies such as the bootstrap and permutations, as well as Bayesian estimation, which relies on a completely different paradigm than the more common frequentist approaches that underpin the other methods examined in this chapter.

Sampling Distributions

In many research scenarios, researchers are interested in making inferences about a population of interest. For example, a pharmaceutical scientist may be interested in assessing the effectiveness of a new drug for treating colon cancer. She hypothesizes that the shrinkage in tumor size over a 6-month period will be greater for the new drug as compared to the standard treatment that is currently in use. In order to address this research question, the

DOI: 10.1201/9781003379324-2

researcher randomly samples 100 colon cancer patients from the hospital where she works. Half (50) of the patients are randomly assigned to be treated with the new drug and the other half are given the traditional treatment. An ultrasound is used to measure the size of the tumor in each patient at the beginning of the study and then again at the end. The outcome of interest is the change in tumor size for each individual. Treatment effectiveness can be assessed by comparing the mean change in tumor size reduction between the groups, which we will calculate as

$$\bar{D} = \bar{X}_{\text{NT}} - \bar{X}_{\text{OT}} \tag{2.1}$$

where

$\bar{X}_{\text{NT}} = $ Average tumor size reduction for those receiving the new treatment
$\bar{X}_{\text{OT}} = $ Average tumor size reduction for those receiving the old treatment.

If $\bar{D}$ is 0, then we conclude that the two drugs were similarly effective in treating colon cancer *for our sample.* On the other hand, if $\bar{D}$ is greater than 0, we would have evidence that the new treatment yielded a larger decrease in tumor size, on average, than did the old treatment. Again, this conclusion would only be applicable to our sample. To take the example one step further, let's assume that for our sample of 100 patients, $\bar{D} = 1.5$ cm, indicating that the average tumor shrinkage for those receiving the new treatment was 1.5 cm greater than that for individuals receiving the old treatment.

This example illustrates several important concepts in statistics. First, our researcher is interested in the entire population of colon cancer patients. In other words, she would like to know whether the new drug will be more beneficial for all individuals with colon cancer, not simply those in her sample of 100 individuals. However, it is typically not possible to collect data from the entire population because populations are usually too large to be fully measured. Furthermore, in most instances, populations are not fully extant at any one point in time. As an example, in the colon cancer study, there are individuals who have not yet been diagnosed with colon cancer but who will belong to the population of interest when their cancer is discovered. Given these problems with obtaining data from the full population, the researcher will need to use a sample instead. For our example, the sample consists of the 100 individuals who were randomly selected to participate in the study. More generally, the sample is the group of individuals upon whom measurements are made in a study, as we discussed in Chapter 1.

A second implication of our example is that the value of $\bar{D}$ can only tell us about our sample. In other words, we can conclude that for the 100 individuals who were included in our study, the new treatment yielded a 1.5 cm larger increase in tumor shrinkage than did the old treatment. However, as we noted above, we are really interested in the effectiveness of the treatment for all patients who have been diagnosed with colon cancer. Indeed, in an ideal world, our researcher would be able to compare the two cancer treatments across the entire population of interest, in which case she would have a definitive answer to the question of which one led to the larger reduction in tumor size. Given this disconnect between what we would like to know (treatment effectiveness in the population) and what the sample statistics can tell us about (treatment effectiveness for the individuals in our sample), we will need to consider how the sample might be used to gain insights into the impact of the treatment in the broader population. This is the focus of inferential statistics.

In order to understand how inferential statistics work in practice, let's consider a thought experiment in which the cancer researcher could repeatedly draw random samples of 100 individuals from the full population of those with colon cancer and repeat her experiment on each sample in turn. In that case, each sample would yield a value for $\bar{D}$. If we were able to draw all possible samples of 100 patients from the population of those diagnosed with colon cancer, we would, of course, have all possible values of $\bar{D}$. This is the sampling distribution of $\bar{D}$ for a sample of size n (100 in our example). In that case, we could calculate $\mu_{\bar{D}}$, or the population average of the mean difference in tumor shrinkage between treatments for all possible samples of size 100. In addition, we could also calculate the standard deviation ($\sigma_{\bar{D}}$) of these values, which would provide us with an estimate of how spread out the sample means are when $n = 100$. More generally, the standard error is denoted as $\sigma_{\bar{x}}$ and we learned how to calculate the sample estimate of $\sigma_{\bar{x}}$ ($s_{\bar{x}}$) in Chapter 1. Together, $\mu_{\bar{D}}$ and $\sigma_{\bar{x}}$ (also known as the population standard error) provide us with a great deal of information about the distribution of sample means for our parameter of interest. From these values, we would know the average difference in treatment effectiveness as well as how different this value is from one sample to another. Larger values of $\sigma_{\bar{D}}$ suggest that the treatment effectiveness is more variable across the samples. Finally, because $\mu_{\bar{D}}$ involves all possible samples of size 100 that can be obtained from the population, it is equal to μ, the population mean. In other words, if we were able to obtain all possible samples of size n and calculate the mean for the variable of interest (e.g., tumor shrinkage), that value would equal the mean of the variable for the population as a whole.

Of course, in the real world, we cannot obtain all possible samples of a given size from the population. We are almost always limited to a single sample and therefore must use it to make inferences about the population of interest. How then can we make inferences about the population if we are unable to obtain all possible samples from the population? The answer to this question lies in the use of mathematical stand-in distributions in place of the sampling distribution of interest. In cases where we can make specific assumptions about the distribution of the population from which we draw our sample, we can use a mathematical sampling distribution in place of the one about which we are interested. Statistical methods based on these statistical proxy distributions are collectively referred to as parametric methods. In the following chapters, we will provide details about specific inferential statistics, the sampling distributions that underlie them, and the assumptions that need to be made in order for researchers to use them. We will also discuss alternatives for constructing such sampling distributions when these assumptions do not hold for our data and the parametric methods are not appropriate. *At this point, the key idea for the reader to take away from our discussion is that in practice, we can make use of a stand-in sampling distribution for the population sampling distribution of interest in order to make inferences about the population when specific assumptions can be made about the population distribution.* We will introduce several of these sampling distributions in subsequent chapters, including the Z, F, t, and chi-square distributions. When the assumptions underlying these parametric statistical methods are not met, they may not provide accurate results regarding population inference. In such cases, we will need to consider alternative techniques that yield accurate information in the face of parametric assumption violations. Collectively, these methods are known as nonparametric or robust statistical methods. The remaining chapters in this book focus on the specifics of using such methods for making inferences about the population using samples when parametric assumptions are met. In addition, we will also describe approaches for making inferences about the population when these parametric assumptions are not met by the data.

Example Data

In order to demonstrate how the methods that we will be discussing work, let's consider a small dataset containing height (measured in inches) for ten students in a college class. These values appear in the first column of Table 2.1. We will discuss the data appearing in the remaining columns in the remainder of this chapter.

TABLE 2.1

Heights for Ten College Students

Original Sample	Ranked Sample	Trimmed Sample	Winsorized Sample	M
65	1	RM	68	RM
67	2	RM	69	RM
68	3	68	68	68
69	4	69	69	69
70	5	70	70	70
71	6	71	71	71
72	7	72	72	72
79	8	79	79	79
82	9	RM	72	RM
83	10	RM	79	RM

TABLE 2.2

Central Tendency and Variability for Data in Table 2.1

Statistic	Central Tendency	Variation	Standard Error
Parametric	72.6	6.41	2.03
Ranked	72	4.45	
Trimmed	71.5		2.53
Winsorized	71.7	4.81	2.74
M	71.5		2.64
Bootstrap	72.6		1.86

Using the equation for the sample mean that we discussed in Chapter 1, we calculate the average height of the original sample to be 72.6 inches and the standard deviation to be 6.41. The reader is encouraged to replicate these calculations as a way of practicing the methods that appear in Chapter 1. The values of central tendency and variation appear in Table 2.2.

Rank-Based Methods

In Chapter 1, we discussed the problem that outliers can present in the calculation of statistics such as the mean and variance. Indeed, outliers can present problems to parametric statistical methods more broadly by leading to violations of the assumptions underlying them, an issue that we will discuss in great detail in Chapter 3. One family of approaches to statistical inference that have been suggested for use when outliers are present is based on ranks of the original data. The goal behind using such ranks is to remove the impact of outliers and/or skewness in the data without actually removing any datapoints. As we will see in subsequent chapters, there exist rank-based statistical techniques that can be used in a wide variety of settings. Generally speaking, the ranked values are used in lieu of the original data although some adjustments to calculations for variances and correlations must be made to accommodate the ranks. Collectively, the rank-based approaches are sometimes referred to as R estimators.

A common measure of location that is associated with rank-based estimators is the Hodges–Lehmann (HL) statistic, which is simply the median of all pairwise averages of the data. The equation for this statistic is

$$\text{HL} = \text{median}\left\{\frac{x_i + x_j}{2}\right\} \text{ for all pairs, } i \text{ and } j \qquad (2.2)$$

where
x_i = Score for case i
x_j = Score for case j.

In order to calculate this statistic, every possible pair of scores in the first column of Table 2.1 are used to calculate the pairwise averages, and then, the median of those average values serves as the HL estimate of central tendency. A common measure of variability is the median absolute deviation (MAD), which is calculated as

$$MAD = median|x_i - M|$$ (2.3)

where
x_i = Score for case i
M = Sample median.

For this sample, MAD=4.45. We should note here that there are a number of other measures of variability associated with traditional nonparametric-based methods based on ranks, and we will address those in Chapter 3.

Trimmed and Winsorized Estimators

Another approach for dealing with long tails in the data distribution, which are commonly associated with outliers and/or high skewness/kurtosis, is simply to trim datapoints. This trimming involves removing a set proportion (e.g., 0.2) from each end of the distribution and then calculating statistics of interest, such as the mean. For example, if we trim 20% of the data in Table 2.1, we are left with the values that appear in the trimmed sample column. We can then use these to calculate the trimmed mean, which is 71.5 (Table 2.2).

Winsorizing is very similar to trimming, with the difference being that rather than simply removing the most extreme values on each end of the sample, we replace them with the next most extreme values. For example, if we Winsorize 0.2 of the sample in Table 2.1, we replace the 65 and 67 with the next lowest 20% of the data, 68 and 69. Likewise, we replace the largest 20% of the sample (82 and 83) with the next highest 20% (72 and 79). The Winsorized estimate of central tendency is the mean of these data, which is presented in the Winsorized column of Table 2.1 and takes the value 71.7 for this example.

The Winsorized variance is very similar to the standard sample variance in that it is based on the difference between individual cases in the Winsorized sample and the Winsorized mean.

$$S_W^2 = \frac{\sum (x_{Wi} - \bar{x}_W)^2}{n-1}$$ (2.4)

where
x_{Wi} = Value for individual i in the Winsorized sample
$\bar{x}_W$ = Winsorized mean
n = Sample size.

The Winsorized standard deviation (S_w) is simply the square root of S_W^2. For this sample, $S_w = 4.81$.

We can use S_w to calculate the standard errors for both the trimmed and Winsorized data. The standard error for the trimmed mean is

$$s_{\bar{x}_T} = \frac{S_w}{(1-2\gamma)\sqrt{n}} \tag{2.5}$$

where
  γ = Proportion of trimming applied to the sample.

In turn, the Winsorized standard error takes the form

$$\left(\frac{n-1}{n-2\gamma n-1}\right)\left(\frac{S_w}{\sqrt{n}}\right) \tag{2.6}$$

For our sample, the trimmed standard error is 2.53 and the Winsorized standard error is 2.74.

M Estimators

M estimators are similar to trimmed means in spirit, but rather than removing a predetermined amount of the data (e.g., $g=0.2$), we use the sample data to determine how many cases (if any) should be removed. This approach involves several steps to identify the observations that are outliers and thus are removed, using the steps outlined below. Readers should not feel overwhelmed by the details of this method. The key idea to take away from this part of the discussion is that M estimators, like trimmed statistics, are designed to identify and remove outlying observations from either end of the data distribution. In the case of trimming, the decision about which values are outliers is made by the researcher a priori by setting the proportion of the sample to be removed. In contrast, the M-estimation approach uses the variability in the data to identify outliers that should be removed. Don't worry about the nitty-gritty details but rather focus on the broader idea underlying M estimators. The M-estimator steps are as follows:

1. Create a weight function $\psi = \text{Max}\left[-K, \text{Min}\left(K, x_i\right)\right]$
   - K is typically set to 1.28,
   - x_i = standardized value (z score) of the variable for subject i.

2. $A = \displaystyle\sum_{i=1}^{n} \psi\left(\frac{x_i - M_K}{\text{MADN}}\right)$ \hfill (2.7)

- M_K=Sample median after weight function is applied to the data.
- $\text{MADN} = \dfrac{\text{MAD}}{0.6745}$ (2.8)

3. $B = \sum_{i=1}^{n} \psi'\left(\dfrac{x_i - M_K}{\text{MADN}}\right)$ (2.9)

4. $\psi' = 1$ if $-1.28 \le x_i \le K$; 0   otherwise

5. Update sample median (M) as $M_{K+1} = M_K + \dfrac{\text{MADN}(A)}{B}$

6. Repeat steps 2–4 until $\lfloor M_{K+1} - M_K \rfloor < 0.0001$

7. $\bar{x}_M = M_{K+1}$

The standard error for the M estimate involves application of ψ to $\dfrac{x_i - M}{\text{MADN}}$ along with the MADN and the proportion of cases identified as outliers. We will not go into the technical details here but do refer the interested reader to the book website (https://holmesfinch.substack.com/) for a full description. For the example data, the data used in the calculation of the M estimators are presented in Table 2.1. The M and trimmed samples are identical in this case (such will not always be true, of course) yielding a mean of 71.5. The M standard error is 2.64.

Research has shown that in many cases, using the first value from the M-estimator algorithm provides reasonably accurate estimates of central tendency, particularly when n is large.

$$\bar{x}_{\text{MOM}} = \frac{1.28(\text{MADN})(i_2 - i_1) + \sum_{i_1+1}^{n-i_2} x_i}{n - i_1 - i_2}$$ (2.10)

where
i_1 =Number of observations< -1.28
i_2 =Number of observations> 1.28.

The bootstrap

The bootstrap is a very general and widely used technique in statistics. It is particularly useful for estimating measures of variability (e.g., standard error) and for constructing confidence intervals in cases where assumptions underlying standard statistical methods don't hold. The basics of the bootstrap are very straightforward, involving creating a large number (B)

of bootstrap samples by sampling the original group with replacement. The bootstrap involves the following steps:

1. Calculate sample statistic of interest (e.g., $\bar{x}$) for the original sample.
2. Randomly sample n individuals from the original sample of size n, with replacement; i.e., individuals can appear multiple times in the bootstrap sample, while others may not appear at all.
3. Calculate the mean, $\bar{x}_B^*$ for the bootstrap sample.
4. Repeat steps 2 and 3 many times (e.g., $B = 10{,}000$) in order to create a sampling distribution for the statistic of interest.
5. Calculate the Bootstrap Standard Error:

$$S_B = \sqrt{\frac{\sum_1^B \left(\bar{x}_B^* - \bar{x}^*\right)^2}{B-1}} \tag{2.11}$$

For the data in Table 2.1, we used the bootstrap with 1,000 samples to calculate the standard error, 1.86. One important point to note here is that because the bootstrap involves randomly resampling the original data, each time we use it the value will differ slightly. However, assuming a relatively large number of bootstrap samples (e.g., 1,000), these values should be very similar each time we use the bootstrap.

Permutation Methods

Another resampling approach that can be used when the parametric assumptions are not met are permutation techniques. This approach is used for statistical hypothesis testing, and as such will be our focus when we discuss statistical inference, beginning in Chapter 4. However, we will review it briefly here. Let's consider a situation in which we would like to compare the mean height for two groups (the general student body and the school's basketball teams). The statistic of interest is the difference in the two groups' mean heights, D. The permutation test approach to making this comparison involves the following steps:

1. Compute $D = \bar{x}_1 - \bar{x}_2$ for the original data.
2. Permute the data by swapping one individual in group 1 with an individual in group 2.
3. Compute $D^* = \bar{x}_1^* - \bar{x}_2^*$ for the permutation created in step 2.
4. Repeat steps 2 and 3 for every possible permutation in the data; i.e., all ways to combine the individuals in the sample into two groups.

5. Create a distribution of the D^* values, which represents the distribution of mean differences when the null hypothesis of no group mean differences is true.

6. Reject the null hypothesis of no group mean differences if $D \geq 97.5$th or $D \leq 2.5$th percentile of the distribution created in step 5.

The estimate of the population parameter of interest (group mean difference in our example) is the value from step 1. The hypothesis test and confidence interval for this statistic are obtained using the permutation steps outlined in steps 2 through 6. We will revisit permutations in the coming chapters.

Bayesian Estimation

The final data analysis framework that we will discuss in this chapter is Bayesian estimation. There is a very broad literature describing the Bayesian paradigm in great detail (Kaplan, 2014; Kruschke, 2015; Gelman, Hill, & Vehtari, 2021). Our goal here is not to replicate these excellent works, but rather to present a brief and (hopefully) easily digestible summary of the Bayesian approach. Readers interested in a more thorough discussion of Bayesian methodology are encouraged to read the works cited above.

The Bayesian approach represents a marked departure in both philosophy and method from the paradigm that underlies the branch of statistics that the other methods described here, known as the frequentist approach. In frequentist statistics, which underlies all of the estimation methods that we have studied in this book to this point, population parameters (e.g., the actual difficulty of a particular item) are viewed as single values. In order to estimate these parameters, we typically draw a random sample from the population and calculate a statistic (e.g., the sample mean). This statistic serves as the single-point estimate of the population parameter. This basic approach of taking a sample from the population and calculating one or more statistics from it to estimate population parameter values can be extended from simple cases, such as the mean, to very complex scenarios involving systems of equations.

In contrast to the frequentist approach, which assumes that the population parameter is a single point, we can conceptualize Bayesian statistics as involving two pieces of information germane to our understanding of the population parameter: The first piece of information is the data that we obtain from the sample (often referred to in the Bayesian literature as the likelihood). We will use the data to calculate a statistic of interest, such as the mean, just as in the frequentist case. The second piece of information that comes into play with Bayes is a prior assumption that we make about the distribution of the population parameter. For example, we might assume that in the population, a math test score is normally distributed with a mean of 80 and a standard deviation of 10. This prior distribution might come from the previous research or from some theoretical considerations. As we will see later in this book, the prior can be informative, as in the math test example, or noninformative where we place a large variance on the distribution. We will discuss this in some detail shortly. The likelihood (observed data) and the prior (assumption about the nature of the population parameter) are combined to form the distribution of the parameter of interest (e.g., mean) in what is typically referred to as the posterior distribution. It is important to note that in Bayesian statistics, the parameter is not a single distinct value unlike in the frequentist domain. Rather, it is a distribution centered around a mean (or median or mode).

Mathematically, we can describe the Bayesian estimate for a parameter of interest, such as the mean, as

$$f(\Omega \mid X) = f(X \mid \Omega) * f(\Omega) \Big/ \left[\int_{\Omega} \big(f(X \mid \Omega) * f(\Omega) d\Omega \big) \right] \qquad (2.12)$$

where
   X = Observed data
   Ω = Parameters of interest
   $f(\Omega \mid X)$ = Posterior probability of a particular parameter value given the data
   $f(\Omega)$ = Prior probability of a particular set of model parameter values
   $f(X \mid \Omega)$ = Probability of the observed data given a particular set of model parameters.

The key idea here is that the Bayesian approach provides us with a probability of the final parameter estimate (e.g., the mean) given the observed data and the prior probability of the parameter value. The value of the parameter estimate with the highest probability is the final estimate. From our prior discussion, $f(X \mid \Omega)$ comes from the data (e.g., mean value calculated from the sample) and $f(\Omega)$ is our prior assumption about the parameter (e.g., the population mean is 80 with a variance of 10).

The focus of Bayesian estimation is the posterior distribution of the model parameter $f(\Omega \mid X)$, which is typically too complex to calculate directly. Statisticians have developed a way around this problem by simulating values from the best estimate of the distribution of interest for our parameter in order to create the posterior distribution. The methodology used to simulate the posterior distribution is known as the Markov Chain Monte Carlo (MCMC). We will not describe the technical aspects of the MCMC here, but the reader is referred to Kaplan (2014) for an excellent and user-friendly discussion of this technique. We will, however, provide a very brief outline of the MCMC here. Using the observed sample, the likelihood estimate of the parameter (e.g., the mean) is calculated and combined with the prior distribution for the mean. The result is a mathematical expression of the distribution for the parameter of interest. The MCMC algorithm then draws a sample from this distribution, which is an estimate of the mean. The posterior distribution is then updated based on the value drawn in this step. Next, a value is drawn from this updated sample yielding a second element to the posterior distribution. Once again, the distribution is updated based on this second random draw. These steps are repeated a very large (e.g., 10,000) number of times, and the set of sampled values forms the posterior distribution for the parameter of interest. This posterior distribution is essentially our estimate of the population distribution for the parameter of interest.

There are a few final details that we need to cover before turning to an example using the Bayesian approach. First, in order to ensure that the posterior distribution is statistically stable, the researcher will need to examine the random samples sequentially to ensure that they are centered around a single value, meaning that the algorithm has converged. This can be done using a trace plot. When convergence has occurred, the trace plot will be centered around a single value. Convergence can also be assessed through the comparison of parameter values taken across multiple sampled distributions, known as chains. In other words, we can simultaneously fit multiple MCMC chains to our data in order to obtain several estimated distributions for the parameter of interest. We can then compare the variance for the parameter estimate within chains to the variance for the same parameter between the chains. The square root of the ratio of these variances can then be used as an indicator of the convergence of the model parameters. If this square root of the ratio is approximately 1, we conclude that convergence has been achieved. In other words, if the variability between estimates obtained from different chains is about the same as the variability within the chains, we can assume that the chains have all converged on the same distribution for our parameter estimate. Furthermore, we will typically discard the first portion of the set of MCMC-sampled values when creating the posterior distribution under the assumption that it takes several iterations before the algorithm converges on a single-parameter estimate. For example, if we have 10,000 total samples, we might discard the first 2,000 (referred to as the burn-in sample) and construct the posterior using the remaining 8,000. Discarding the burn-in portion of the sample helps to ensure that the posterior distribution of the parameter estimate has converged on a single value.

In addition to convergence, the researcher must also take account of the fact that adjacent random samples are correlated with one another because the sampling distribution for one random draw is updated based on the value in the previous random draw. This autocorrelation can be monitored using an autocorrelation plot and can be dealt with by thinning the final posterior distribution. Thinning simply refers to only retaining every kth value in the set of sampled values in order to create the posterior distribution. For example, we may retain every 10th value of the 8,000 remaining after removing the burn-in portion. This final set of 800 sample values serves as the posterior distribution. If we want a single value for a parameter estimate, we can use the mean or median of this posterior. The Bayesian standard error estimate is the standard deviation of the posterior distribution. And, the Bayesian analog to the frequentist confidence interval is known as the credibility interval. The 95% credibility interval is obtained by taking the 2.5th and 97.5th values of the posterior distribution.

Finally, before turning to an example of Bayesian estimation of the mean, we need to consider more closely the issue of priors. As we discussed earlier, a key component of Bayesian estimation is the use of a prior distribution for the parameter of interest. Priors can be used to inform the analysis of prior research. In the height example, we might use results from previous studies regarding the typical heights of American college students. On the other hand, if no such data exist, we can use noninformative priors, which will have little impact on the form of the posterior distribution. Common priors for the mean of some value would be the normal distribution with a given mean and variance. The mean might be the average value (or a set of values) from prior studies on college student height. The variance is expressed through the variance of the prior. If prior studies have consistently shown that the average height of American college students is about 70 inches, then we might set this variance to be very small, reflecting our confidence in the likelihood that the population height is close to 70 inches. On the other hand, if prior research has reported a wide array of average heights for American students, then we can set the variance to be relatively large, reflecting our uncertainty regarding what prior research has found.

Now let us use the Bayesian approach to estimate the mean and standard error for the height of American college students based on our sample. First, let's use a normally distributed prior with a mean of 70 and a standard deviation of 10, reflecting the fact that we are not particularly confident that the mean height of college students is 70 inches. Figure 2.1 displays the posterior (blue) and prior (red) distributions.

The dotted red curve in Figure 2.1 represents the prior distribution, and the solid blue curve represents the posterior distribution. Note how much more diffuse the prior is than the posterior, reflecting our relative uncertainty about the distribution of student heights prior to collecting data. The mean and standard error of the posterior distribution are 72.26 and 0.23, respectively.

Now, let's assume that we are more confident that the mean student height in the population is 70. We reflect this greater confidence by setting the standard deviation of the prior distribution to be 1. The prior and posterior distributions are presented in Figure 2.2, again with the prior being the dotted red curve and the posterior being the solid blue one. We can see that the center of the prior is somewhat less than that of the posterior, whereas for the less-informative prior displayed in Figure 2.1, the two distributions' centers were closer together. The mean and standard deviation of the posterior distribution using the more informative prior were 73.36 and 0.30, respectively. Clearly, the use of an informative prior can impact the results.

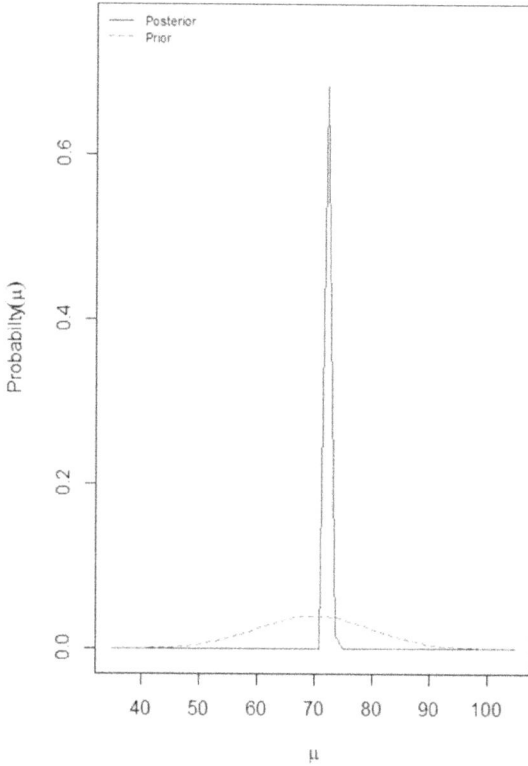

FIGURE 2.1
Posterior and prior distributions for student height for a normal prior with mean=70 and standard deviation of 10.

Finally, we could use a set of means, rather than a single value, when setting the prior distribution. Table 2.3 displays a set of mean student heights obtained from 11 prior studies, along with probabilities associated with them. These probabilities were derived by the researcher and reflect their subjective opinion about the likelihood of each value being the true mean for the population.

The posterior and prior distributions based on the priors in Table 2.3 are presented in Figure 2.3, again with the dotted red line representing the prior distribution and the solid blue line representing the posterior distribution.

The mean and standard deviation of the posterior distribution are 72.73 and 0.44, respectively. Taken together, we can see that the means of the posterior distributions were all quite similar, hovering between 72 and 73 inches. However, the standard deviations of the posterior were much larger when we used more informative prior distributions. We will explore these and other issues in Bayesian estimation more fully in subsequent chapters.

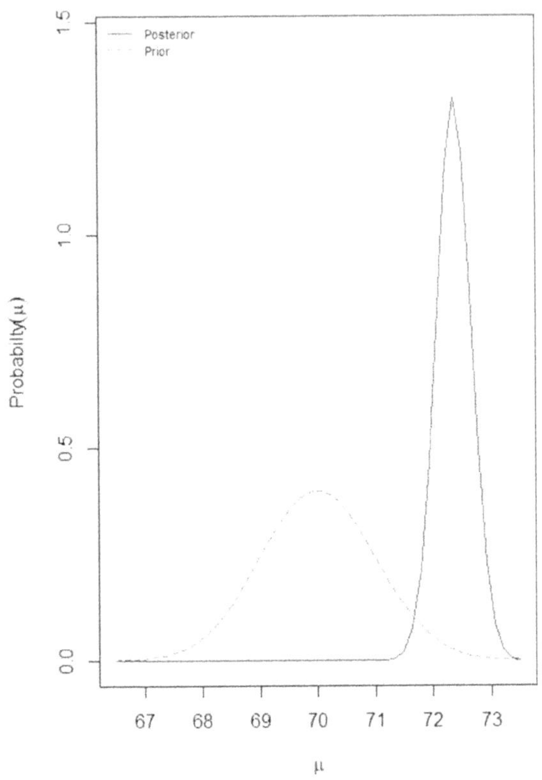

FIGURE 2.2
Posterior and prior distributions for student height for a normal prior with mean=70 and standard deviation of 1.

TABLE 2.3

Prior Distribution of Mean Heights
and Associated Probabilities

Height	Probability
65	0.05
66	0.10
67	0.10
68	0.10
69	0.20
70	0.20
71	0.10
72	0.05
73	0.05
74	0.025
75	0.025

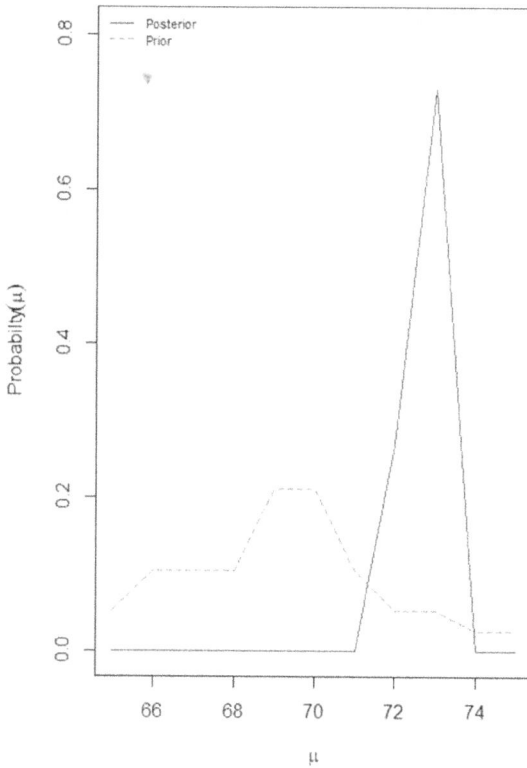

FIGURE 2.3
Posterior and prior distributions for student height for prior distribution in Table 2.2.

Summary

The purpose of this chapter was to introduce some core estimation methods that we will use throughout the remainder of this book. These approaches each have their place in the data analyst's toolbox, and one of the primary goals of the remaining chapters is to provide you with information about which of these techniques are most appropriate for which situations. As we saw in this chapter, there are many approaches that we can use to solve the same problem. For example, if we know that outliers are present in our sample and we would like to reduce their impact on the final estimate of our population parameter, we could make use of a method based on ranks, or we could remove the outliers using a trimmed or M-estimation approach. On the other hand, if we want to reduce the impact of outliers or skewness in our sample but don't want to reduce the sample size, we could use Winsorization. We also learned about three other approaches that provide us with estimates of entire distributions, as opposed to a single-point estimate. Both the bootstrap and permutation use a type of resampling to derive estimates of a distribution either to represent the population from which the sample was drawn (bootstrap) or when the null hypothesis is true (permutation). Bayesian estimation uses the data in conjunction with prior beliefs about the population to derive an estimate of the population parameter distribution (the posterior). We are now ready to move forward with these core concepts and explore estimation and inference for situations in which we have a single sample. We will see how the techniques that were introduced here can be used to gain insights into population parameters such as the mean, variance, and median.

3

One-Sample Parameter Estimation

Our goal in this chapter is to build on the work that we did in Chapters 1 and 2 by examining methods that researchers can use when they are interested in estimating parameters for a single sample. We will apply the methods that we learned about in Chapter 2 to address some of the issues raised in Chapter 1, particularly around the idea of making inferences about population parameters. There are many scenarios that researchers may face in which estimation for a single population is of interest. For example, researchers who gather data about academic performance for a particular group, such as gifted students, may wish to make inferences about performance on an academic assessment for the population of all gifted students. Another very common scenario in which one-sample parameter estimation is likely to be of interest involves the collection of scores on a measure before and after an intervention. For example, a clinical psychologist may want to assess the effectiveness of a particular intervention for anxiety. In order to assess the effectiveness of the intervention, she can give a sample of patients an anxiety measure prior to the intervention and then again once the intervention is completed. The researcher would then calculate the difference between the pre and postintervention scores as a measure of its effectiveness. She could then make inferences about the mean difference score as a way to understand program effectiveness. More specifically, she could ascertain whether the mean change score was likely to be 0 in the population of all anxiety patients who received the intervention. If 0 is unlikely to be the population mean change score, then the psychologist can conclude that the intervention likely made a difference in the level of anxiety experienced by patients. The one-sample methods that we discuss in this chapter can be brought to be in addressing such questions.

In order to provide context for our discussion, we'll be working with a dataset containing standardized reading test scores for a sample of 30 elementary school students who have been randomly selected from a school district program designed to improve reading. These students have participated in the program for 6 months and were then given the exam. School personnel are interested in assessing whether the students in the accelerated reading program are reading above the typical level for their grade. The test has been normed so that a score of 10 equates to typical performance at that grade level. In other words, students with a score at or close to 10 are viewed as reading at a typical level. Our job as data analysts is to use the scores from the 30 students to help the district leaders assess whether students are indeed performing at the typical level.

DOI: 10.1201/9781003379324-3

Exploration of the Data

To begin our analysis, we can use graphs to gain insights into the data. We can start with a histogram of the test scores in order to get a sense of how they are distributed across the 30 students (Figure 3.1).

The distribution of test scores is positively skewed, characterized by a long tail in the graph heading to the right. We would interpret this skewness as evidence that while the largest set of scores is around 10, some students scored quite high on the test. In other words, although most students read slightly below or above the standard score of 10, some read well above what is typical based on the national norms.

A boxplot is another useful tool that we can use to explore the data and in particular to gain insights regarding where the midpoint of the values lies (Figure 3.2).

Boxplots provide several very useful pieces of information about the variable's distribution. The upper and lower limits of the box itself represent the third quartile (75th percentile) and the first quartile (25th percentile), respectively. The dark line in the box shows where the median lies and the width of the box is the interquartile range (IQR). The whisker that ranges below the box represents the first quartile minus 1.5 X IQR. A similar whisker is often present at the top of the box and represents the third quartile plus 1.5 X IQR. However, for this example, the upper whisker does not exist due to the skewed nature of the data. Finally, data points with scores less than the lower whisker or with scores greater than the upper whisker are plotted individually and represent outlying, or unusual values. For this example, there are no low outlying scores, but there are four individuals with reading scores that are unusually high when compared to the rest of the sample.

A final plot that we will use here to examine the data is a violin plot. Violin plots are similar to boxplots but with additional information about the density of the data. What we mean by density is simply where most of the data values for the variable lie. The violin plot will be wider where there are more scores present. Here's the violin plot for our reading test data (Figure 3.3).

The plot shows us where the greatest density of students is with respect to scores. For this example, we see that most of the scores lie between approximately 9 and 11, given that this is where the violin is widest. There are also pockets of data with scores greater than 12.5, reflecting the skewness that was evident in both the histogram and boxplot.

In addition to graphs, we can also summarize the data using several statistical tools that we discussed in Chapter 1. The mean and median of the reading test scores for our sample are 10.73 and 10, respectively. Thus, the center of our sample distribution of reading test scores lies very close to what is typical for the norming sample. The standard deviation for our sample is 2, and the IQR is 1. We may also be interested in the degree of skewness in our scores. In particular, symmetric data have a skewness of 0 and it may be

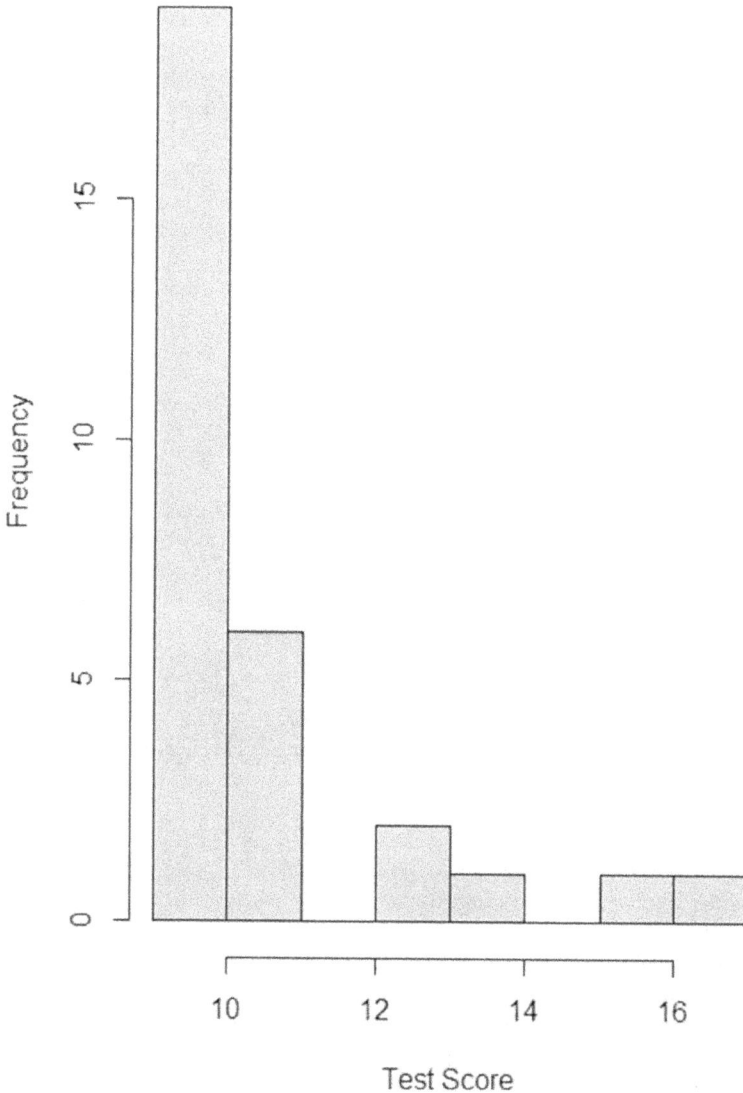

FIGURE 3.1
Histogram of reading test scores.

of interest to see how our sample's skewness compares to that. The skewness for this sample is 1.71, reflecting the positive skewness that we saw in the graphs. A negative skewness value would be indicative of a tail to the left of the center of the score distribution.

Boxplot of Test Scores

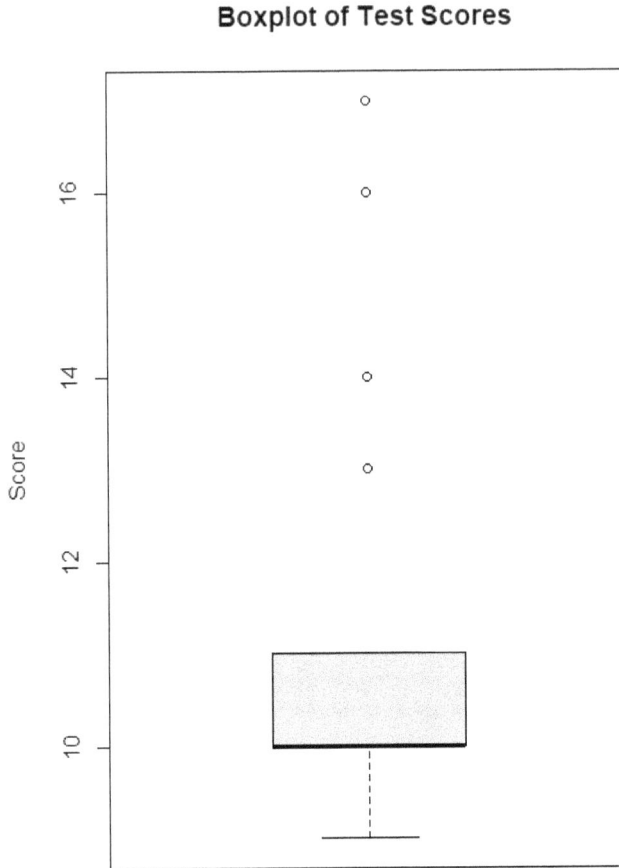

FIGURE 3.2
Boxplot of reading test scores.

Considering these results together, we come away with the impression that most of the reading scores for our sample lie around a value of 10, reflecting typical reading performance. However, there are also several individuals in the sample with scores well above 10. A natural question to ask is whether our sample conform to what we would expect if it were drawn from the same population that produced the norming results. In other words, is our sample from a "typical" population? It is important for us to say here that results from this single sample, by themselves, cannot answer this question. The fact that our sample mean reading score is 10.73, which is higher than the norming sample mean of 10, does not indicate that these 30 individuals come from a population of stronger readers than is typical. The difference in means could simply be due to sampling variability. In other words, we may have obtained a sample of readers from the typical population who just happen to be particularly good readers. On the other hand, it is also possible that

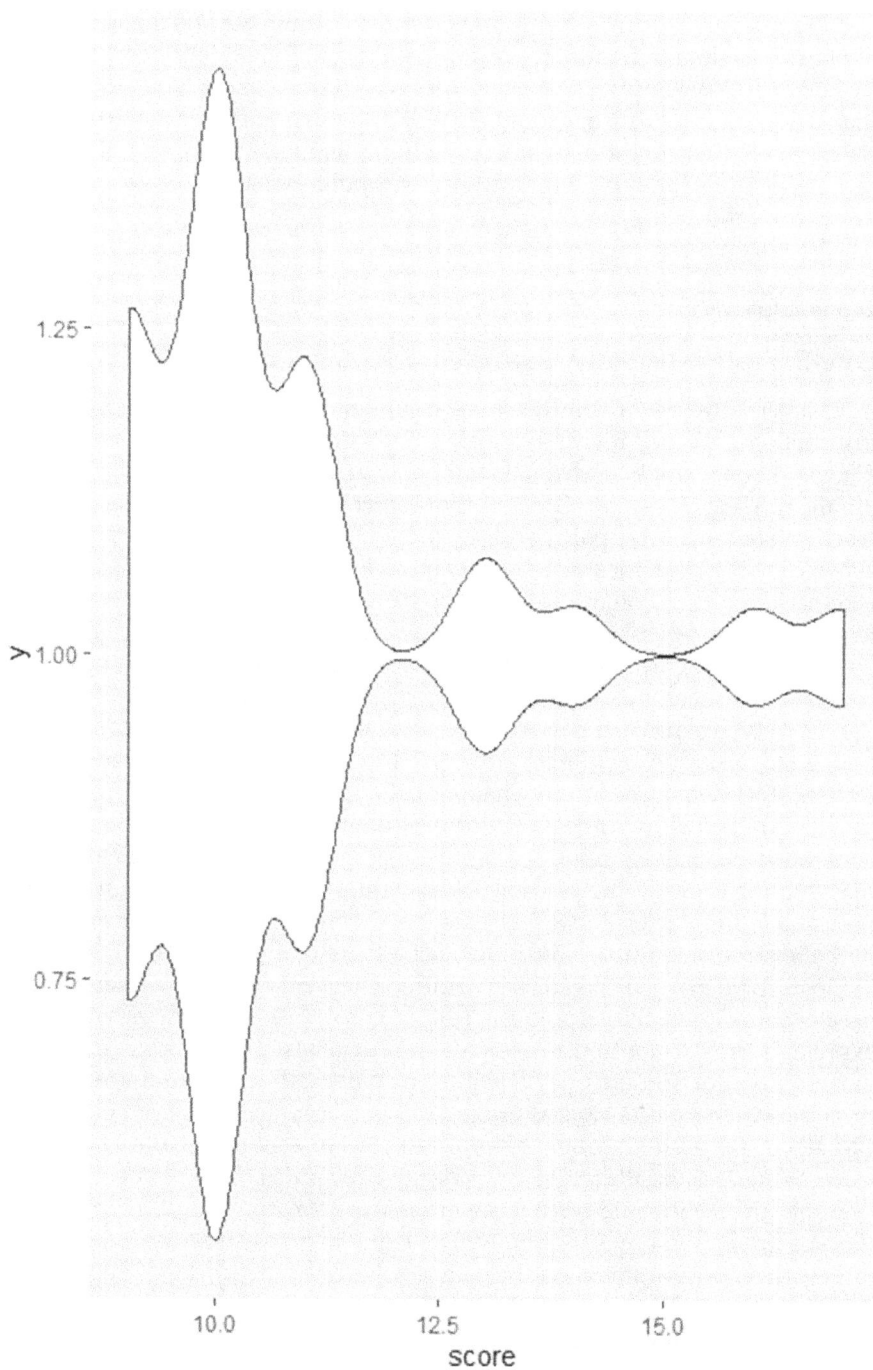

FIGURE 3.3
Violin plot of reading test scores.

our 30 students come from a different population than the one that produced the norming sample and that our population consists of very good readers. At this point, we simply don't know which scenario is most likely to be true. In order to address this question, we will now turn our attention to statistical methods that might be useful in helping us make inferences about the sample vis-à-vis the typical (or any other) population.

One-Sample *t*-Test and Confidence Interval for the Mean

In Chapter 1, we discussed the concepts of hypothesis testing and sampling distributions. We will revisit and expand upon those notions here in order to discuss how we might address the school district's research question regarding the reading test scores. Specifically, district personnel would like to know whether students who participate in the accelerated reading program read above what is typical for students in their grade level. This is identical to asking whether the population of students who have received the intervention is different from the general population of readers. In order to explore this question, the district randomly selected 30 students who had been participating in the program for 6 months and gave them a standardized test where a mean score of 10 reflects typical reading performance, all of which we outlined above.

How can we translate this research goal to a question that can be addressed statistically? Remember from Chapter 1 that we can phrase research questions as statistical hypotheses. In this case, we want to know whether the students participating in the accelerated reading program perform at higher than typical levels on the reading assessment. For the test that we are using, typical reading performance translates to a score of 10, based on prior work with the norming sample. Given that fact, we are really interested in whether students who have received the reading intervention have a mean score greater than 10. In the hypothesis testing context, this statement would be expressed as $H_A : \mu > 10$. In other words, our alternative hypothesis is that mean scores for students participating in the reading program are greater than 10, which reflects typical grade-level performance. As we discussed in Chapter 1, for each alternative hypothesis, we also need a null hypothesis, *where the null reflects all outcomes that are not contained in the alternative*. For this example, the null would be that the mean score for students participating in the accelerated reading program will be equal to or less than 10: $H_0 : \mu \leq 10$.

Now that we have stated our hypotheses, we need to ascertain which one is likely to be true given our sample data. In an ideal world, we would know the sampling distribution for reading test score means when $N = 30$ from the general population. If we had this information, we could very easily determine whether our sample is unusual (or not) when compared to samples of typical students. This would be evident simply by plotting the reading test means using a histogram or boxplot and then checking to see where the mean for our sample falls. If the mean reading test score for our sample exceeds the mean for most of the values in the sampling distribution of means when $N = 30$, then we would conclude that students participating in the accelerated program perform better on the reading assessment than typical students at their grade level. Otherwise, we would conclude that being in the accelerated reading program is not associated with higher performance than is typical.

Unfortunately, although the logic of our approach to solving the problem is sound, we don't have access to the sampling distribution of interest. However, we can potentially make use of mathematical sampling distributions if our sample has certain characteristics. (It's these characteristics that we make assumptions about when conducting many of the procedures described in this book.) In particular, Student's t-distribution can serve as a useful proxy for the distribution of interest if the population distribution of scores is normal and if the individual scores in the sample are independent of one another. *Put another way, if the reading test scores are normally distributed in the population, and the individual scores are independent, then we can use the t-distribution as a stand-in for the sampling distribution that we are actually interested in, namely that for reading test score means for samples of 30 individuals.* We will discuss how to check the normality assumption in a moment, but first, let's take a look at how the t-statistic is calculated.

For a single sample, t is calculated as

$$t = \frac{\bar{x} - \mu_0}{\frac{s}{\sqrt{n}}} \tag{3.1}$$

where
  $\bar{x}$ = Sample mean
  μ_0 = Mean under the null hypothesis
  s = Sample standard deviation
  n = Sample size.

The denominator of the t-statistic is the standard error of the mean and reflects variability in the mean for the sampling distribution. The t-distribution is centered on 0, making it a good proxy for the sampling distribution of reading score means when the null hypothesis is true, as it reflects no difference between the mean for accelerated reading program participants and the norming population mean of 10. Using the mean and standard deviation for the sample that we reported above, the value of t is calculated as

$$t = \frac{\bar{x} - \mu_0}{\frac{s}{\sqrt{n}}} = \frac{10.73 - 10}{\frac{2}{\sqrt{30}}} = \frac{0.73}{0.36} = 2.03$$

Quite literally, this value of t tells us that the difference between the sample mean reading test score and the value that represents typical reading performance in the population is 2.03 standard errors of the mean. But what does that indicate about the real world, and how can we use it to address our question? Because the *t*-distribution is well understood mathematically, we can use the value that we calculated based on our sample data to obtain a probability (the *p*-value from Chapter 1) that if the null hypothesis is indeed true, we would obtain the sample that we did in fact get. Put another way, the *p*-value represents the probability that if we drew our sample of 30 students from the general population of readers, we would have obtained the sample statistics that we got. This probability comes from comparing our *t* value to the *t* sampling distribution and is given by statistical software programs.

There is one additional piece of information that we need to calculate before we can obtain our *p*-value, which is the degrees of freedom. Rather than there being a single *t*-distribution, there are actually an infinite number of them, each referenced by its degrees of freedom. The degrees of freedom simply reflect the number of independent pieces of information available in the sample for us to use in making our estimates. Recall from Chapter 1 that the sample standard deviation is calculated as

$$s = \sqrt{\frac{\sum_{i=1}^{n}\left(x_i - \bar{x}\right)^2}{n-1}} \tag{3.2}$$

where
  x_i = Score for individual *i*
  n = Sample size.

Thus, when estimating the sample standard deviation, we have $n-1$ independent pieces of information with which to work. Given that the standard error for the mean is based upon the sample standard deviation, the *t* value derived from the standard error must also be based on $n-1$ independent pieces of information, making the degrees of freedom for the *t*-distribution equal to $n-1$. For our problem, the degree of freedom is 29; $df = 30 - 1 = 29$.

Now that we have our values of *t* and *df*, we are ready to assess the likelihood that the null hypothesis holds true in light of our sample data. Figure 3.4 shows the *t*-distribution for 29 df, and with our *t* value of 2.03 represented by the vertical line.

We can see that the *t* value for our sample is relatively large when compared to the sampling distribution of *t* values for $df = 29$. How large is it? Well, we can characterize its relative magnitude in terms of the probability that we would obtain a value of 2.03 or larger. This probability corresponds to the portion of the curve that lies right of the vertical line, which is 0.027. In other words, approximately 2.7% of values in the *t*-distribution with 29 *df* are 2.03 or larger. This probability is small, suggesting that it is unlikely that we would obtain a *t* value of this magnitude or larger. Because the *t*-distribution is standing in for the sampling distribution of interest (the difference between the intervention group reading mean and the mean for the general

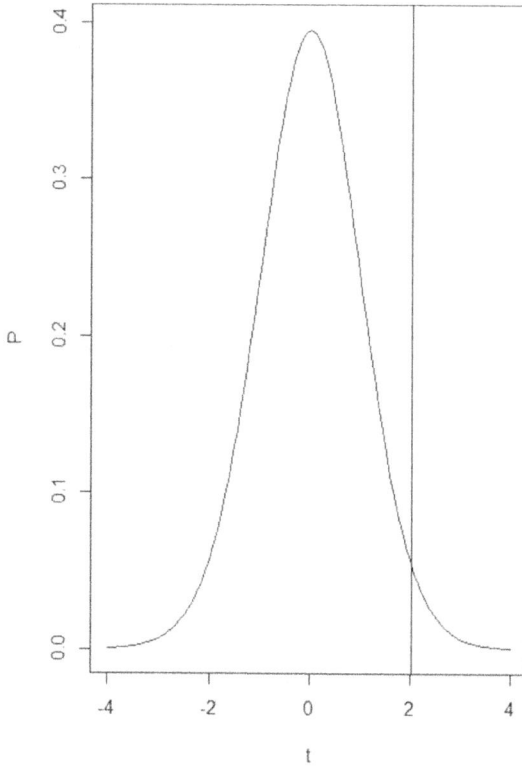

FIGURE 3.4
t-Distribution with 29 degrees of freedom and sample *t* value.

population), we can also conclude that obtaining a sample of 30 students from the general population who have a reading score mean of 10.73 is also unlikely.

Before moving on with this example, let's consider what we have discovered thus far. By making the assumption of a normal population and independent reading test scores, we were able to use the *t*-distribution as a proxy for the sampling distribution that we actually care about, the distribution of reading test means for samples of 30 examinees from the general population of students. We converted our observed reading score mean of 10.73 to this *t*-distribution using equation (3.1) and obtained a value of 2.03. We then compared this rescaled mean difference to the *t*-distribution with 29 *df* and found that it is relatively rare, with values of this size or larger occurring in only 2.7% of cases.

Referring back to our discussion of hypothesis testing from Chapter 1, we will compare this *p*-value of 0.027 to a predetermined threshold of 0.05, which we call α. Given that the *p*-value is less than 0.05, we can reject the null hypothesis using the standard hypothesis testing framework described in Chapter 1. A couple of points should be noted here. First, we framed our research question as a directional hypothesis. In other words, we were interested in assessing whether being in the accelerated reading program was associated with higher reading test scores than what we see in the norming population. It would certainly be possible to set the hypotheses up as nondirectional, where $H_0 : \mu \neq 10$. In the real world, this would correspond to a research question in which we want to know whether participants in the reading program perform differently, on average, than students in the general population. However, given that the program is designed to improve reading performance, this does not seem like a reasonable null hypothesis. The second point that we need to make here is that null hypothesis testing provides us with somewhat limited information. In this case, we were able to conclude that the mean reading test performance for individuals in the accelerated program is likely to be higher than that of the general population. But that's it. Hypothesis testing doesn't provide us with any information about whether this is a large difference or what range of differences we might expect to see at the population level. We will address these issues next.

Confidence Interval

A number of authors have suggested that researchers use confidence intervals as a way to gain insights into how a sample mean compares to that from the population, rather than using hypothesis testing (see Cumming, 2011 and Harlow, Mulaik, & Steiger, 1997 for an excellent discussion of this issue). In the context of inference about the mean for a variable of interest (e.g., reading test performance), confidence intervals reflect a set of reasonable values that it might take in the population. Confidence intervals are typically described in terms of the percent coverage of the reasonable mean values in the population. Perhaps the most common such coverage rate is 95%, which we will use as the default throughout this book. The 95% confidence interval for the population mean is calculated as

$$\bar{x} \pm t_{cv}\left(\frac{s}{\sqrt{n}}\right) \tag{3.3}$$

where
t_{cv} = Value of the t-distribution with $n-1$ degrees of freedom corresponding to the 95% coverage level.

Using the mean and standard deviation for our sample (Table 3.1), the 95% confidence interval for the accelerated reading program is

$$10.73 \pm 2.045\left(\frac{2}{\sqrt{30}}\right)$$

$$(9.98, 11.48)$$

TABLE 3.1

Raw Scores, Difference from
Mean, and Sign

Score	Difference from 10	Sign
10	0	0
9	−1	−1
10	0	0
10	0	0
11	1	1
10	0	0
11	1	1
11	1	1
11	1	1
10	0	0
10	0	0
9	−1	−1
10	0	0
9	−1	−1
10	0	0
10	0	0
9	−1	−1
10	0	0
11	1	1
9	−1	−1
10	0	0
9	−1	−1
11	1	1
10	0	0
9	−1	−1
17	7	1
14	4	1
13	3	1
16	6	1
13	3	1

Thus, we are 95% confident that the mean reading test score for the population of students in the accelerated reading program is between 9.98 and 11.48.

Given this result, we cannot rule out that participation in the reading program is not associated with higher mean test scores than what we would find in the general population, given that 10 falls in our interval. A natural question that you might have is: why is there a discrepancy in the conclusions we would reach based on the *t*-test versus the confidence interval? This difference comes down to the fact that the *t*-test was one directional (one tailed) and the confidence interval is bidirectional (two tailed). Remember that our alternative hypothesis for the *t*-test was $H_A : \mu > 10$, reflecting the fact that we entered the study believing that the accelerated program would improve reading test performance. When the research question implies a unidirectional hypothesis, as ours does, we could construct a one-tailed confidence interval as

$$\left(10.73 - 1.699 \left(\frac{2}{\sqrt{30}} \right), \infty \right)$$

$$(10.11, \infty)$$

Thus, if we conform the confidence interval to the research question of interest (i.e., is the reading program associated with higher mean test scores than is seen in the general population?), we do conclude that 10 is an unlikely value for the mean test score for a sample of 30 students.

We will conclude this section by recommending that researchers rely more on confidence intervals than they do on hypothesis tests because the former yields more information than does the latter. In this case, we see that a "reasonable" lower bound for the mean reading test score for students in the accelerated program is 10.11. In other words, it is unlikely (though certainly possible) that participation in the accelerated program will result in a reading test performance of less than 10.11. We can then go to the school district personnel and ask whether such a difference would be educationally meaningful. The hypothesis test by itself does not provide us with this higher degree of information, but rather simply indicates that participation in the program will yield mean performance above what would be expected in the general population. At the end of this chapter, we will learn about effect sizes, which provide yet more context for differences that we might see between our sample and a hypothesized population.

Assessment of Assumptions

We have mentioned previously that the *t*-test relies on the assumptions that the data come from a normally distributed population and that the individual scores in the sample are independent of one another. We can assess the normality assumption using a QQ plot, which shows us how closely the observed data conform to what they would be if the data were in fact normally distributed. The QQ plot for the reading test data is shown in Figure 3.5.

The solid line represents the data distribution under the assumption of normality, and the points show the observed data values. The shaded region around the line is the 95% confidence interval for the normal

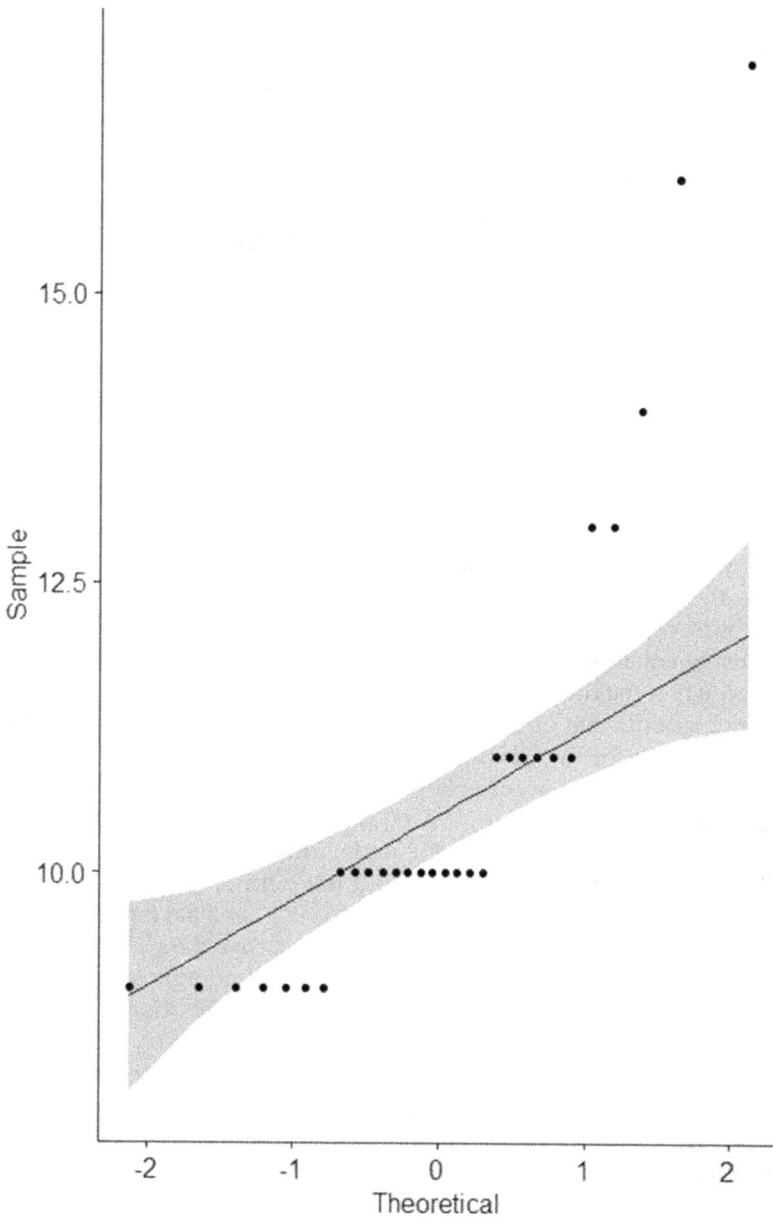

FIGURE 3.5
QQ plot for the reading test data.

distribution. The closer that the observed data conform to the line (i.e., lie within the confidence region), the more closely the reading test data conform to the normal distribution. In this case, we see that there are 5 data points that lie well above the line (corresponding to the positive skewness of the score distribution that we saw in both the histogram and boxplot) and several other data points lie below the line and outside of the confidence region. We would interpret this to mean that the data may not be normally distributed.

A second approach for assessing the normality assumption involves using the Shapiro–Wilks test for which the null hypothesis is that the data are normally distributed (H_0: Data are normally distributed). Thus, if the p-value for this test is less than α (e.g., 0.05), we would conclude that the data are not likely to be normally distributed. For our example, the p-value is 0.000008, leading us to reject H_0 and conclude that the data are probably not normally distributed. Considering both the QQ plot and the results from the Shapiro–Wilks test, we would conclude that the reading test data do not likely come from a normally distributed population.

In Chapter 1, we discussed the central limit theorem (CLT), which says that as sample sizes increase, the distribution of the sample mean converges to the normal distribution. This is an extremely important theorem because it allows us to use sampling distributions relying on the normality assumption (such as the t-test outlined above) if the sample size is sufficiently large. In other words, when n is large enough, we can convert the mean difference to a t value using equation (3.1) and obtain accurate p-values and confidence intervals even when the data are not normally distributed. On the other hand, if we cannot assume that our data are normally distributed and our sample is not large enough for the CLT to apply, then equation (3.1) is not appropriate and the results will not accurately reflect the t-distribution. A natural question is: how large does the sample need to be in order for us to apply the CLT. The answer to this question is not simple, unfortunately. If the underlying distribution of the data is symmetric but not normal (think something like the uniform distribution, which is symmetric but has a flat top), a sample size as small as 25 or 30 might be sufficient for us to use the CLT (Zhang et al., 2023). On the other hand, if the population distribution is skewed, we will likely need a much larger sample in order for the distribution of the mean to converge to the normal (Zhang et al., 2023).

Our results above showed that the reading test data are fairly skewed. Considering this fact in conjunction with the sample size of 30 and the results presented by Zhang et al., we should be wary of applying the CLT to our problem. Therefore, the t-test and parametric confidence interval may not be appropriate for our dataset. For this reason, let's consider some alternative statistical approaches to address the research question regarding whether the reading test performance of students in the accelerated program is better than that of the general population. These methods are based on the approaches described in Chapter 2. In the following pages of the chapter, we will see how statistical techniques based on these methods can be applied to cases like ours, where the sample has potential outliers and the population does not appear to be normally distributed.

Sign Test and Wilcoxon Signed Rank Test

Recall that for the reading intervention study, the alternative hypothesis is that the mean of the reading test scores for students participating in the accelerated program will be higher than 10, which is what we would expect in the general population. The corresponding null hypothesis was $H_0 : \mu \leq 10$. We found that the assumption of normality may not be tenable for the data, which would lead us to consider the use of alternatives that do not require this assumption. There exist several such alternatives to the one-sample t-test and associated confidence interval for the mean. Perhaps, the simplest of these techniques is the sign test, which relies on the binomial sampling distribution in order to provide a test of the null hypothesis. The only assumption that we need to make in order to use the sign test is that the individual scores in the sample are independent of one another.

In order to conduct the sign test, we need to convert each score into a sign (+ or –) by subtracting it from the mean if the null hypothesis is true. The frequency of positive signs (scores greater than 10) is then compared to the total number of possible positive signs, which corresponds to the sample size. If the null hypothesis is true, then the probability of a score being greater than 10 is 0.5; i.e., a student is just as likely to have a score less than 10 as they are to have a score greater than 10. On the other hand, if the null is false, we would expect there to be more + or – signs than 0.5. The data, differences, and signs are presented in Table 3.1.

There were 11 of the 30 cases with positive scores and 7 individuals with negative scores. The remaining 12 students had test scores equal to the null mean of 10. For the purposes of the sign test, the individuals with a score equal to the null mean are excluded from the total sample size, leaving us with 18 individuals to consider. We can now use the binomial distribution to obtain the probability of having 11 individuals with scores greater than 10 (i.e., 11 + values) out of a sample of 18 scores if the probability of having a higher score is actually 0.5. Why do we use 0.5 as our comparison probability? Because when the null hypothesis is true, we would expect to have equal proportions of scores above and below 10. The p-value for our sample is 0.2403, leading us not to reject the null hypothesis ($H_0 : \mu \leq 10$). We would thus conclude that the accelerated reading intervention did not yield a statistically significant proportion of scores greater than 10, meaning that it does not seem to yield improved performance vis-à-vis the general population.

Perhaps the greatest strengths of the sign test are its flexibility, simplicity, and robustness to the distribution underlying the observed data. It can be used in any situation in which data can be ordered. However, a major weakness of the sign test is its relatively low power when compared with other methods (Sprent, 1989). The Wilcoxon signed rank test extends the sign test, yielding power that is close to that of the t-test when the data are normally distributed and higher for data with heavier tails (Sprent). Use of the Wilcoxon signed rank test requires an assumption that the data are continuous and symmetrically distributed. With this test, we will be testing hypotheses about the population median (θ), rather than μ. For our example, the null hypothesis is, therefore, $H_0 : \theta \leq 10$ and the alternative is $H_A : \theta > 10$.

The Wilcoxon signed rank test is based upon the ranks of the scores, rather than the raw data themselves. Thus, if we denote the raw test scores for the N individuals in our sample as $x_1, x_2, \ldots, x_i$, then the ranks of these scores are $R_1, R_2, \ldots, R_i$. The Wilcoxon signed rank statistic is then calculated as

$$W = \sum_{i=1}^{N} \text{sign}(x_i) R_i \tag{3.4}$$

where
   $\text{sign}(x_i) = $ Sign of the difference between x_i and μ_0.

For the purposes of testing the null hypothesis, the statistic is the sum of the ranks for the scores with positive differences:

$$W^+ = R_i^+ = \frac{1}{2} W + \frac{N(N+1)}{4} \tag{3.5}$$

The exact p-value for W^+ can be obtained for any sample size, given the fact that there are 2^N possibilities for the value of W^+. Most software packages limit the computation of the exact p-value to cases where $N \leq 50$, due to the computation burden for larger samples. For sample sizes greater than 50, a normal approximation is used to obtain the p-value. Given our earlier discussion of the CLT, this is usually a reasonable approach when the data are symmetrically distributed. Finally, when there are ties in the score of interest, the exact p-value cannot be obtained regardless of the sample size, meaning that the normal approximation will be used in that case as well. Finally, it is possible to obtain a confidence interval for θ based on the binomial distribution.

For our example, the ranks and signs for the test data are presented in Table 3.2.

Notice that we have a number of tied scores/ranks, meaning that the normal approximation will be used to obtain the p-value even though our sample size is less than 50. The p-value for this sample is 0.0505, leading us to not reject $H_0 : \theta \leq 10$, given that $\alpha = 0.05$. The one-sided 95% confidence interval for the median is (9.99, ∞). Given that 10 lies within this interval (albeit barely), we conclude that 10 is a possible value for the population median. In other words, using the Wilcoxon signed rank test leads us to conclude that the population of students participating in the accelerated reading program does not perform better on the reading test than does the general population.

Finally, in addition to the standard median, researchers using the Wilcoxon signed rank test often make use of the Hodges–Lehman (HL) statistic to characterize the center of the data. Recall from Chapter 2 that

$$\text{HL} = \text{median} \left\{ \frac{x_i + x_j}{2} \right\} \tag{3.6}$$

for all pairs, i and j. For our example, the median is 10 and HL is 10.5 with a 95% confidence interval of (9.99, 11.00). This latter result reinforces our finding

TABLE 3.2

Raw Scores, Difference from Mean, Sign, and Rank

Score	Difference from 10	Sign	Rank
10	0	0	12
9	−1	−1	24
10	0	0	12
10	0	0	12
11	1	1	6
10	0	0	12
11	1	1	6
11	1	1	6
11	1	1	6
10	0	0	12
10	0	0	12
9	−1	−1	24
10	0	0	12
9	−1	−1	24
10	0	0	12
10	0	0	12
9	−1	−1	24
10	0	0	12
11	1	1	6
9	−1	−1	24
10	0	0	12
9	−1	−1	24
11	1	1	6
10	0	0	12
9	−1	−1	24
17	7	1	1
14	4	1	3
13	3	1	4
16	6	1	2
13	3	1	4

with the Wilcoxon test that the center of the score distribution for students in the accelerated reading program may be 10.

One-Sample Confidence Interval for the Median

In Chapter 2, we discussed the robustness of the sample median relative to the mean. This robustness makes the median an excellent tool to use when the data are not normally distributed, particularly when they are skewed. There are a number of approaches for calculating standard errors and confidence intervals for the sample median, thereby allowing us to make inferences about the population median. One such approach is the Harrell-Davis (HD) confidence interval for the median. The HD standard error for the median (S_{HD}) is obtained using the bootstrap, which we described in Chapter 2. In this case, a large number of bootstrap samples (e.g., $B = 1,000$) are drawn, and for each, the median is calculated. The standard error of the median is estimated by the standard deviation of this bootstrap distribution. The 95% confidence interval is then constructed as

$$M \pm z_{CV}(S_{HD}) \tag{3.7}$$

where
  M = Sample median
  z_{CV} = Critical value from the standard normal distribution, e.g., 2 for a 95% confidence interval.

A second approach for calculating the confidence interval of the median was described by Maritz and Jarrett (1978). The Maritz–Jarrett standard error (S_{MJ}) is calculated using the Beta distribution and is described in detail by Johnson and Kotz (1970). The confidence interval for the median is then calculated as

$$M \pm z_{CV}(S_{MJ}) \tag{3.8}$$

McKean and Schrader (1984) introduced a third approach to calculating a confidence interval for the median:

$$M \pm z_{CV}(S_{MS}) \tag{3.9}$$

where

$$S_{MS} = \sqrt{\frac{\left(x_{n-k-1} - x_k\right)^2}{2z_{0.995}}}$$

$$k = \frac{n+1}{2} - Z_{0.995}\sqrt{\frac{n}{4}}$$

For the reading test example, the 95% confidence intervals for the median, using the three methods outlined above are

  HD: (9.88, 10.66)
  MJ: (9.56, 10.38)
  MS: (9.62, 10.38).

Regardless of the approach, these confidence intervals support the conclusion that 10 is a reasonable value for the population median of reading test scores for students in the accelerated reading program. In other words, there is not sufficient evidence for us to conclude that participation in the program is associated with reading performance above that in the general population.

One-Sample Confidence Interval
for Trimmed and Winsorized Means

Trimmed and Winsorized means were described in Chapter 2. Recall that trimming involves removing a predetermined proportion (e.g., $\gamma=0.2$) of the data from each end of the sample distribution. The mean is then calculated using the remaining data points. For Winsorizing, the extreme values on each end of the sample distribution are replaced by the next most extreme values in the sample, as we described in Chapter 2. Confidence intervals for both the trimmed and Winsorized means can be used to make inferences about the central location for the bulk of the population values. Note that when using these approaches, we are estimating the Winsorized and trimmed population means, rather than the full mean. However, if our interest is in gaining insights about the center of the largest portion of the data, these approaches may be more appropriate than the standard mean when there are outliers and/or skewed sample data.

One approach for calculating confidence intervals for the Winsorized ($\bar{X}_W$) and trimmed ($\bar{X}_T$) means involves the Winsorized variance (S_W^2), which we discussed in Chapter 2. The standard errors for the Winsorized and trimmed means are calculated as

$$S_{\bar{x}_W} = \frac{\sqrt{S_W^2}}{\sqrt{n}} \quad \text{and} \tag{3.10}$$

$$S_{\bar{x}_T} = \frac{\sqrt{S_W^2}}{(1-2\gamma)\sqrt{n}} \tag{3.11}$$

The confidence intervals for the trimmed and Winsorized means are then

$$\bar{x}_T \pm t_{cv}\left(\frac{S_{\bar{x}_W}}{(1-2\gamma)\sqrt{n}}\right) \quad \text{and} \tag{3.12}$$

$$\bar{x}_W \pm t_{cv}\left(\frac{S_{\bar{x}_W}}{\sqrt{n}}\right) \tag{3.13}$$

An alternative approach to finding the confidence interval for the trimmed and Winsorized means is to use the bootstrap. For example, the 95% confidence interval for the trimmed mean can be determined by taking B bootstrap samples, calculating the trimmed mean for each and then using the 2.5th and 97.5th percentiles of the resulting bootstrap distribution as the lower and upper bounds of the confidence interval. A similar approach can be used with the Winsorized mean. There is no strong evidence to support one approach over the other when it comes to obtaining confidence intervals for the Winsorized and trimmed means.

For the reading intervention data, the 95% confidence intervals for the Winsorized and trimmed means appear below.

Winsorized:	(9.62, 10.64)
Trimmed:	(9.72, 10.72)
Bootstrap Winsorized:	(9.72, 10.89)
Bootstrap Trimmed:	(9.78, 10.83)

As with the median, these results suggest that the norming population reading test mean of 10 is also a reasonable possibility of the population mean for students participating in the accelerated reading program.

Confidence Intervals for M-Estimators
Measures of Central Tendency

We discussed M estimation in Chapter 2 and learned that it is similar in spirit (though not in method) to the trimmed approach. The reader is encouraged to review this description in the previous chapter prior to our discussion of confidence intervals. The standard error for the M estimator of central tendency takes the form

$$S_{M \text{ est}} = \sqrt{\frac{1}{n(n-1)} \sum_{i=1}^{n} U_i^2} \qquad (3.14)$$

$$U_i = \frac{\left[\left(\text{MADN} * \psi(\gamma_i) \right) - V_i \bar{C} \right]}{\bar{D}}$$

$$\text{MADN} = \frac{\text{Median of } |x_i - M|}{0.6745}$$

$$\gamma_i = \frac{x_i - M}{\text{MADN}}$$

$$D_i = 1 \quad \text{if } \gamma_i < K, \quad 0 \quad \text{otherwise}$$

$\bar{D}$ = Mean of D_i; proportion of cases with $\gamma_i < K$
  K is typically set to 1.28:

$$\bar{C} = \frac{\sum_{i=1}^{n} D_i \gamma_i}{n}$$

$$V_i = \frac{C(x_i)}{2(0.6745)\left[(M+\text{MAD})+(M-\text{MAD})\right]}$$

$$C(x_i) = A(x_i) - \frac{B(x_i)}{M}\left[(M+\text{MAD})+(M-\text{MAD})\right]$$

$$A(x_i) = \text{sign}\left((x_i - M) - \text{MAD}\right)$$

$$A(x_i) = \text{sign}(x_i - M)$$

The confidence interval for the M estimator is calculated as

$$\bar{x}_M \pm t_{cv} * S_{M\ est} \tag{3.15}$$

where
t_{cv} = Critical value from the *t*-distribution.

The approach to estimating the standard error outlined above is appropriate when the distribution is symmetric. When the distribution is skewed, the bootstrap is recommended for estimating the standard error of the M estimator (Wilcox, 2012). The 95% confidence interval using $S_{M\ est}$ is (9.79, 10.89). If, instead, we use the bootstrap approach, the 95% confidence interval for the M estimator is (10.0842, 10.5949). The analytic standard error was larger than that of the bootstrap (0.27 versus 0.13), yielding a wider interval using the former. Given that our data are skewed (see Figure 3.1), the bootstrap approach to estimating the standard error is probably preferred. Using the confidence interval for the M estimator, we would, therefore, conclude that 10 is not a reasonable value for the central tendency of the reading test scores for students in the accelerated program.

Bayesian Credibility Intervals for Measures of Central Tendency

The final approach that we will consider here involves use of the Bayesian paradigm. We will not review the technical details of Bayesian estimation here, but rather we will focus on how it can be used as an alternative to obtain the standard error and confidence intervals for the mean reading test score. The reader is encouraged to review the discussion of Bayesian estimation basics in Chapter 2. Given that the mean of the reading test is 10 for the general population (based on the norming sample), we will use this as the prior value for the Bayesian estimation algorithm. In addition, we will set the prior variance for this mean to a very small value (0.0001), thereby making it an informative prior. This means that, based on prior evidence (the norming studies), we are fairly confident that the population mean of the reading assessment is 10. An alternative to using such an informative prior would be to set no values and use software default values, which correspond to a mean of 0 and an infinite variance. In summary, the informative prior distribution will be a normal with a mean of 10 and a variance of 0.0001: $N(10, 0.0001)$ and the noninformative prior will be $N(0, \infty)$.

When using Bayesian estimation, we need to determine the total number of iterations for the MCMC algorithm, as well as its burn-in value. For the reading test, we will set the total number of iterations to 100,000 and the burn-in value to 5,000. The data are thinned so that every 10th value is retained, yielding a final posterior distribution estimate containing 9,500 data points. For the reading test example, the one-sided 95% confidence interval for the mean based on the informative prior analysis is (10.13, ∞) and the p-value for the test of the null hypothesis is 0.023. For the noninformative prior, the 95% confidence interval is (10.11, ∞) with a p-value of 0.027. As we can see, the results are nearly identical, regardless of the prior distribution that we select. The conclusion we reach is essentially the same as for the standard t-test: We reject the null hypothesis and conclude that the mean reading score for the students participating in the accelerated reading program is greater than 10.

We will conclude our discussion of Bayesian analysis in the one-sample context with an examination of using the Bayes factor (BF) to investigate the reading test performance of students in the accelerated reading program. Of particular interest is the likelihood that the mean score for the population of accelerated program participants is greater than that of the norming population (10). Heretofore, we have investigated this question using hypothesis testing and confidence intervals. The BF approach differs from these techniques in that it focuses on the likelihood of specific hypotheses about the population, in light of the data and prior information. Thus, rather than making probabilistic statements about obtaining the data that we have if the null hypothesis is true, or creating a set of reasonable values for the population mean, we will instead focus on the likelihood of one or more hypotheses of interest holding in the population, given the observed data and the prior distribution for the variable of interest. Prior to discussing this approach for our specific example, we will first review the concept of the BF in general for the single sample case.

The BF is particularly useful for comparing the likelihood of two hypotheses about the population given the data and prior distribution (Gu, 2021; 2022). For example, let's consider three hypotheses that might be associated with the reading test problem upon which we've been focused. The three possibilities of interest are

$$H_1 : \mu > 10$$

$$H_2 : \mu < 10 \tag{3.16}$$

$$H_3 : \mu = 10$$

The first hypothesis corresponds to what we might expect: participants in an accelerated reading program perform better than the general population on the reading test. The second hypothesis states the converse, intervention participants do worse on the reading test than does the general population. Finally, the third hypothesis corresponds to accelerated reading program participants performing equivalently to students in the general population.

We can assess these hypotheses using the testing approach that has been a focus of much of this chapter. While potentially useful for determining group mean differences, this hypothesis test-based approach will not provide us with much direct information about the likelihoods of the various hypotheses that might be of interest. In addition, null hypothesis testing doesn't allow for comparisons involving more than two hypotheses. Confidence intervals help to address this problem by providing us with a range of reasonable values for the population mean, but they don't provide any information about the relative likelihoods of the various hypotheses. Rather, the values contained in the interval are simply considered to be equally likely possibilities for the population mean. The BF can be quite useful in helping researchers ascertain the relative likelihood of each hypothesis in light of the data and prior information.

The BF is based on the marginal likelihood associated with each hypothesis, where marginal likelihoods reflect the probability of a specific hypothesis given the data at hand (Gu et al., 2021). The BF is calculated as

$$BF_{ia} = \frac{f_i}{c_i} \qquad (3.17)$$

where
f_i = Fit of H_i
c_i = Complexity of H_i.

The fit of hypothesis H_i refers to how well the hypothesis conforms to patterns seen in the data with larger values indicating a closer match. In Bayesian modeling terms, this proximity of the hypothesis to the data (f_i) is referred to as the posterior density of the data. For example, we can obtain the correspondence of the posterior distribution of the mean with H_1 by examining the posterior distribution of reading test scores obtained using the MCMC algorithm. If most of the values in the posterior exceed 10, then we conclude that H_1 provides a good fit to the data; i.e., it matches up well with what we see in the observed data. The denominator of the BF, c_i, is a number assessing the complexity of the hypothesis. More informative hypotheses (i.e., those that make more detailed statements about the population) have higher levels of complexity. In Bayesian terms, c_i is the prior density of H_i. Thus, the BF for a hypothesis is the ratio of how closely it conforms to the observed data divided by how specific it is about the population. Larger values of the BF indicate that the hypothesis is more likely to be true in the population.

Once we have the BF_{ia} for each hypothesis, the likelihoods of competing hypotheses vis-à-vis one another can be derived using a ratio of their BFs (Gu et al., 2018):

$$\text{BF}_{ii'} = \frac{\text{BF}_{ia}}{\text{BF}_{i'a}} \tag{3.18}$$

where
  BF_{ia} = BF for H_i
  $\text{BF}_{i'a}$ = BF for $H_{i'}$.

Thus, in order to compare H_1 and H_2 above, we would calculate the BF for each and then take their ratio as

$$\text{BF}_{12} = \frac{\text{BF}_1}{\text{BF}_2} \tag{3.19}$$

One standard used by many researchers to identify support for one hypothesis over another is a BF of 3 or more (Kaplan, 2014). Thus, values of this statistic between 1/3 and 3 fall into what is termed the region of indecision, meaning that it is not possible to say that one hypothesis definitively fits with the data better than the other. In addition to the BF ratio, we can also directly interpret the posterior probabilities associated with each hypothesis. The posterior probability reflects the likelihood of an individual hypothesis holding in the population, given the data at hand and the prior distribution. Hypotheses with larger posterior probabilities are considered more likely than those with lower such probabilities. These posterior probabilities can be used to select from among the considered hypotheses. The posterior probabilities are simply the ratio of the likelihood and the prior probability to the prior.

   Prior to examining the BF results, we should note some caveats to consider when using this technique. Schad et al. (2023) provide an excellent workflow for data analysts to consider as they employ BF to address their research questions. We will not review each of these here, but we will highlight some of the more salient points. Perhaps one of the core recommendations from this work is that BF results should be seen as continuous evidence regarding the hypotheses of interest. This means that rather than focusing on a single dichotomous decision (e.g., reject the null hypothesis or not), we should consider the degree of evidence regarding specific hypotheses. In other words, how likely is a particular hypothesis given the data and the priors? A higher likelihood for a hypothesis leads to greater confidence regarding its correspondence to the population. However, we are never able to reach absolute conclusions regarding the nature of the population using BFs because they are subject to sampling variability in exactly the same way as any other inferential technique. This does not render BFs useless but simply means that they are evidentiary and not definitive in nature. Such is true, however, for any statistical method we use to gain insights into the population.

   A second major caveat to consider when using BFs is that they are sensitive to the priors and to the number of MCMC iterations we use. Having longer chains is more likely to yield accurate results. Likewise, the use of informative priors should be done very carefully. If the researcher is not fairly certain

TABLE 3.3

Bayes Factors and Likelihoods Associated with
H_1, H_2, and H_3

Hypothesis	Bayes factor	Probability
$H_1 : \mu > 10$	0.978	0.72
$H_2 : \mu < 10$	0.145	0.27
$H_3 : \mu = 10$	0.022	0.01

regarding prior information, it would be best to use noninformative priors with large variances so as not to bias the final posterior distribution. BFs are also sensitive to sample size such that results obtained using smaller samples should be interpreted with more caution than results based on large samples. Again, this is true for any other inferential statistical method.

For the reading example, the BF values for hypotheses H_1, H_2, and H_3 above are presented in Table 3.3.

The fit value is largest for H_1 as is the posterior probability of the hypothesis, whereas the smallest of these values were associated with H_3. The ratios for the various BFs appear below:

$$H_1 \text{ vs } H_2 : \frac{0.978}{0.145} = 6.74$$

$$H_1 \text{ vs } H_3 : \frac{0.978}{0.022} = 44.45 \text{ and}$$

$$H_2 \text{ vs } H_3 : \frac{0.145}{0.022} = 6.59$$

Based on these results, H_1 is more likely to be true than either of the other hypotheses. The BF ratios involving it exceed the threshold of 3 when compared to the other hypotheses of interest. Although H_2 was over 6 times more likely than H_3, the fact that H_1 was over six times more likely than H_2 leads us to conclude that H_1 is the most likely to be true in the population, given the data and priors. In addition, we would conclude that H_1 has a probability of 0.72 of holding in the population.

This set of Bayesian results provides us with a bit more information than either the hypothesis test or the confidence intervals associated with the other methods discussed in this chapter. The hypothesis that the mean reading test score for the population of students participating in the accelerated reading program exceeds 10 is over 6 times more likely to be true than is the next most likely hypothesis about the population. And, the hypothesis of the reading test mean exceeding 10 has an overall posterior probability of 0.72. Thus, although it is not a certainty, it is by far the most likely to be true, given the sample data and the prior distribution.

Effect Size Estimation

In Chapter 1, we discussed effect sizes, which can be used to provide information about the magnitude of the statistical effects present in the data. For example, based on the hypothesis test results presented above, we concluded that students participating in the accelerated reading program had higher mean test scores than those in the norming population. However, by itself, this result does not provide us with information regarding how large this difference might be. Do the reading program participants perform much better than the general population or just slightly better? It is this question that an effect size can help to answer.

Cohen's *d* Effect Size

For comparisons involving means, one of the most popular effect size statistics is Cohen's *d*, which standardizes the difference between the sample mean and the mean associated with the null hypothesis. Cohen's *d* is calculated as

$$d = \frac{\bar{x} - \mu}{s} \tag{3.20}$$

For our example, $d = \dfrac{10.73 - 10}{2} = \dfrac{0.73}{2} = 0.37$. What does this value tell us? It reflects the fact that the difference between the sample reading test mean for students participating in the reading program and the mean for the general population (as represented by the norming group) was approximately 1/3 of a standard deviation. Another approach for interpreting *d* is to convert it to a probability based on the standard normal distribution. We can treat our *d* value (0.37) as a z-score, given that it represents the standardized difference between the observed and null population means. We can then get the probability of a value being smaller than this z-score from the standard normal distribution. This value can be interpreted as the probability that a randomly selected individual from the accelerated reading program will have a higher score than a randomly selected individual from the general population. For this example, the probability associated with a z of 0.37 is 0.64. Thus, we can interpret this to mean that the probability of a randomly selected student from the accelerated reading program having a higher reading test score than a randomly selected student from the general population is 0.64.

There exist some guidelines for interpreting *d* from Cohen (1988) that are widely used in the social sciences: 0.2–0.5=Small effect, 0.5–0.8=Moderate effect, and 0.8+=Large effect. There are also subject-specific sets of guidelines for interpreting Cohen's *d* (e.g., Lovakov & Agadullina, 2019). Although many researchers default to interpreting *d* based on readily available guidelines, we would encourage a more nuanced approach to using effect sizes. Referring to the common guidelines but also using an alternative approach to interpretation such as that based on the *z*-score transformation is recommended. For our example, using the standard guidelines for interpreting Cohen's *d* suggests that there is a small difference between the accelerated reading program and the norming population means. In addition, we would expect that a randomly selected program participant would have a higher reading test score than approximately 64% of individuals in the general population. In short, reading test scores appear to be higher for those in the accelerated program, but this difference is not particularly large.

It is also possible to obtain a confidence interval for *d*, which can be used in much the same manner as the confidence interval for the mean. The confidence interval for *d* is calculated as

$$d \pm d_{SE} z_{cv} \tag{3.21}$$

where

z_{cv} = Critical value for the standard normal distribution; e.g., 2 for 95% confidence.

$$d_{SE} = \sqrt{\frac{1}{n} + \frac{d^2}{2n}}$$

For the reading example, the 95% confidence interval for *d* is (−0.01, 0.73). Thus, the difference between the mean reading test score of students participating in the accelerated reading program and 10 (the mean for the general population) could be essentially 0 or as large as moderate.

Hedges' *g*

Hedges (1981) proposed an alternative effect size measure for mean differences that represents a slight adjustment of Cohen's *d*. Hedges showed that *g* is somewhat less biased than *d* for small samples (i.e., less than approximately 20). It is calculated as

$$g = d * \left(1 - \frac{3}{4(n-1)-1} \right) \tag{3.22}$$

Interpretation and guidelines for magnitude are the same as with *d*. For the reading test example, Hedges' *g* is $0.37 * \left(1 - \dfrac{3}{4(30-1)-1} \right) = 0.36$. In general, the two effect size estimates will be quite similar for samples greater than 20. The confidence interval for g is

$$g \pm g_{SE} z_{cv} \qquad (3.23)$$

where

$$g_{SE} = \sqrt{\frac{1}{n} + \frac{g^2}{2n}}$$

The 95% confidence interval for g is (−0.01, 0.71), ranging from no effect to a moderate effect.

Nonparametric Effect Sizes

Researchers have found that Cohen's d is not robust when data are not normally distributed, particularly for skewed data (see Wilcox, 2012 for a discussion). One alternative effect size that has been recommended for use with single sample problems when the data are not normally distributed is calculated as

$$r = \frac{Z}{\sqrt{n}} \qquad (3.24)$$

where

$$Z = \frac{U - m_u}{s_u} \quad \text{and}$$

$$U = R - \frac{n(n+1)}{2}.$$

R = Sum of ranks for cases with positive signs, as in the Wilcoxon signed rank test

$$m_u = \frac{n}{2}$$

$$s_u = \sqrt{\frac{n(n+1)}{12}}$$

When there are ties in the score, s_u is calculated as

$$s_{ties} = \sqrt{\frac{n(n+1)}{12} - \frac{n \sum_{k=1}^{K} \left(t_k^3 - t_k \right)}{12n(n-1)}}$$

where
t_k = Number of ties for the kth ranked value.

The r effect size for the reading test data is 0.238, which can be categorized as of small magnitude based on the guidelines described in Fritz et al. (2012). More generally, these guidelines are 0.1–0.3 (small), 0.3–0.5 (medium), and 0.5+ (large). It is important to note that as with Cohen's d, these guidelines are suggestions and should not be treated as hard and fast rules.

The confidence interval for r can be obtained using the percentile bootstrap. We can take B (e.g., 1,000) bootstrap samples and for each calculate the value of r. The 95% confidence interval corresponds to the 2.5th and 97.5th percentiles of the bootstrap distribution. This approach is very similar to the bootstrap approach for calculating confidence intervals for the Winsorized and trimmed means. For the reading test problem, the 95% confidence interval for r is (0.02, 0.54), suggesting a wide range of possible effects from negligible to large.

Summary

In this chapter, we have learned about a number of approaches for estimating the central tendency and variability of a single population and making inferences about that population. A natural question to ask is: how do we know which approach is most appropriate in a given situation? The answer to this question is, as you might imagine, not a simple one. In some instances, the choice is clear, but in many other real-world cases, there will not be an obvious optimal technique to use. And, as we saw with our example, quite often the various methods will yield similar results to one another. With all of that in mind, we offer a few pieces of advice here. In addition, we also recommend that when faced with a nonstandard situation (or one not discussed in this book), the reader do some independent investigation in order to identify the best approach for their particular problem.

When the data conform to the assumptions underlying the parametric methods (i.e., the t-test and associated confidence interval), these methods are optimal. In other words, if you assess the assumptions for the parametric *t*-test and they are met, then you should use that approach. In addition, when the sample is sufficiently large, the parametric approach is also probably the best bet. Of course, as we discussed a bit in this chapter, the idea of sufficiently large is not absolutely defined. Certainly, 200 or more would be large enough to use the parametric methods, in most cases, even when the data are not normally distributed. On the other hand, a sample size of 30 is probably not large enough for us to depend on the CLT. What to do in those cases? For skewed data (particularly with a high degree of skewness), the M and trimmed estimators seem to be most optimal. Likewise, for heavy-tailed distributions, these methods are also likely to yield the most accurate results. On the other hand, when there are a number of outliers, the researcher should consider the rank-based approach, or even the sign test, as these techniques are particularly resistant to the influence of unusual datapoints. An additional advantage of the rank-based method compared to the M and trimmed techniques is that it does not remove observations from the data. This last point is particularly salient when the sample size is not particularly large.

We should say a word about Bayesian estimation in particular. The Bayes paradigm is fundamentally different from the frequentist approach that is the basis for the other methods described in this chapter. Indeed, it may be best to view it as another way in which to conceive of population inference and not as an alternative to be used for misbehaving data. Bayes allows us to incorporate prior information and expertise into the estimation and inference framework. In this chapter, we saw how both informative and noninformative priors can be readily employed in the data analysis framework. In addition, we saw how the BF can provide us with information about the relative likelihoods and probabilities of competing hypotheses holding true in the population. These are unique aspects of Bayesian estimation that are not available in the frequentist paradigm. Given all of these advantages, we would recommend using Bayesian estimation when you have prior information that you would like to incorporate into the analysis and/or when you have multiple hypotheses of interest and you would like to compare their likelihoods. And in truth, some statisticians would argue that the Bayesian paradigm is always the way to go.

Finally, we recommend that researchers always make use of confidence intervals and effect sizes when working with statistical estimation. Confidence intervals provide us with the same information as the hypothesis test, as well as so much more. We can assess whether a specific value is reasonable for the mean of our population (e.g., 10 for the reading test example). But with confidence intervals, we can also develop a full set of reasonable values for the population mean, so that we're not limited to simply making a dichotomous reject or not reject about a single hypothesis. Effect sizes are also crucial additional pieces of information about how our sample mean fits into the broader question about the nature of the population. Statistics such as d and g tell us something about how different (or not) our sample mean is from the hypothesized mean. Again providing additional information beyond the simple dichotomous reject or not that is inherent in standard hypothesis testing.

4

Comparing Measures of Central Tendency between Two Independent Groups

In Chapter 3, we discussed a wide variety of analysis options that researchers can use when working with a single sample. In particular, we examined multiple approaches to estimate central tendency, including the mean, the median, trimmed and Winsorized means, and M estimators, among others. We also saw that there are multiple alternatives available for estimating sampling variability and for constructing confidence intervals, such as the bootstrap, Bayes, and various techniques for the median. We concluded our description of the single sample problem by discussing the utility of Bayes Factors (BF) and effect sizes to more fully describe the likelihood of specific hypotheses as well as the magnitude of statistical results. In Chapter 4, we will extend these ideas to situations in which we want to compare means across two independent samples. More formally, we will be interested in comparing the mean of a variable for two independent populations, using data that we obtain from samples drawn from those populations. As we will see throughout Chapter 4, most of the techniques and principles that we outlined in Chapter 3 for the one-sample problem will also be available to us when working with two samples. When the underlying techniques are fundamentally different than those described in Chapter 3, we will delve into them in some detail. However, for many of the methods in Chapter 4, the basic mathematics that we learned about in the previous chapter still apply.

Motivating Example

We will continue working with data from the reading intervention program that we used in Chapter 3. In Chapter 4, we expand this example to include a dataset of 100 students half of whom were randomly assigned to the reading intervention and half of whom were assigned to a control group in which the usual instruction was used. After 9 weeks, students in both conditions were given a reading test. The research question of interest is whether there are differences in mean reading performance between the two groups after the 9 weeks of instruction. Prior research suggests that participation in the accelerated reading program should yield higher reading test scores than

DOI: 10.1201/9781003379324-4

the standard approach. We can express the statistical hypotheses to be addressed as

$$H_0 : \mu_C \geq \mu_T \quad \text{and}$$

$$H_A : \mu_C < \mu_T$$

where

μ_C = Population mean reading score for the control (standard instruction) group

μ_T = Population mean reading score for the treatment (accelerated reading) group.

Exploration of the Data

It is always a good idea to start our exploration of the data with visualization and calculation of descriptive statistics. We saw in Chapter 3 that boxplots provide us with useful information about the overall distribution of the target variable (reading test score). In the two-sample case, they also provide us with this information, thereby allowing for a visual comparison of the location and spread of scores across the two groups. The reading test score boxplots by treatment condition are shown in Figure 4.1.

Several interesting data features are made apparent by the boxplots. First, the medians of the two groups (dark lines in the boxes) appear to be quite similar, suggesting that there may not be a large treatment effect *for this sample*. Until we explore the data more fully using inferential techniques, we cannot say anything about population differences. Second, there was greater variability in the treatment group, which also included several potential outliers with unusually high test scores. The box associated with the treatment group was also wider than that of the control. Descriptively, then, it would appear that the accelerated reading treatment was associated with a wider range of possible scores than the control group, but not necessarily with a higher central tendency value.

The descriptive statistics for the two groups, including the mean, the median, the standard deviation (SD), and the mean absolute deviation (MAD), are presented in Table 4.1.

For our sample, the treatment group had a higher mean than the control group, although their medians were identical. This divergence in differences for the measures of central tendency is likely due to the outliers present in the treatment group. Recall from Chapter 2 that the median is less sensitive to outliers than is the mean and we saw in the boxplots that there were multiple possible outliers in the treatment condition. The SD for reading test scores was also larger for the treatment group than for the control. But as with the medians, the MAD values for the two groups were the same. *It is important*

Boxplot of Test Scores by Group

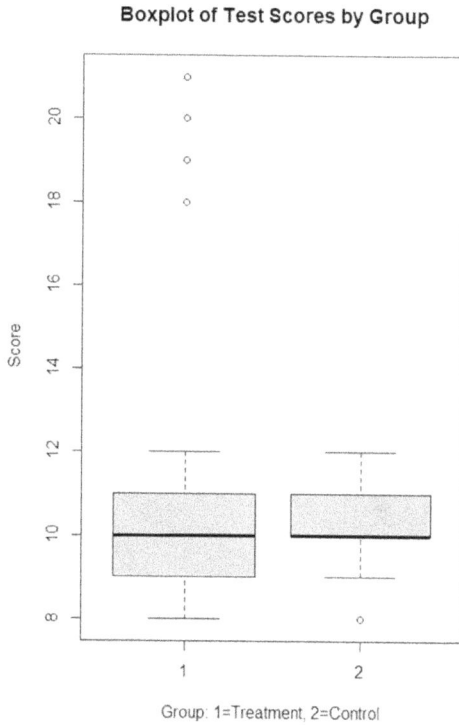

Group: 1=Treatment, 2=Control

FIGURE 4.1
Boxplots of reading test scores by the treatment group.

TABLE 4.1

Descriptive Statistics for Reading Test Scores by Groups

Group	Mean	Median	SD	MAD
Treatment	11.8	10	4.13	1.48
Control	10.2	10	0.97	1.48

to emphasize again that these observations are made only for the sample data and we cannot yet infer anything about differences in scores for the population. Population inference will come in the following pages of the chapter.

Before moving to inference, however, we should first consider the distribution of test scores for the full sample of 100 students, as well as for the two samples separately. The histogram of reading test scores is shown in Figure 4.2.

The positive skewness of the sample is very evident in the histogram. Indeed, the estimate of skewness for the full sample is 2.24, whereas, for the normal distribution, skewness is 0. Thus, we can conclude that *for this sample,* the data are strongly positively skewed. Figure 4.3 displays the histograms for reading test scores separately for each group.

Histogram of Test Scores

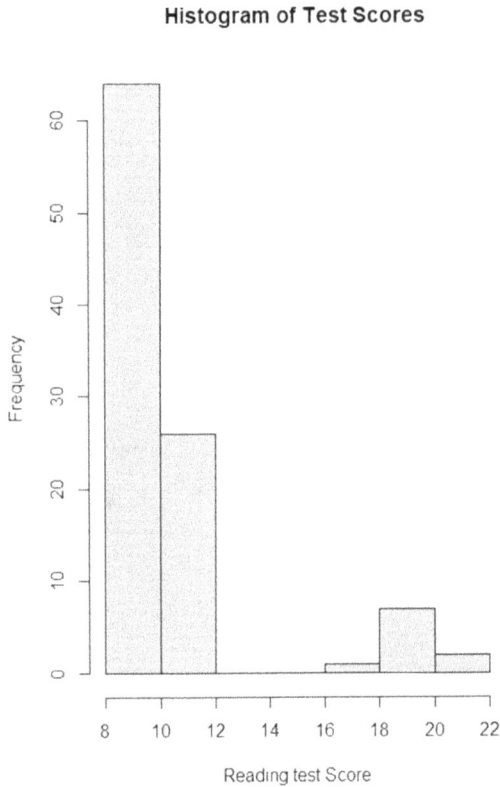

FIGURE 4.2
Histogram of reading test scores.

The separate histograms show that the positive skewness is totally associated with group 1 (accelerated reading treatment). Indeed, the skewness value for group 1 is 1.34, whereas for group 2 it is 0, providing further evidence that the treatment condition was associated with positively skewed scores in the sample, whereas there was essentially no skewness for the control group. We should keep the positive skewness of the overall sample in mind as we move forward with statistical inference, as it may influence the statistical method that we ultimately choose to address our research question.

Independent Sample *t*-Test

The parametric approach for comparing means between two groups is the independent samples *t*-test, which is closely associated with the one-sample *t*-test that we discussed in Chapter 3. As in the one-sample case, the test can

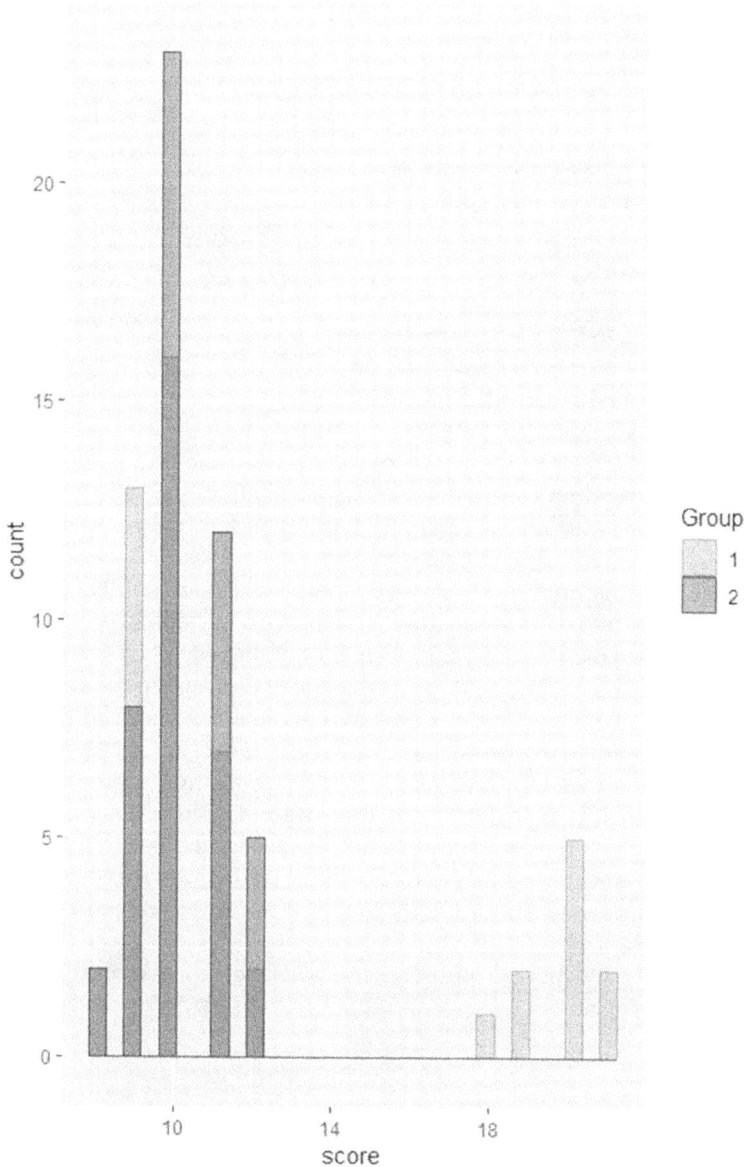

FIGURE 4.3
Histogram of reading test scores by groups.

be used to test both two-sided (i.e., $H_0 : \mu_C \neq \mu_T$) and one-sided ($H_0 : \mu_C \leq \mu_T$ or $H_0 : \mu_C \geq \mu_T$) null hypotheses. The logic underlying the two-sample t-test, with respect to making inferences about the population using sample data, is the same as what we described for the one-sample statistic in Chapter 3. The two-sample t-test statistic is calculated as

$$t = \frac{\bar{x}_T - \bar{x}_C}{S_P \sqrt{\dfrac{1}{n_T} + \dfrac{1}{n_C}}} \qquad (4.1)$$

where

$\bar{x}_T$ = Sample mean for the treatment group
$\bar{x}_C$ = Sample mean for the control group

S_P = Pooled SD = $\sqrt{\dfrac{(n_T - 1)S_T^2 + (n_C - 1)S_C^2}{n_T + n_C - 2}}$

n_T = Sample size for the treatment group
n_C = Sample size for the control group
S_T^2 = Sample variance for the treatment group
S_C^2 = Sample variance for the control group.

The degree of freedom for the two-sample t-test is $n_T + n_C - 2$, reflecting the fact that for each sample, there are $n-1$ independent pieces of information with which to estimate the statistics of interest.

The assumptions underlying the two-sample t-test are similar to those for the one-sample test, including independence of the observations once we account for group membership and normality of the outcome variable (reading test score in this case). For the two-sample t-test, we must make an additional assumption that the variances of the two groups are equal in the population. When we examine the equation for S_P, the necessity of this assumption holding true is readily apparent. In order to use a single value to represent the two groups' SDs, it makes sense that the individual SDs (or variances) are equal in the population. If this is not the case, then pooling them into a single value would not be a reasonable thing to do. We will discuss how to compare the variances, and what to do if this assumption is not met, in the following pages.

As a reminder, for our example the null and alternative hypotheses are

$$H_0 : \mu_C \geq \mu_T \quad \text{and}$$

$$H_A : \mu_C < \mu_T$$

This corresponds to our expectation, based on prior research, that the students participating in the accelerated reading program will have higher mean reading test scores than those assigned to the standard instruction. Let's calculate the t-statistic and use it to test the null hypothesis. The pooled SD is calculated as

$$S_P = \sqrt{\frac{(n_T - 1)S_T^2 + (n_C - 1)S_C^2}{n_T + n_C - 2}} = \sqrt{\frac{(50 - 1)17.06 + (50 - 1)0.94}{50 + 50 - 2}}$$

$$= \sqrt{\frac{836.94 + 46.06}{98}} = 3.00$$

The *t*-statistic can then be calculated as

$$t = \frac{11.84 - 10.20}{3.00\sqrt{\dfrac{1}{50} + \dfrac{1}{50}}} = \frac{1.64}{0.6} = 2.73$$

The degrees of freedom for this statistic are $50 + 50 - 2 = 98$.

We are testing a directional hypothesis (i.e., $H_0 : \mu_C \geq \mu_T$) so that we are only interested in the upper portion of the *t*-distribution. The t with 98 degrees of freedom, including a vertical line for our *t*-value, is graphed in Figure 4.4.

It is clear that the *t*-value for our sample is large when compared to the full *t*-distribution with 98 degrees of freedom. In fact, the probability of having a *t*-value greater than 2.73 with 98 degrees of freedom is 0.004. Because this *p*-value is less than our predetermined α of 0.05, we will reject the null hypothesis $H_0 : \mu_C \geq \mu_T$. Therefore, we conclude that the mean reading test score for the treatment group is larger than the mean reading test score for the control group.

Perhaps, more important than the hypothesis testing result is the confidence interval for the difference between the groups' means. Similar to the one-sample scenario from Chapter 3, this confidence interval will include the range of values that we might consider "reasonable" estimates of the difference between the reading test means in the population as a whole. And just as for the single sample case, the confidence interval for the groups' mean difference provides us with a full set of values for the mean difference, rather than simply indicating whether the means are likely to be different or not. The confidence interval for the difference in means is calculated as

$$(\bar{x}_T - \bar{x}_C) \pm t_{cv} S_P \sqrt{\frac{1}{n_T} + \frac{1}{n_C}} \qquad (4.2)$$

The terms in equation (4.2) are as in equation (4.1), with the addition of t_{cv}, which is the value from the *t*-distribution with $n_T + n_C - 2$ degrees of freedom that corresponds to 95% coverage. We can create a one-sided confidence interval by selecting the appropriate t_{cv} and only adding (for alternative hypotheses in which the treatment group has the lower mean) or subtracting (for alternative hypotheses in which the treatment group has the larger mean) the standard error for the difference, just as we did in Chapter 3.

For our example, the one-sided confidence interval would be

$$(11.84 - 10.20) - 1.661 * 0.6 = (0.64, \infty)$$

This result shows that the lowest reasonable value for the group mean difference is 0.64 points on the reading test. Given that 0 is not in this interval,

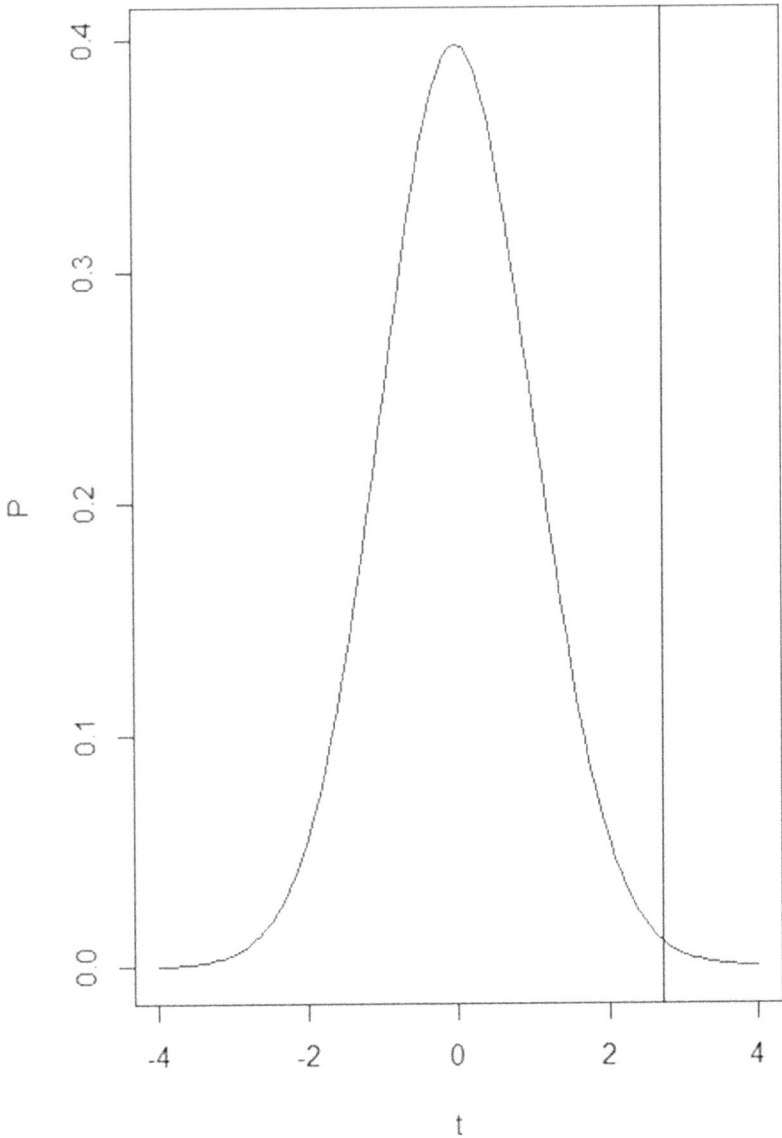

FIGURE 4.4
t-distribution with 98 degrees of freedom and sample *t* values.

we can conclude that the treatment group is likely to perform better on the reading test than does the control group, with the smallest likely value of the difference in means being just over half of a point. The two-sided confidence interval, corresponding to $H_0 : \mu_C \neq \mu_T$, is (0.45, 2.83).

Assessment of Assumptions

As we noted above, the accuracy of the parametric two-sample *t*-test rests on key assumptions about the population distribution of the dependent variable. First, scores for the individuals in our sample should be independent of one another, once we account for their group membership. Second, the dependent variable should be normally distributed, and third, the dependent variable variance should be equal across the groups. As in the single sample case, we can use the QQ plot and the Shapiro–Wilks test to assess the normality assumption. The QQ plot for the current example is shown in Figure 4.5.

Remember from Chapter 3 that the graph reflects the actual datapoints on the *y*-axis and the theoretical values if the data are normally distributed on the *x*-axis. The individual data points are represented by dots, the normal distribution appears as the solid line, and the shaded region reflects the 95% confidence interval for the normal distribution line. Normally distributed data will fall along the line and within the confidence bounds. The further from this pattern the individual points lie, the further from normal are the data. For our reading test example, it seems clear that the data are not normally distributed. The outlying values that we saw in the histogram and box plots are represented in the QQ plot by the dots lying well above the confidence band at the upper end of the distribution. This graph would suggest that the normality assumption may not be viable. The results of the Shapiro–Wilks test also suggest that we reject the null hypothesis that the data were normally distributed, given the *p*-value of 0.00001.

There exist several tests for the assumption regarding equality of group variances. We'll consider two of these, both of which can be used with two or more groups, which will be helpful when we get to the analysis of variance in the next chapter. Levene's test (Levene, 1960) is certainly one of the most widely used techniques for assessing the equality of group variances. For our study, the null hypothesis is $H_0 : \sigma_T^2 = \sigma_C^2$; i.e., the population variances for the treatment and control groups are equal. This statistic essentially compares the within-group variation across the groups and is calculated as

$$W = \frac{(N-k)}{(k-1)} \left(\frac{\sum_{j=1}^{k} N_j \left(\bar{Z}_j - \bar{Z}_. \right)^2}{\sum_{j=1}^{k} \sum_{k=1}^{N_j} \left(Z_{ij} - \bar{Z}_j \right)^2} \right) \tag{4.3}$$

where

$$Z_{ij} = \left| Y_{ij} - \bar{Y}_j \right|$$

Y_{ij} = Value on the dependent variable for person *i* in group *j*
$\bar{Y}_j$ = Group *j* mean or median
$\bar{Z}_j$ = Group *j* mean for Z_{ij}

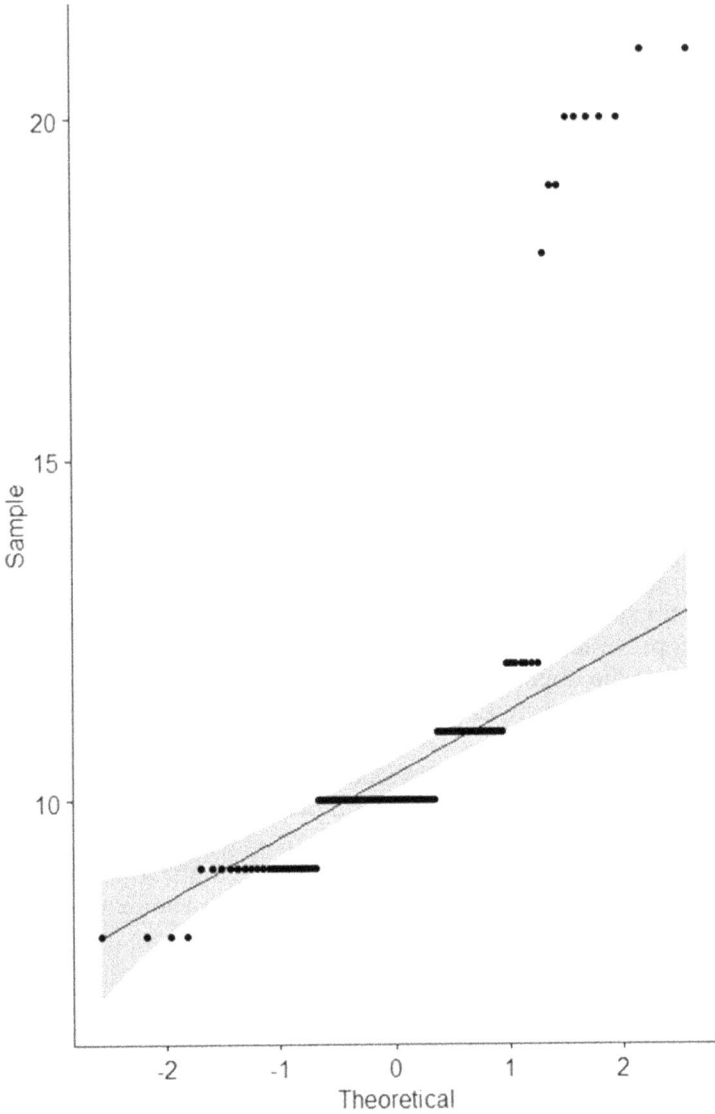

FIGURE 4.5
QQ plot for the reading data in the two-sample case.

$\bar{Z}_{.}$ = Overall mean for Z_{ij}
N_j = Sample size for group j
k = Number of groups.

Levene's test has been shown to be more accurate with normally distributed data, though it is somewhat more robust when the median is used in the calculation of Z_{ij} (Brown & Forsythe, 1974). When the dependent variable is not normally distributed, a potentially more robust alternative to Levene's test is the Fligner–Killeen test, which is based on ranks (Fligner & Killeen, 1976). For this test, Z_{ij} are calculated as for W in equation (4.3) and these values are then ranked. In turn, the ranks are normalized to yield a_{ij} and the Fligner–Killeen statistic is calculated as

$$\text{FK} = \frac{\sum_{j=1}^{k} N_j \left(\bar{a}_j - \bar{a}_. \right)^2}{S_a^2} \tag{4.4}$$

where
$\bar{a}_j$ = Mean normalized rank difference for group j
$\bar{a}_.$ = Overall mean normalized rank difference
S_a^2 = Variance of all normalized values.

The null hypothesis for the Fligner–Killeen test is identical to that of Levene's, namely that the population variances of the groups are equal.

For our example, the p-value for Levene's test was 0.001, whereas for the Fligner–Killeen test, it was 0.01. In either case, we would reject the null hypothesis that the population variances of the treatment and control groups are equal, assuming that we set $\alpha = 0.05$. These results indicate that the assumption of equal variance is not viable. Taken together with the lack of normality, it appears that the standard t-test is probably not the appropriate tool to address our research question. Thus, we should be very careful when interpreting the statistically significant t-test results for mean differences in the reading test scores between the groups. Next, we will consider some alternatives to the t-test for comparing the central tendency for two groups that are more robust to violations of the normality and equal group variances assumptions.

Welch's *t*-Test

One option for data analysts to use when the groups' variances are not equal is Welch's t-test. This approach to calculating the t-statistic simply does not pool the SDs when calculating the standard error for the mean difference. The calculation takes the form

$$t = \frac{\bar{x}_T - \bar{x}_C}{\sqrt{\dfrac{S_T^2}{n_T} + \dfrac{S_C^2}{n_C}}} \tag{4.5}$$

In addition to a different equation for the standard error, the degree of freedom for Welch's t-statistic is also different than $n_T + n_C - 2$ for the standard t. The Welch degrees of freedom are calculated as

$$df_w = \frac{\left(\dfrac{S_T^2}{n_T} + \dfrac{S_C^2}{n_C}\right)^2}{\left(\left(\dfrac{S_T^2}{n_T}\right)^2 /(n_T - 1)\right) + \left(\left(\dfrac{S_C^2}{n_C}\right)^2 /(n_C - 1)\right)} \tag{4.6}$$

For our reading test example, Welch's t would be

$$t = \frac{11.84 - 10.20}{\sqrt{\dfrac{17.06}{50} + \dfrac{0.94}{50}}} = \frac{1.64}{\sqrt{0.34 + 0.02}} = \frac{1.64}{0.6} = 2.73$$

This value is very close to that of the t-statistic under the assumption of equal variances. We will not demonstrate the calculation of the Welch degrees of freedom here (the motivated and interested reader is encouraged to make this calculation!), but the value for our example is 54.38. The p-value associated with a t-value of 2.73 with 54.38 degrees of freedom is 0.004.

Because the p-value is less than the α of 0.05, we reject the null hypothesis and conclude that the test score mean for students in the accelerated reading program is greater than that for students in the business as usual condition. The one-sided Welch 95% confidence interval for the difference between the means is $(11.84 - 10.20) - 1.661 * 0.6 = (0.64, \infty)$. Therefore, we are confident that the actual mean difference in the population is 0.64 or larger, favoring the accelerated reading program participants.

Wilcoxon Test and Mann–Whitney U

Although Welch's t-test is an effective tool to use when the groups' variances are unequal, it is not robust to non-normality in the dependent variable. One approach that has been shown to produce accurate hypothesis tests and confidence intervals when the data are not normally distributed is based on the ranks of the data, as opposed to the raw values. In Chapter 3, we discussed the Wilcoxon signed rank test for the one-sample location problem. In that case, the raw values of the dependent variable were ranked and the test statistic was the sum of ranks for values that were larger than the mean under the null hypothesis. For the scenario in which we want to compare the locations (e.g., medians) of two groups, we start by ranking all of the raw data values across both groups. We then sum the ranks for the group with the smaller sample size (or either of the groups if they are of equal size). This sum serves as the test statistic of the null hypothesis that the groups' locations are equal in the population. Traditionally, this difference is referred to as the shift in location between the two groups and can be estimated by the difference in sample means or medians. The null and alternative hypotheses regarding the median (M) for our reading test example would be

$$H_0 : M_C \geq M_T \text{ and}$$

$$H_A : M_C < M_T$$

Formally, the Wilcoxon test can be expressed as

$$T = \sum_{i=1}^{n_j} R_{ji} \tag{4.7}$$

where
R_{ji} = Rank of the raw score for individual i in group j.

The p-value for this statistic is then obtained using a permutation-based approach (see Chapter 2 for the description of the permutation methodology). For each possible permutation of the ranks, T is calculated. This set of all possible T values for our sample are then used to create a sampling distribution for T when the null hypothesis of no group difference in location holds true. The actual value of T for our sample is then compared against this null distribution, and the p-value is the proportion of the distribution greater than the observed value of T.

When the overall sample size is greater than 50 or there are ties in the data, the permutation-based approach outlined above is replaced by a large sample approximation in order to obtain the p-value. The approximation is

$$z = \frac{T - \left[n_c (n+1)/2 \right]}{\sqrt{n_T n_c (n+1)/12}} \tag{4.8}$$

The quantity $n_c (n+1)/2$ is an estimate for the mean of T under the null hypothesis of no group difference, and $n_T n_c (n+1)/12$ is an estimate of its variance. The p-value for this statistic is then obtained by comparing z with the standard normal distribution. This comparison is valid when the sample is sufficiently large for the Central Limit theorem to apply (Hettmensperger & McKean, 2011).

Finally, before looking at how these tests can be applied to our reading test data, we should note that there exists an alternative approach to calculate this rank-based estimate of the test statistic for the differences in location for the two groups. It is based on the set of all possible differences in scores between the two groups, which we call T^+. This statistic can also be calculated as

$$T^+ = T - \frac{n_c (n_c + 1)}{2} \tag{4.9}$$

This alternative test statistic is called the Mann–Whitney after the statisticians who developed it (Mann & Whitney, 1947). It will yield identical results to the Wilcoxon test statistic.

For our reading test example, the raw scores and ranks for the control group are presented in Table 4.2.

TABLE 4.2

Raw Reading Test Scores and Ranks
for the Control Group

Raw Reading Test Score	Rank of the Reading Test Score
9	15
10	45
10	45
9	15
8	2.5
10	45
11	74
10	45
10	45
10	45
10	45
9	15
11	74
9	15
12	87
9	15
10	45
9	15
9	15
9	15
11	74
9	15
9	15
10	45
10	45
9	15
10	45
10	45
10	45
9	15
8	2.5
10	45
11	74
11	74
9	15
11	74
12	87
11	74
10	45

(*Continued*)

TABLE 4.2 (*Continued*)

Raw Reading Test Scores and Ranks
for the Control Group

Raw Reading Test Score	Rank of the Reading Test Score
10	45
19	92.5
20	96
20	96
21	99.5
19	92.5
18	91
20	96
20	96
21	99.5
20	96

Given the sample size and the presence of tied ranks, we will use the large sample approximation test statistic. The sum of the ranks is

$$T = \sum_{i=1}^{n_j} R_{ji} = 2,567$$

The mean rank for the large sample approximation when the null is true is calculated as

$$\frac{n_c(n+1)}{2} = \frac{50(100+1)}{2} = 2,525$$

The standard error for this statistic is

$$\sqrt{n_T n_c (n+1)/12} = \sqrt{50 * 50(100+1)/12} = 145.06$$

Finally, the test statistic is calculated as

$$z = \frac{T - [n_c(n+1)/2]}{\sqrt{n_T n_c (n+1)/12}} = z = \frac{2,567 - 2,525}{145.06} = 0.29$$

This value of z yields a p-value of 0.38. Given $\alpha = 0.05$, we can't reject the null hypothesis meaning that there is not sufficient evidence to conclude that the groups' reading test locations differ. The sample medians for the two groups are both 10 (Table 4.1), and the Hodges–Lehmann estimate of the difference between the groups is 0.00003. The one-sided 95% confidence interval for this difference is $(-0.00007, \infty)$. The fact that 0 is included in this interval further reinforces our conclusion that there is no difference in group medians on the reading test.

When interpreting these results, it is important to note that the Wilcoxon/ Mann–Whitney U statistics are sensitive to differences in distributions between the two groups being compared. Therefore, any differences in distributions between the two groups (e.g., location or spread) could be detected by the Wilcoxon test statistic and reflected in a statistically significant result. In the current example, the nonstatistically significant result would suggest that when we base our inference on ranks, the distributions of the groups (including location) are unlikely to be different in the population.

Finally, the reader should note that there are several variants of these nonparametric test statistics that might be particularly useful for specific types of data distributions, such as those with heavy skewness or with light tails. Hogg (1974) described a simple adaptive testing approach that the data analyst could employ so that the most appropriate statistical test is used given the shape of the data. In Hogg's technique, two selector statistics are calculated based on descriptive information in the data:

$$Q_1 = \frac{\bar{U}_{0.05} - \bar{M}_{0.5}}{\bar{M}_{0.5} - \bar{L}_{0.05}} \quad \text{and} \tag{4.10}$$

$$Q_2 = \frac{\bar{U}_{0.05} - \bar{L}_{00.5}}{\bar{U}_{0.5} - \bar{L}_{0.5}}$$

where
$\bar{U}_{0.05}$ = Mean for the upper 5% of the data
$\bar{M}_{0.5}$ = Mean for the middle 50% of the data
$\bar{L}_{0.05}$ = Mean for the lower 5% of the data
$\bar{U}_{0.5}$ = Mean for the upper 50% of the data
$\bar{L}_{0.5}$ = Mean for the lower 50% of the data.

Hogg viewed Q_1 as a measure of skewness in the data and Q_2 as a measure of tail heaviness. Based on these two statistics, Hogg's adaptive selection algorithm makes the following decision as to which test statistic to use (Table 4.3).

TABLE 4.3

Hogg's Adaptive Test Statistic Selection Algorithm

Selector Statistics Values	Test Statistic to Use
$Q_2 > 7$	Sign test statistic
$Q_1 > 1$ and $Q_2 < 7$	Bent test statistic for right-skewed data
$Q_1 \leq 1$ and $Q_2 \leq 7$	Light-tailed test statistic
Otherwise	Wilcoxon test statistic

For the reading test data, we have the following values:

$$\bar{U}_{0.05} = 20.4$$
$$\bar{M}_{0.5} = 10.22$$
$$\bar{L}_{0.05} = 8.2$$
$$\bar{U}_{0.5} = 12.62 \text{ and}$$
$$\bar{L}_{0.5} = 9.42$$

Using these values, we then obtain the following for the selector statistics:

$$Q_1 = \frac{20.4 - 10.22}{10.22 - 8.2} = \frac{10.18}{2.02} = 5.03 \quad \text{and}$$
$$Q_2 = \frac{20.4 - 8.2}{12.62 - 9.42} = \frac{12.2}{3.2} = 3.81$$

Now applying the algorithm in Table 4.2 to our values for Q_1 and Q_2, we find ourselves in the second row where the Bent statistic for right-skewed data will be selected. This makes sense given our earlier exploration of the data, which showed that the reading test scores were indeed right skewed. The value of the Bent statistic for our sample is −0.56 with a p-value of 0.37. This result agrees with that of the Wilcoxon and Mann–Whitney U tests and leads us to conclude that the location of the two groups' reading test scores is not likely to be different in the population.

Bootstrap-Based Test for Comparing Central Tendency for Two Groups

We discussed the basics of the bootstrap in Chapter 2 and then described its use in making inference about the mean in the one-sample case in Chapter 3. The bootstrap can be easily extended to making inference about the difference in the means for two groups. The percentile bootstrap technique in this context is carried out as follows:

1. Create B (e.g., 10,000) bootstrap samples from the original data.
2. For each of the B samples, calculate $D^* = \bar{x}_T^* - \bar{x}_C^*$.
3. Calculate 1-proportion of B samples for which $\bar{x}_T^* > \bar{x}_C^*$, and call this P^*.
4. Reject $H_0 : \bar{x}_T^* = \bar{x}_C^*$ when $P^* \leq 0.05$ for a one-tailed test, or $P^* \leq 0.025$ for a two-tailed test.

For the reading test data, P^* is 0.011. In other words, in only 1.1% of the bootstrap samples was the mean for the treatment group *not* larger than that of the control. The 95% percentile bootstrap confidence interval for the mean difference is (−0.94, −0.14). Taking these results together, we would conclude that the mean reading score for the treatment group is likely to be larger than that of the control in the population.

The percentile bootstrap can also be applied to comparisons of the median or the M estimator of location as well. The four steps outlined above would be applied as described, but in step 2, the difference in medians or difference in M estimators would replace the difference in means. Note that M estimation would be applied to the complete sample prior to the application of the bootstrap. For our example, the bootstrap p-value for the difference in medians is 0.809 with a 95% confidence interval of (−1, 0). This result would lead us to the conclusion that the population medians are not different between the two groups. The p-value for the M estimator is 0.0.025 and the 95% confidence interval of (−0.96, −0.08). Thus, as for the mean, we would reject the null hypothesis of equal reading test M estimators for the two groups.

Yuen–Welch Test for Comparing Trimmed Means

We saw in Chapters 2 and 3 that techniques based on trimmed means can be particularly effective in situations where there are outliers. Trimmed means can also be applied to the case where we would like to compare locations for two groups. The approach for doing this was developed by Yuen (1974) and tests hypotheses about the trimmed mean (μ_T), as opposed to the mean for the full sample. More formally, for our example, the null and alternative hypotheses are

$$H_0 : \mu_{TT} > \mu_{TC} \quad \text{and}$$
$$H_A : \mu_{TT} \leq \mu_{TC}$$

The statistic used to test the null hypothesis is then calculated as

$$t_{Yuen} = \frac{\bar{x}_{TT} - \bar{x}_{TC}}{\sqrt{d_T + d_C}} \tag{4.11}$$

where

$$d_T = \frac{(n_T - 1)s_{WT}^2}{h_T(h_T - 1)} \quad \text{and}$$

$$d_C = \frac{(n_C - 1)s_{WC}^2}{h_C(h_C - 1)}$$

n_T = Number in the treatment group
n_C = Number in the control group
h_T = Number retained (not trimmed) in the treatment group
h_C = Number retained (not trimmed) in the control group
s_{WT}^2 = Winsorized variance for the treatment group
s_{WC}^2 = Winsorized variance for the control group
$\bar{x}_{TT}$ = Trimmed mean for the treatment group
$\bar{x}_{TC}$ = Trimmed mean for the control group.

The degree of freedom for the Yuen–Welch test is

$$df_y = \frac{(d_T + d_C)^2}{\dfrac{d_T^2}{h_T - 1} + \dfrac{d_C^2}{h_C - 1}} \qquad (4.12)$$

To obtain the p-value for our test, we compare the value of t_{Yuen} with the t-distribution and df_y degrees of freedom.

For the reading test data, $t_{\text{Yuen}} = 0.1127$ with a p-value of 0.91. Thus, we would not reject the null hypothesis that the treatment group's trimmed mean is less than or equal to that of the control group. In addition, the 95% confidence interval for the difference in trimmed means is (−0.6317, 0.565). Therefore, because 0 lies within the interval, we would conclude that it is possible that the groups' population trimmed means are not different from one another. In other words, when we remove the most extreme values on either end of the score distribution for each group, the mean reading performance is unlikely to be different between the groups in the population.

Permutation Test to Compare Means for Two Groups

The permutation approach has been shown to be useful in cases where the assumption of normality is not viable (Brown & Rothery, 1993). We described the basics of this methodology for comparing two groups' means in Chapter 2 and saw that it essentially involves repeatedly swapping dependent variable values between groups and calculating the statistic of interest, i.e., the difference in group means. For the reading test example, this approach takes the form:

1. Compute $D = \bar{x}_T - \bar{x}_C$ for the original data.
2. Permute the data by swapping one individual in the treatment group with an individual in the control group.
3. Compute $D^* = \bar{x}_T^* - \bar{x}_C^*$ for the permutation created in step 2.

4. Repeat steps 2 and 3 for every possible permutation in the data; i.e. all ways to combine the individuals in the sample into two groups.

5. Create a distribution of the D^* values, which represents the distribution of mean differences when the null hypothesis of no group mean differences is true.

6. Reject the null hypothesis of no group mean differences if $D \geq 97.5$th or $D \leq 2.5$th percentile of the distribution created in step 5.

For our example, the p-value of the permutation test is 0.01, leading us to reject $H_0 : \mu_C \geq \mu_T$.

Bayesian Comparison of Central Tendency for Two Groups

The reader is surely seeing the pattern of this book where the methods described in Chapter 2 can be applied to a wide variety of problems. Bayesian estimation is no exception to this pattern. As we will see in this chapter, Bayesian estimation can be applied to the problem of comparing central tendency for two groups. In addition, we will also see how the BF can be used to assess the likelihood of various hypotheses of interest. The technical aspects of Bayesian estimation in the two-sample means comparison context, including the use of priors, the burn-in period, the number of replications, and the number of chains, all remain essentially the same as in Chapters 2 and 3. The use of the BF is also quite similar in the two-sample case as it was for a single sample (see Chapter 3), with appropriate extensions for the multiple groups, as we will see below. Given this high degree of similarity in so many of these aspects of Bayesian estimation, we will not review them again here.

We will first apply the Bayesian estimation approach to the comparison of the reading test means for the accelerated reading and control groups using an uninformative normal prior with a mean of 0 and variance of ∞. We will use 100,000 iterations with a burn-in of 5,000, a thinning rate of 1 out of 10, and two Markov Chain Monte Carlo (MCMC) chains. The resulting t-statistic is 2.74 with 98 degrees of freedom, yielding a p-value of 0.004. Given that the p-value is less than our α of 0.05, we will reject the null hypothesis and conclude that the treatment group mean is likely to be larger than the control group mean in the population. The one-sided 95% confidence interval for the difference in group means is (0.64, ∞). These results are essentially an identical result to that for the standard t-test. This similarity in results between the standard t-test and the Bayesian version should not be particularly surprising for two reasons. First, we used an uninformative prior with a large variance, thereby placing much greater emphasis on the observed data than on any prior information we have. Second, the sample size is not particularly small (100), giving it more weight in the final estimate than would be the case for a small sample size.

Next, let's apply an informative prior distribution that the population mean reading test score for each group is 10. We can justify this prior by the fact that the test is normed so that the population mean should indeed be 10, as discussed in Chapter 3. The prior variances associated with these prior means of 10 were set to 0.0001 for each group, which yields a very tightly concentrated distribution for each group, as seen in Figure 4.6.

The t for the informative Bayesian estimator was 2.78 and $p = 0.003$. The one-sided 95% confidence interval for the difference was (0.66, ∞). Thus, using informative priors did not yield appreciably different results from the Bayesian estimator with noninformative priors, nor from the standard t-test for that matter.

We can use the BF to assess the probability of each of several hypotheses given the sample data, just as we did for the one-sample case in Chapter 3. A natural set of hypotheses for the accelerated reading program study would be

$$H_1 : \mu_C = \mu_T$$

$$H_2 : \mu_C < \mu_T \quad \text{and}$$

$$H_3 : \mu_C > \mu_T$$

We reviewed the calculation of BF and the BF ratios in Chapter 3, and so we would refer the reader there for a reminder as to how this is done. We would also refer the reader to the caveats regarding the limitations of the BF (Schad et al., 2023), including the need to avoid discrete (e.g., reject, don't reject) decisions based on the BF as well as its sensitivity to the choice of priors and the number of MCMC iterations that are used. For this example, we used noninformative priors for the groups' reading test means and 100,000 iterations with 5,000 iterations for the burn-in and a thinning rate of 1 out of 10. The resulting BF and hypothesis likelihoods are presented in Table 4.4.

The ratio of BFs for the hypotheses are

$$H_1 \text{ vs } H_2 : \frac{0.237}{1.994} = 0.12$$

$$H_1 \text{ vs } H_3 : \frac{0.237}{0.006} = 39.50 \quad \text{and}$$

$$H_2 \text{ vs } H_3 : \frac{1.994}{0.006} = 332.33$$

The most likely hypothesis is clearly $H_2 : \mu_C < \mu_T$, which has a probability of 0.891. It is 8.41 (1/0.12) times more likely to be true than the hypothesis of equal group means. Likewise, it is more than 332 times more likely than the

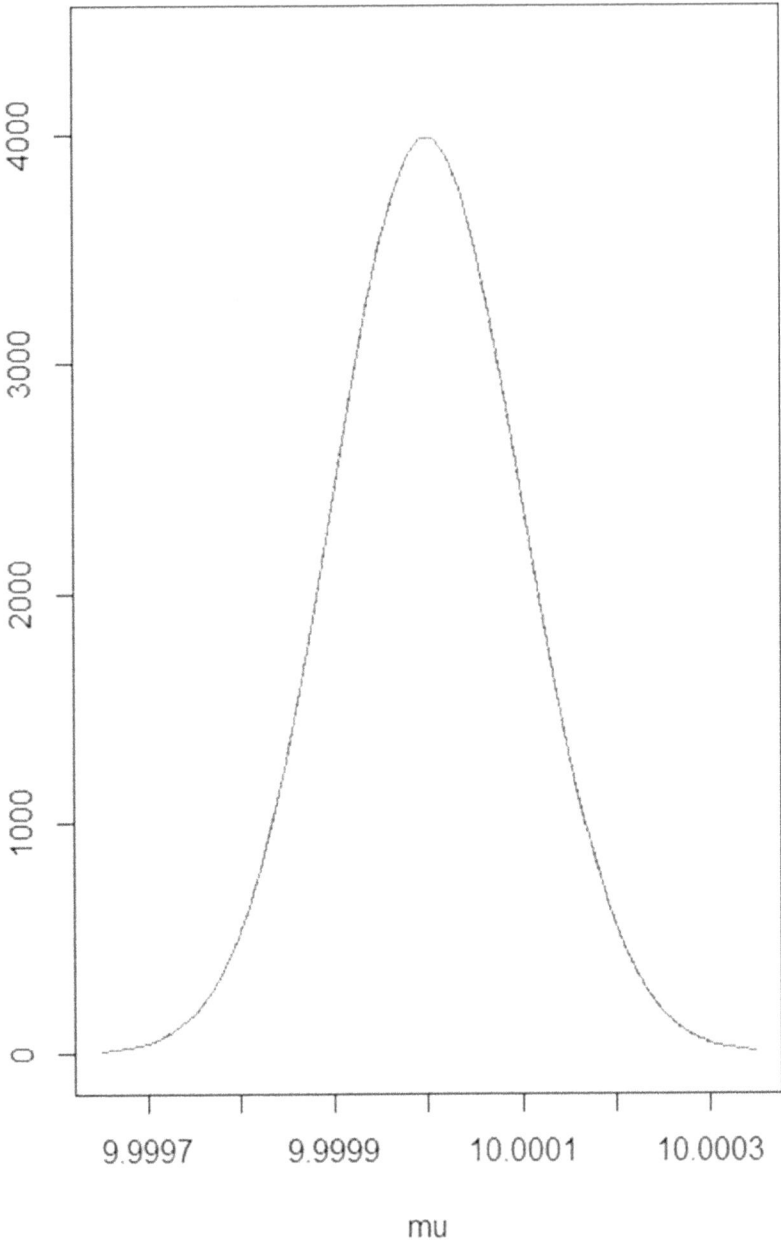

FIGURE 4.6
Prior distribution with a mean of 10 and a variance of 0.0001.

hypothesis that the control group has a higher reading test mean than the treatment group. Taken together with the results of the Bayesian hypothesis test and confidence interval, the BFs in Table 4.3 and the associated ratios

TABLE 4.4

Bayes Factors and Likelihoods Associated
with H_1, H_2, and H_3

Hypothesis	Bayes factor	Probability
$H_1 : \mu_C = \mu_T$	0.237	0.106
$H_2 : \mu_C < \mu_T$	1.994	0.891
$H_3 : \mu_C > \mu_T$	0.006	0.003

provide strong evidence that the mean reading test score for the treatment group was higher than that of the control.

Finally, prior to completing our discussion of comparing group means in a Bayesian framework, we will turn our attention to the concept of a region of practical equivalence (ROPE) as described by Kruschke (2015), among others. The philosophy underlying the use of ROPE is somewhat different than the standard hypothesis testing that underlies most of the approaches discussed in this chapter and differs from the application of BF as well. We will not go into great detail regarding the mathematical underpinnings of ROPE but instead will consider it from a conceptual standpoint. In the context of comparing central tendency for two groups, ROPE involves comparing the posterior distribution of the observed difference between group means with a predetermined definition of what constitutes equivalence between the groups in terms of the mean difference. For example, if we determine that a standardized mean difference (δ) of less than −0.1 or greater than 0.1 constitutes a "meaningful" difference (Kruschke, 2015), then we can ascertain how much of the posterior distribution falls outside of the range between $-0.1*s_y$ and $0.1*s_y$. If most of the Bayesian posterior distribution for the group mean difference lies outside of this range, we have evidence for a difference in means between the groups. Conversely, if a substantial portion of the posterior distribution lies within this range, we have little evidence for a difference in the groups' means.

The determination of what should constitute the ROPE is ultimately up to the researcher and should be based on both prior research and conceptual considerations. Kruschke (2018) suggested $0.1*s_y$ as a reasonable ROPE value based on some very general guidelines for interpreting effect sizes (which we discuss later in this chapter). However, there may be instances where the context of a specific problem might suggest different rules. If we apply the $0.1*s_y$ rule to our problem, then the ROPE would lie between −0.31 (−1*3.09) and 0.31 (1*3.09). Another approach for characterizing the ROPE results is to calculate the standardized difference between the groups' means and then compare them with a commonly used heuristic for interpreting the resulting value. This standardized value is similar to Cohen's d effect size, which we will discuss in more detail in the Effect Size Estimation section of this chapter. In the context of ROPE, this standardized difference in means is known as δ and is calculated as

$$\delta = \frac{\overline{X}_T - \overline{X}_C}{S} \qquad\qquad (4.13)$$

where
    $\overline{X}_T$ = Mean reading score for the treatment group
    $\overline{X}_C$ = Mean reading score for the control group
    S = SD of the reading score for the full sample

The results of the Bayesian analysis based on ROPE are shown in Figure 4.7.
    For our example, the posterior distribution of the standardized difference in group means was centered at −0.545, as seen in the top panel of Figure 4.7. In addition, we can see that the vast majority of the posterior distribution lies above the ROPE (denoted by the vertical dashed line at the left of the histogram). Of the posterior distribution, 65.72% was in the medium effect range, and 34.28% was in the small effect range (bottom panel of Figure 4.7). None of the posterior lay in the no effect range. Thus, because the distribution of δ values lies fully outside of the region of equivalence, we would conclude that there is in fact a difference in reading test means for the two groups.

Effect Size Estimation

In addition to the hypothesis test results, it is very useful for us to consider the magnitude of the group difference in the means. In Chapter 3, we learned about Cohen's d, which is the standardized difference between the sample mean and the population value if the null hypothesis were true. This statistic can easily be extended to the two-sample case:

$$d = \frac{\overline{X}_T - \overline{X}_C}{S_P} \qquad\qquad (4.14)$$

For our example, $d = \dfrac{11.84 - 10.20}{3.00} = \dfrac{1.64}{3.00} = 0.55$. The interpretive guidelines for d outlined by Cohen (1988) are applicable in the two-sample case, although as we discussed in Chapter 3, the researcher should use these as very rough guidelines, and not as hard and fast rules. Cohen's guidelines for interpreting d are 0.2–0.5=Small effect, 0.5–0.8=Moderate effect, and 0.8+=Large effect. Using this heuristic, the difference in reading score means between the two groups is slightly more than half of a standard deviation, placing it in Cohen's moderate range.

    One problem with using d in our case is that it relies on the pooled standard deviation, and we know the groups' variances are not equal. So, as with the t-test, this calls into question our use of d. Fortunately, there are

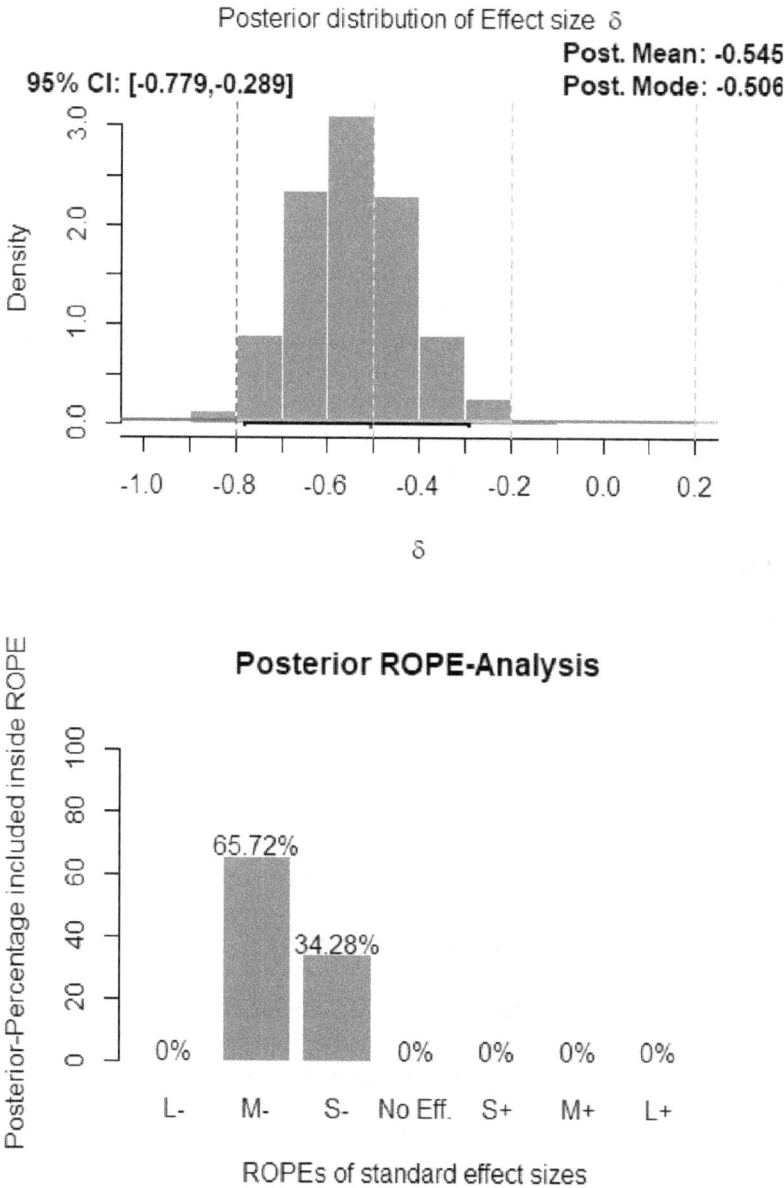

FIGURE 4.7
Posterior distribution of δ and classification of δ magnitudes.

alternative statistics that can provide us with information about the magnitude of group mean difference. Recent research has identified Hedges' g_s^* as the best performer among these options across a variety of real-world sampling conditions (Baguley, 2009). It is based on a version of d that does not pool the variances and is corrected so as to remove bias for smaller samples, as in the case of Hedges' g for a single sample (see Chapter 3). Hedges' g_s^* is based on a version of Cohen's d that does not pool the standard deviations:

$$d_s^* = \frac{\bar{X}_T - \bar{X}_C}{\sqrt{\dfrac{S_T^2 + S_C^2}{2}}} \tag{4.15}$$

This value is then corrected in much the same way as we saw in Chapter 3 for Hedges' g in the one-sample case. There is a very helpful web application for calculating g_s^* in the two-sample case: https://effectsize.shinyapps.io/deffsize/. Using this app for our reading test example, we obtained a value of Hedges' g_s^* of 0.539. We can interpret this just as we do Cohen's d, leading us to conclude that the groups' means are roughly half of a standard deviation apart and thus moderately different. The 95% confidence interval for g_s^* is 0.201, ∞. Thus, the effect on the population could be small in magnitude, or quite large.

Algina et al. (2005) recommended a second nonparametric analog for Cohen's d, which uses the trimmed mean and Winsorized variances in equation (4.13). This alternative is calculated as

$$d_{\text{TR}} = 0.642 \frac{\bar{X}_{\text{TTR}} - \bar{X}_{\text{CTR}}}{\sqrt{\dfrac{(n_T - 1)S_{\text{WT}}^2 + (n_C - 1)S_{\text{WC}}^2}{n_T + n_C - 2}}} \tag{4.16}$$

where
$\bar{X}_{\text{TTR}}$ = Trimmed mean for the treatment group
$\bar{X}_{\text{CTR}}$ = Trimmed mean for the control group
S_{WT}^2 = Winsorized variance for the treatment group
S_{WC}^2 = Winsorized variance for the control group.

The value 0.642 is a correction factor that is included to account for the use of Winsorized variances as opposed to the standard variance. When the data are normally distributed and the groups' variances are equal, $d_{\text{TR}} = d$. Guidelines for interpreting d_{TR} are the same as for d.

For the reading intervention example, d_{TR} is

$$d_{\text{TR}} = 0.642 \frac{10.2 - 10.23}{\sqrt{\dfrac{(50 - 1)1.32 + (50 - 1)0.23}{50 + 50 - 2}}} = 0.642 \frac{-0.03}{0.88} = -0.02$$

Based on Cohen's guidelines, we would conclude that the difference between the trimmed means is negligible. In other words, when the most extreme values are removed from the sample, there is no evidence of a difference in the groups' performance on the reading test.

Wilcox and Tian (2011) proposed an effect size estimate that reflects the proportion of variability in the dependent variable associated with group membership. Conceptually, this effect size, ξ^2, is quite similar to the R^2 from linear regression, which we will discuss in Chapter 9. A robust sample estimate based on Winsorized variances is calculated as

$$\hat{\xi}^2 = \frac{S_W^2\left(\hat{Y}\right)}{S_W^2\left(Y\right)}$$

(4.17)

where
$S_W^2\left(\hat{Y}\right)$ = Winsorized variance for the dependent variable predicted by group membership
$S_W^2\left(Y\right)$ = Winsorized variance for the observed values of the dependent variable.

Wilcox and Tian referred to the square root of ξ^2 as the explanatory measure of effect size and showed that Cohen's guidelines for d correspond to the following for ξ: 0.15–0.35=Small, 0.35–0.50=Medium, and 0.50+=Large. We can also use the percentile bootstrap to obtain a confidence interval for ξ. For our example with 20% trimming, $\xi^2 = 0.0004$ and $\xi = 0.02$. The 95% confidence interval for ξ is (0, 0.84). Taken together, these results mirror those based on d_{TR}; namely, the difference between the trimmed means was negligible in size. Again, it appears that when we remove the most extreme scores in the sample, there is no evidence for a difference in group performance on the reading test.

Cliff's delta offers a very different way in which we can compare the distributions of reading scores for our two groups. The previous effect size statistics expressed this difference in terms of the standardized difference in group means. In contrast, Cliff's delta describes this difference in terms of the proportion of cases for which the dependent variable value for one group exceeds that of the other group. More specifically, this technique compares the reading test score for each person in the treatment group with that of each person in the control group and delta reflects the proportion of these comparisons for which the treatment group reading scores were higher than those of the control group. Put another way, Cliff's delta is the probability that a randomly chosen individual from the treatment group would have a higher reading test mean than a randomly chosen person from the control group. For our example, the delta is 0.03, meaning that the probability that a randomly selected individual in the accelerated reading program will have a higher reading test score than a randomly selected individual in the business as usual control group is 0.03. This result suggests that the likelihood of an individual in the reading treatment performing better on the test than a business as usual student is not very large. The 95% confidence interval for this effect size value is −0.16 to 1, meaning that 0 is a possible value of delta; i.e., it would be reasonable to conclude that there may be little practical difference in the performance of the groups on the reading exam.

Finally, we can use the nonparametric effect size based on ranks (Fritz et al., 2012) that we first introduced in Chapter 3. Extended to the two-sample case, it is calculated as

$$r = \frac{2\left(\bar{R}_T - \bar{R}_C\right)}{n_T + n_C} \tag{4.18}$$

where
$\bar{R}_T$ = Mean rank for the treatment group
$\bar{R}_C$ = Mean rank for the control group
n_T = Treatment group sample size
n_C = Control group sample size.

For the reading treatment example, $r=0.0301$ with a 95% bootstrap confidence interval of (0.003, 0.24). Based on the guidelines provided by Friz et al. (0.1–0.3 (small), 0.3–0.5 (medium), and 0.5+ (large)), this value would fall into the negligible (less than small) range.

Summary

As was true for the one-sample case in Chapter 3, there are a large number of options that we can use to compare central tendency for two groups. Furthermore, as we saw with our example data, the choice of method can lead to apparently different conclusions for the same sample. Thus, it is very important that we carefully consider which method we should use. It is also important to note that although our ultimate conclusions might differ depending on the method used, these differences are typically down to differences in the nature of the null hypotheses being tested. For example, the parametric *t*-test involves inference about differences in the population means, whereas the Wilcoxon test provides inference about differences in the population median and the *t*-test for trimmed means yields inference about differences in the population trimmed means. In short, these tests are looking at the same problem in different ways. In order to determine which of these approaches to use, we need to consider carefully both the distribution of the data at hand and the research question that we would like to address.

In light of the caveats outlined in the previous paragraph, we can provide some general guidelines regarding when to use which approach. First of all, when the data are normally distributed and the groups' variances are equal, then the standard *t*-test is always optimal. It has the highest power of the methods discussed in this chapter and maintains good control over the Type I error rate. If the data are non-normal, but the non-normality is not severe, then *t* also controls the Type I error rate well, particularly with large N. However, power will be low for t when data are non-normal at all. The Wilcoxon test tends to have higher power than the parametric t for heavy-tailed and skewed distributions, but lower power for normal and light-tailed distributions. Generally speaking, the permutation test is similar to t in terms of power and Type I error rate when the data are normally distributed but has superior power for detecting group mean differences when the data are not normal. In addition, it is also important to keep in mind that the Wilcoxon and the permutation both actually test the null hypothesis of equal distribution shape (location and dispersion) between groups and not simply differences in location. Thus, when group variances are not equivalent, we are more likely to obtain statistically significant results, even in cases where the means do not differ. Indeed, when the groups' variances and sample sizes are unequal, the Type I error rate of the t-test will be especially inflated.

In terms of the bootstrap approach to hypothesis testing, for normally distributed data, it performs similarly to the parametric t-test. If the data do not come from a normally distributed population, the bootstrap is generally preferred to the parametric t-test. Having said this, like the parametric approach, the bootstrap-based test can have inflated Type I error rates for some non-normal distributions. This inflation will generally not be as severe, however. For heavily skewed distributions, the Yuen–Welch test performs better than the t-test in terms of both controlling the Type I error rate and maintaining higher power. But for normally distributed data, the Yuen–Welch test has lower power than the parametric t-test. Finally, the trimmed and Winsorized approaches are less negatively impacted by skewness and outliers than the parametric method and should thus be considered worthy options in such cases. However, it is also true that greater trimming of the data generally yields lower power, so care must be taken in considering the tradeoff. As discussed at the end of Chapter 3, the Bayesian paradigm offers a very different and unique way to understand the comparison of central tendency. We would certainly encourage researchers to use this framework, particularly when there is viable prior information about the nature of the group differences that could be brought to bear on the problem. In addition, we also find the application of BF to be quite informative with respect to group differences and recommend that researchers consider its use for many research problems.

5

Comparing Measures of Central Tendency among More Than Two Independent Groups

In Chapter 2, we were introduced to the robust and Bayesian techniques that we then applied in Chapter 3 to the one-sample problem. We extended that work to the comparison of means for two independent groups, which, for the standard parametric approach, is the independent samples t-test. We saw that there are a large number of options available for comparing central tendency between groups and that these alternatives perform better in a number of cases than does the parametric t-test. In particular, methods based on ranks, trimming, and Winsorizing all provide the research with tools that can be employed when the population distribution of the variable in question is unlikely to be normal and/or when the groups' variances are not equal in the population. Finally, we saw that the Bayesian paradigm also provides us with a variety of useful tools for comparing two groups with respect to a dependent variable of interest, including through the use of the Bayes Factor (BF) and the region of practical equivalence (ROPE). We will build on these ideas in Chapter 5 by extending them to the case where we have more than 2 independent groups across which we wish to compare central tendency. As we will see, a number of the techniques available to us in the two-group case will also be useful when we have more than two groups.

Motivating Example

To explore the comparison of central tendency for more than two groups, we will turn to a dataset containing math test scores for a group of 18 high school students involved in an experiment focused on the effectiveness of tutoring. The students were randomly assigned to one of three instructional groups (six in each): 1=one-on-one tutoring in-person, 2=one-on-one tutoring online, and 3=group tutoring (GT) in person. After the group assignment, the students continued with their regular math instruction for 8 weeks, along with the tutoring protocol to which they were assigned. Prior research suggests that one-on-one tutoring should yield greater gains in math performance than GT. However, it is not clear whether tutoring in-person will yield higher test scores than online tutoring (OLT). If the two approaches yield

DOI: 10.1201/9781003379324-5

TABLE 5.1

Descriptive Statistics for Math Test Score by Treatment

Treatment	Mean	Median	SD	MAD
In-person individual	96.5	96.5	9.7	9.6
Online individual	83.8	81	13.8	15.6
Group	64.5	74	28.4	23.7

comparable results, educators could conceivably provide effective tutoring to a larger group of people using an online platform. The math test scores are scaled so that the mean for the population is 100 and the standard deviation is 15.

Exploration of the Data

As with any data analysis, it is always good to begin with an exploration of the data. We can start by examining descriptive statistics of the math tests for each of the groups (Table 5.1).

The pattern of means *for this sample* follows what theory would predict: In-person individual, followed by online individual, with GT having the lowest mean and median math test scores. In addition, *for the sample* the variability in math test scores was much larger for the GT treatment than for the others. We can see this pattern in the boxplots (Figure 5.1).

Based on the results in Figure 5.3, it appears that treatment 3 (GT) exhibited a much broader range of scores than either of the other groups. It is important to remember that these observations are for the sample data only. At this point, we cannot make any inferences about the population. We will get there shortly, however.

Before beginning our discussion of statistical inference, however, we will consider one additional plot of the math test scores. This graph, known as a ridgeline plot, is essentially a set of separate density plots (similar to histograms) of the math test score for each of the treatment conditions. It allows for an easy comparison of the score distributions across the groups and can be a useful supplement to descriptive statistics and the boxplot. The ridgeline plot for the math test scores by treatment condition appears in Figure 5.2.

Perhaps the most notable feature of this graph is the high degree of overlap among math scores, with the exception of the lowest values for treatment 3 (GT). This pattern was also evident in the boxplot, but here we see that there is a clear division between the main body of test scores for group 3 and those with very low values. In other words, most of the students in the GT sample seem to have performed similarly to members of the other two groups.

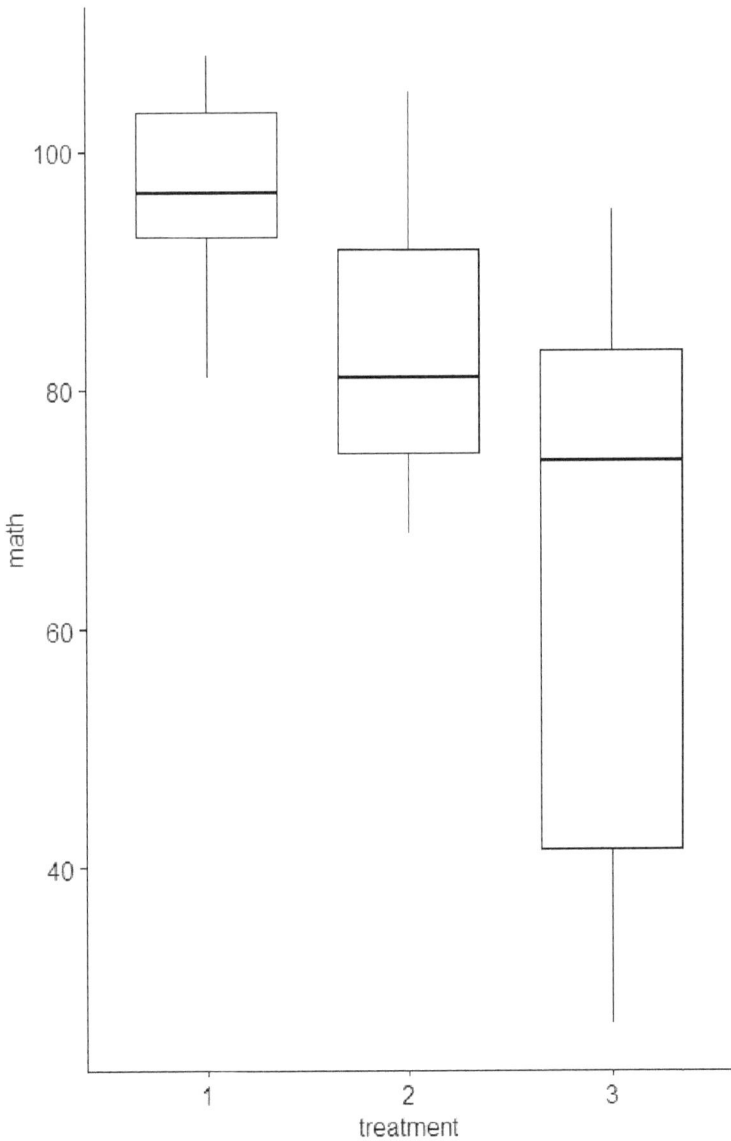

FIGURE 5.1
Boxplots of math test scores by treatment group.

However, there was a subset of these students whose performance on the math test fell below that for the rest of the sample, regardless of treatment. As we work through the approaches for inference in the following pages, this distinct pattern of scores in group 3 vis-à-vis groups 1 and 2 may inform our conclusions regarding the relative effectiveness of the treatments.

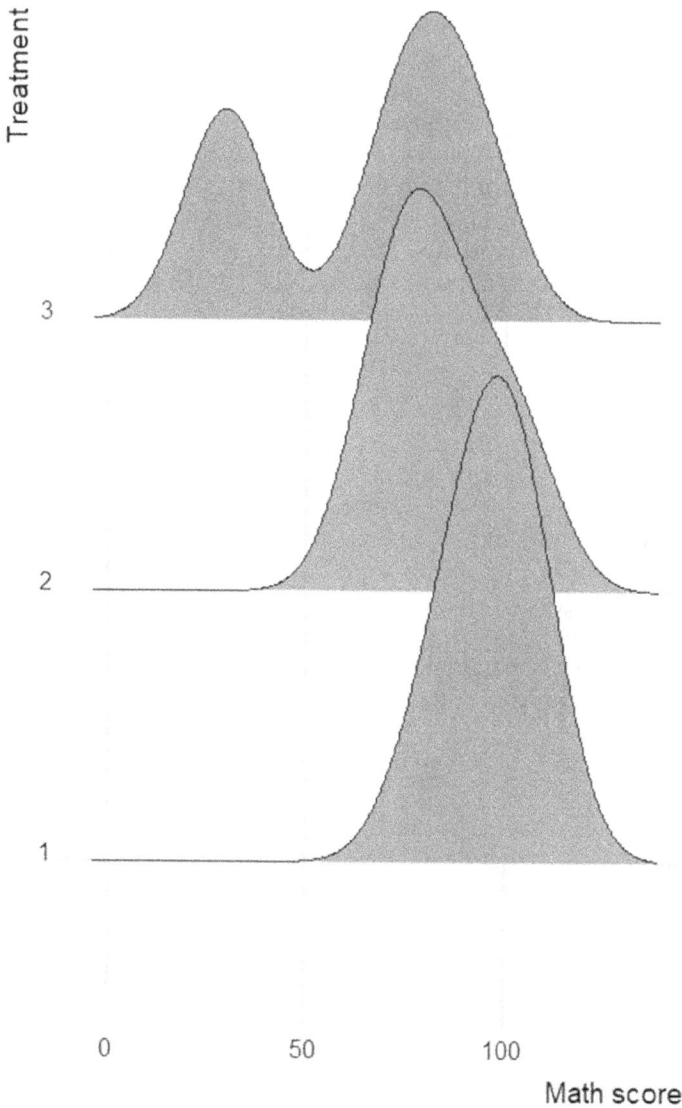

FIGURE 5.2
Ridgeline plot of math test scores by treatment group.

Analysis of Variance

In Chapter 4, our goal was to compare the means of a dependent variable (i.e., reading test scores) between two groups of an independent variable (i.e., treatment and control groups) for which we used a *t*-test. Recall that under the assumption of equal variances, the *t*-test was calculated as the ratio of the difference between the groups' means versus the pooled standard error of the mean. We can very easily extend this idea to the problem of comparing means for a dependent variable across more than two groups, using analysis of variance (ANOVA). The test statistic at the heart of ANOVA is the *F*, which, as we shall see, is a ratio of the variability between the groups' means to the variability within the groups, much like the *t*. Indeed, it is from this ratio of two variances that the name ANOVA comes.

Before we take a look at the *F* statistic, let's consider the null and alternative hypotheses for ANOVA. For our example, the null hypothesis is $H_0 : \mu_{GT} = \mu_{OLT} = \mu_{IPT}$. This null hypothesis states that the population mean math scores are equal for GT, OLT, and in-person tutoring (IPT). Thus, if we reject H_0, we would conclude that there is some difference among the groups' means. Importantly, however, we won't know where that difference lies. Is there a difference between GT and the two one-on-one approaches? Is there a difference among all three methods? A rejection of the ANOVA null will not provide us with insights in this regard. For that reason, we will need to follow a statistically significant ANOVA omnibus test with another procedure designed to determine exactly where the null hypothesis is false.

The one-way ANOVA model can be expressed as a linear equation isolating the individual sources of variability in the dependent variable.

$$y_{ij} = \mu_. + x_j + \varepsilon_{ij}, \tag{5.1}$$

where y_{ij} is the dependent variable value for person *i* in group *j*, $\mu_.$ is the overall mean for the dependent variable, x_j is the effect of being in group *j* of the independent variable *x*, and ε_{ij} is the random error for person *i* in group *j*.

We refer to this model as a one-way ANOVA because it involves a single independent variable. In the next chapter, we will further extend this model to include multiple independent variables, giving us two-way, three-way, and more ANOVA models. Equation (5.1) highlights the fact that the value of the dependent variable for any individual in the population is a function of the overall population mean of the variable ($\mu_.$), the level of the independent variable to which an individual belongs (x_j), and random error that is unique to the individual (ε_{ij}).

We are now ready to consider the *F* statistic, which is central to the ANOVA. It is calculated as

$$F = \frac{\dfrac{\sum_{j=1}^{J}\left(\bar{x}_j - \bar{x}_.\right)^2}{J-1}}{\dfrac{\sum_{i=1}^{N}\left(x_{ij} - \bar{x}_j\right)^2}{N-J}}, \tag{5.2}$$

where $\bar{x}_{.}$ is the overall mean for dependent variable, $\bar{x}_j$ is the group j mean for dependent variable, x_{ij} is the dependent variable value for individual i within group j, N is the total sample size, and J is the number of groups.

The numerator of F, known as the mean square between groups (MSB), is a measure of between group differences in the means, where larger values indicate greater differences among the groups' means. The quantity

$$\sum_{j=1}^{J} n_j\left(\bar{x}_j - \bar{x}_{.}\right)^2 \tag{5.3}$$

is known as the sum of squares between groups (SSB). The denominator contains the mean square within groups (MSW) and reflects differences in the dependent variable among individuals within the same group. The quantity

$$\sum_{i=1}^{N} \left(x_{ij} - \bar{x}_j\right)^2 \tag{5.4}$$

is the sum of squares within groups (SSW). The total sum of squares (SST) is calculated as

$$SST = \sum_{i=1}^{N} \left(x_{ij} - \bar{x}_{.}\right)^2 \tag{5.5}$$

The SST can also be calculated as SSB+SSW.

Large F values indicate that differences among the groups' means are larger than differences among individuals within the same group. As with the t-statistic, we can obtain a probability value (p-value) associated with a given F. The F distribution has two associated degrees of freedom, one for the numerator (df_1) and the other for the denominator (df_2):

$$df_1 = J - 1$$
$$df_2 = N - J \tag{5.6}$$

For our example problem, there are 2 (3−1) numerator degrees of freedom and 15 (18−3) degrees of freedom meaning that the statistic is written as $F_{2,15}$. The distribution for $F_{2,15}$ appears in Figure 5.3.

Now that we have an understanding of the core statistic used in ANOVA, let's calculate it by hand for our data. The group membership and math test scores appear in Table 5.2. In addition, the difference between each individual's score and the mean for their group appears in the Difference Within column, and the difference between the individual scores and the overall mean appears in the Difference Total column. The group and overall means appear at the bottom of the table.

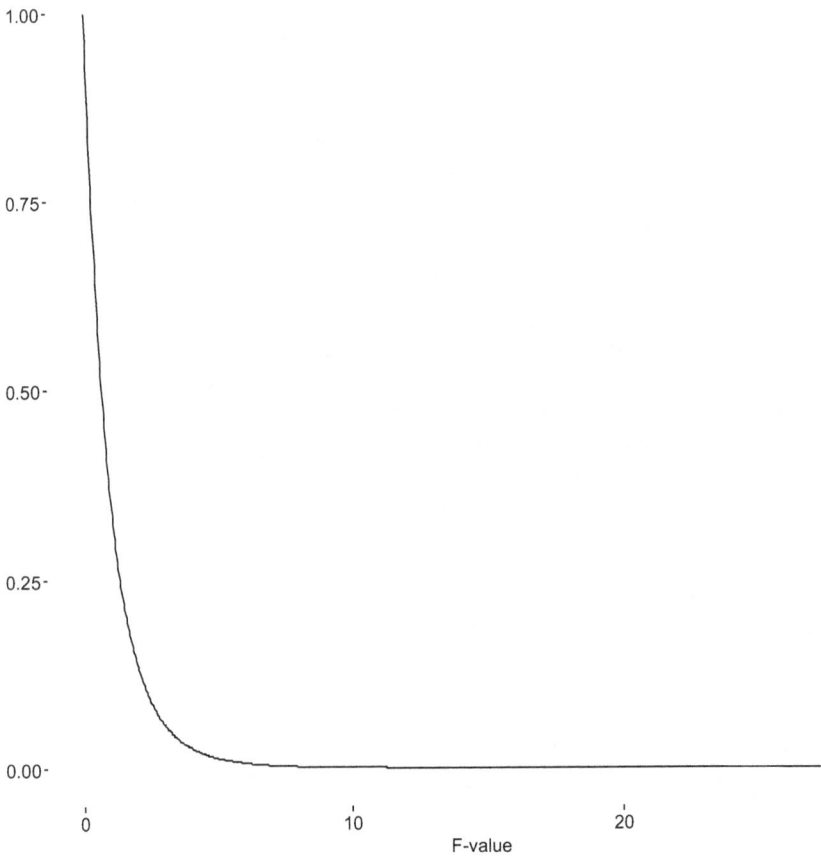

FIGURE 5.3
F distribution for 2 and 15 degrees of freedom.

The SSW is the sum of the Difference Within column, or 5,453.83, and the SST is the sum of the Difference Total column, 8,570.28. We can then calculate SSB as SST − SSW = 8,570.28 − 5,453.83 = 3,116.44. We can use these values to calculate $MSW = \dfrac{SSW}{N-J} = \dfrac{5,453.83}{18-3} = 363.59$ and $MSB = \dfrac{SSB}{J-1} = \dfrac{3,116.44}{3-1} = 1,558.22$. Taking these values, $F = \dfrac{MSB}{MSW} = \dfrac{1,558.22}{363.59} = 4.29$.

Given a particular F with numerator and denominator degrees of freedom, we can obtain the probability of getting such a value when the null hypothesis of no group mean differences is true. Just as for the one- and two-sample t-tests, when this p-value is less than a predetermined cut-off (e.g., $\alpha = 0.05$), we conclude that the null hypothesis is unlikely to be true in the population

TABLE 5.2

Group Membership and Math Test Scores

Treatment	Math Score	Difference Within	Difference Total
1	108	132.25	696.3734568
1	105	72.25	547.0401235
1	98	2.25	268.595679
2	105	448.0277778	547.0401235
2	95	124.6944444	179.2623457
2	73	117.3611111	74.15123457
3	78	182.25	13.04012346
3	85	420.25	11.4845679
3	95	930.25	179.2623457
1	92	20.25	107.9290123
1	95	2.25	179.2623457
1	81	240.25	0.37345679
2	80	14.69444444	2.595679012
2	68	250.6944444	185.2623457
2	82	3.361111111	0.151234568
3	27	1406.25	2982.373457
3	70	30.25	134.8179012
3	32	1056.25	2461.262346
Group 1 mean	96.50		
Group 2 mean	83.83		
Group 3 mean	64.50		
Overall mean	81.61		

and therefore reject it. We can visualize this result as the region greater than the vertical line in Figure 5.4. This region corresponds to 0.0337 (3.37%) of the F distribution with 2 and 15 degrees of freedom. In other words, the probability of obtaining the result that we got, if the null hypothesis of equal group means is true, is 0.0337. Given that this value falls below the predetermined α of 0.05, we will reject $H_0 : \mu_{GT} = \mu_{OLT} = \mu_{IPT}$. This leads to the conclusion that there is some combination of groups that have different math test score means. This may correspond to a pair of groups, two groups versus one other group, or all of the groups' means being different. At this point, we cannot say. We will learn shortly about follow-up procedures that allow us to ascertain how the groups' means differ.

The F of 3.68 is associated with the a of 0.05. Thus, any value larger than this has associated with it a p-value less than 0.05. For our math tutoring example, $F = 4.29$, which lies above 3.68, leading us to reject the null hypothesis of no group mean difference.

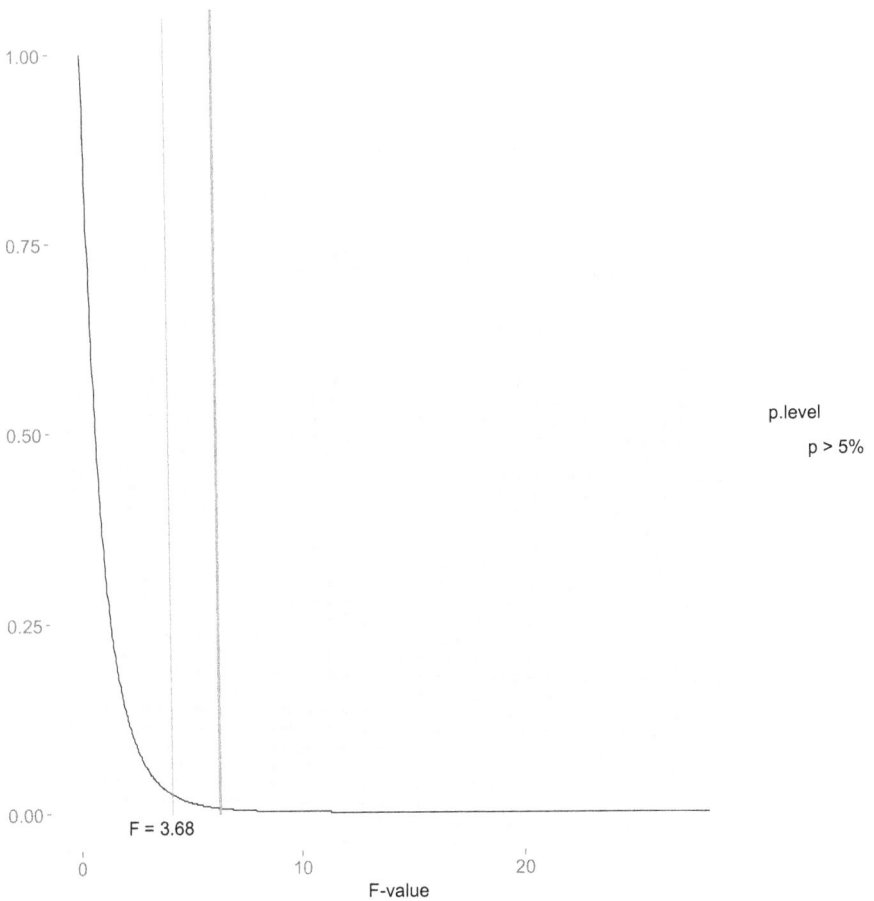

FIGURE 5.4
The F distribution with 2 and 15 degrees of freedom, F associated with $\alpha=0.05$ (3.68), and F associated with 4.29 in orange.

Table 5.3 displays the ANOVA results using the format in which they are typically presented. We see the degrees of freedom, sums of squares, mean squares, F, and p-value.

Assessment of Assumptions

Prior to discussing the pairwise comparisons follow-up procedures, we should first assess the assumptions underlying the use of the F-test. These assumptions are basically identical to those for the independent samples t-test, which we reviewed in Chapter 4. They include independence of the individuals in

TABLE 5.3

Results of One-Way ANOVA for Comparing Math Test
Score Means

Variable	DF	SS	MS	F	p
Tutoring method	2	3116.0	1558.2	4.29	0.03
Error	15	5454.0	363.6		

the sample after accounting for group membership, normality of the outcome variable, and equality of the variances across the groups. The methods used to assess these assumptions are also the same for ANOVA as they were for the *t*-test. For this reason, we will not review them in detail here but instead simply apply them to the current problem. If you need a quick reminder for how these methods work, please take a moment and review those details in Chapter 4. We will assess the normality assumption using the QQ plot and the Shapiro–Wilk test for normality. The QQ plot appears in Figure 5.5.

We can see that most of the data fall along the line that corresponds to the normal distribution and within the 95% confidence bounds. Two points do fall away from the line and outside the shaded confidence region for normality at the lower end of the distribution. We can also see these cases in Figure 5.6, as well as the fact that the data appear to be somewhat negatively skewed.

Indeed, the skewness estimate for this sample is −1.36, and the result of the Shapiro–Wilk test is $W = 0.86$ with $p = 0.01$. Because the *p*-value for the test is less than 0.05, we will reject the null hypothesis that the math test scores follow a normal distribution. Our sample size of 18 is unlikely to be large enough for us to rely on the central limit theorem to assume that the sampling distribution of the mean will behave as if it were normal. Put more succinctly, the data do not support the assumption that the math test scores come from a normally distributed population.

We must also assess the assumption of homogeneity of group variances. For ANOVA, Levene's test is used to test the null hypothesis that the groups' variances are equal in the population: $H_0 : \sigma^2_{GT} = \sigma^2_{OLT} = \sigma^2_{IPT}$. For the math tutoring data, the test statistic value is 2.20 with $p = 0.15$. We therefore don't reject the null hypothesis and conclude that the variances are likely to be equal in the population. We will assume independence of observations given the random selection of our sample.

Post Hoc Testing

As we have already discussed, the rejection of the null hypothesis in the context of ANOVA with more than two groups is not particularly informative. This result does tell us that the population means are likely to be different among the groups. However, we don't know much about what those

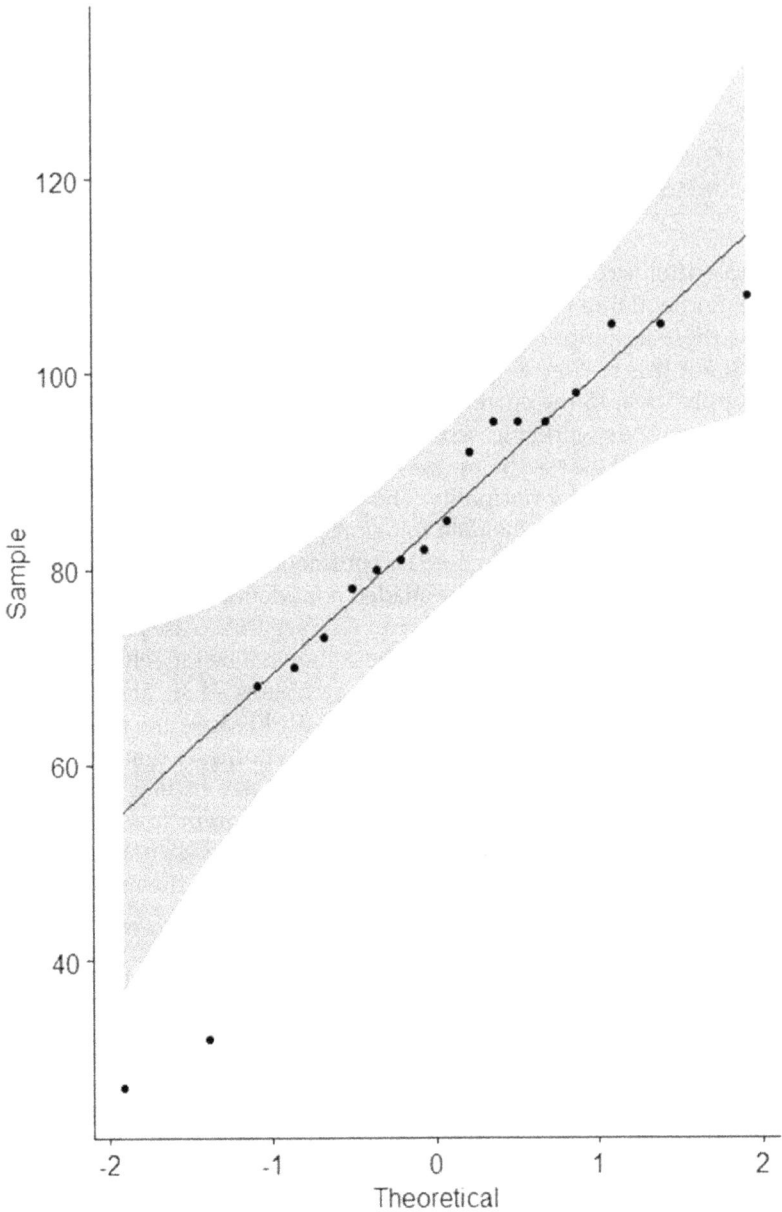

FIGURE 5.5
QQ plot for math test scores.

differences might be. Is the mean math test score higher for the IPT group versus the GT group? Or is the mean higher for the two one-on-one groups compared to the GT condition? Or are all of the groups' means different from one another? The rejection of the null hypothesis in ANOVA provides no

Histogram of Math Test Scores

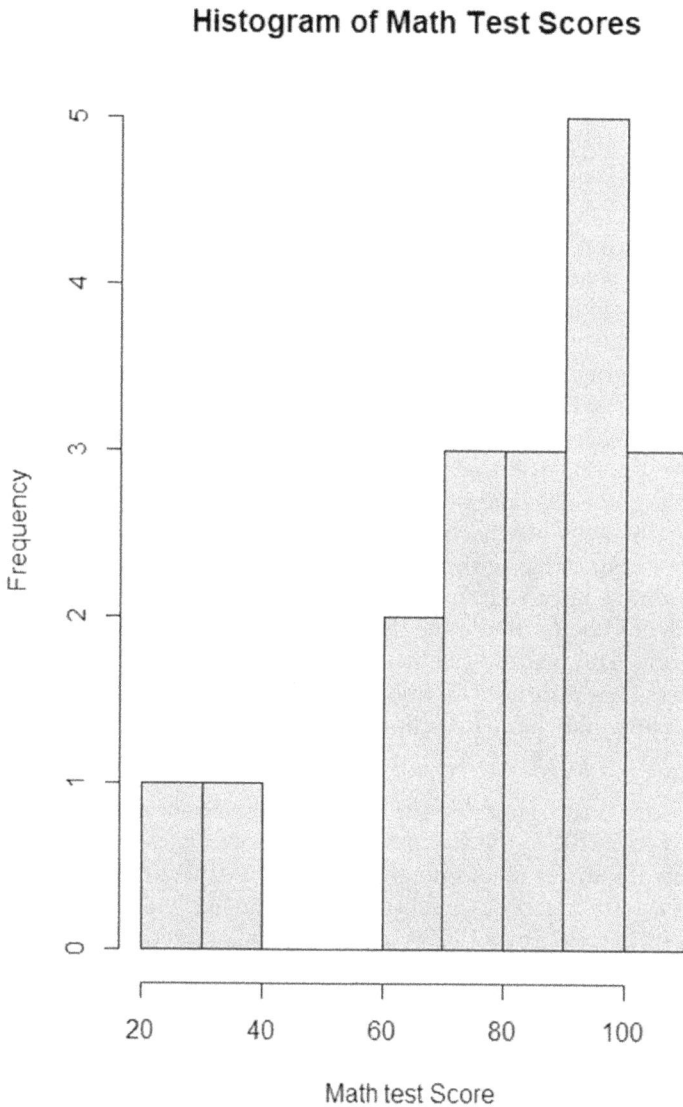

FIGURE 5.6
Histogram of math test scores.

insights into these questions. Therefore, we typically follow a statistically significant ANOVA with a post hoc comparison procedure, of which there are many. We will focus on a small number of them here, featuring those that are the most flexible and most widely used. However, the reader should know that there are many, many more possibilities than those discussed here.

An important issue that we need to be aware of when conducting such procedures is that repeatedly applying statistical tests to the same set of data leads to inflation of the Type I error rate, which we introduced in Chapter 1. Remember that a Type I error occurs when we reject the null hypothesis when in fact it holds in the population. For any given statistical test, we set the Type I error rate to be α, which is most often 0.05. When we conduct a series of c statistical tests on the same sample, however, the probability of having at least one Type I error in the set (sometimes referred to as the familywise error rate) is not 0.05 but rather $1-(1-\alpha)^c$. Thus, if we conducted a t-test to compare every pair of means for the math test example, we would have three such comparisons (group vs online, group vs in-person, and online vs in-person). If we set $\alpha = 0.05$, then the familywise error rate would be $1-(1-0.05)^3 = 1-0.86 = 0.14$. In other words, the probability of making at least one Type I error across the three hypothesis tests is 0.14, which is well above the nominal 0.05 that we use for each individual test. The purpose behind using special post hoc comparisons is to allow the data analyst to explore the reasons behind the rejection of the omnibus null hypothesis in ANOVA without inflating the Type I error rate for the family of tests.

A very simple approach for controlling the Type I error rate in such comparisons would be to divide the desired α level (e.g., 0.05) by the number of comparisons. The resulting value would then serve as the α for each individual test. This approach is known as the Bonferroni correction, named after the mathematician who proposed it. For our example, the Bonferroni α would be $\dfrac{0.05}{3} = 0.017$. We could then use three t-tests (one-on-one in-person vs one-on-one online; one-on-one in-person vs group; one-on-one online vs groups) to make all of the pairwise comparisons necessary to know which groups' means differ from one another, using 0.017 as our α for each test. Although the Bonferroni correction is simple to apply and does indeed control the familywise Type I error rate, it has also been shown to have low power for detecting mean differences when they are present in the population (Jennions & Møller, 2003). For the math test example, the p-values for the pairwise comparisons using the independent samples t-test are 0.10 (GT vs OLT), 0.04 (GT vs IPT), and 0.18 (OLT vs IPT). None of these p-values are less than 0.017, meaning that we cannot conclude that there are any statistically significant pairwise differences when we apply the Bonferroni correction.

Holm (1979) proposed an adjustment to the Bonferroni approach involving sequentially correcting the α values by updating the denominator in the ratio. Here's how it works. We start by obtaining p-values for each of the three t-tests comparing the math test means for our groups. We then order these from the smallest to the largest. We then compare the smallest p-value to 0.017. If this result is statistically significant; i.e., $p \leq 0.017$, we would then compare the second smallest p-value to $\dfrac{0.05}{2} = 0.025$. If the second smallest p-value is less than 0.025, then we would compare the largest of the three

p-values to 0.05 $\left(\dfrac{0.05}{1} = 0.05\right)$. If, at any point in the sequential comparisons, we do not have a statistically significant result, we would stop the procedure. The ordered math tutoring p-values are 0.04, 0.10, and 0.18. The smallest of these is larger than 0.017, again meaning that we cannot reject the first null hypothesis and will not make any further comparisons. This result leads us to conclude that there are no statistically significant pairwise differences when we use the Holm procedure.

The third approach that we will consider here is the excellently named Tukey's honestly significant difference (HSD) procedure, which is based on the studentized range statistic:

$$q = \frac{\bar{x}_j - \bar{x}_k}{\sqrt{\dfrac{1}{2}\text{MSW}\left(\dfrac{1}{N_j} + \dfrac{1}{N_k}\right)}} \tag{5.7}$$

where $\bar{x}_j$ is the mean for group j, $\bar{x}_k$ is the mean for group k, N_j is the number of cases in group j, N_k is the number of cases in group k, and MSW is the mean square within groups from the ANOVA.

The resulting q is then treated as a test statistic in the same way as for the t-test. The value is compared to the sampling distribution for the studentized range statistic with $N - J$ degrees of freedom (equivalent to the within groups df used with the F statistic). Tukey's HSD can be used to compare every possible pair of means in the sample while maintaining the familywise error rate at the desired α (e.g., 0.05).

It is also possible to construct a 95% confidence interval for the difference between a pair of means and use that as a follow-up to a rejection of the omnibus ANOVA null hypothesis. In this case, we check to see whether 0 falls within the interval, and if not, we conclude that the groups' means are likely to be different in the population. The confidence interval associated with HSD is calculated as

$$\left(\bar{x}_j - \bar{x}_k\right) \pm q_{N-J}\sqrt{\dfrac{1}{2}\text{MSW}\left(\dfrac{1}{N_j} + \dfrac{1}{N_k}\right)} \tag{5.8}$$

where q_{N-J} is the critical value for the studentized range statistic with $N - J$ degrees of freedom.

The 95% confidence intervals of the pairwise mean differences based on Tukey's HSD appear below:

$$\mu_{GT} - \mu_{OLT} : (-41.27, 15.94)$$

$$\mu_{GT} - \mu_{IPT} : (-60.61, -3.40)$$

$$\mu_{OLT} - \mu_{IPT} : (-47.94, 9.27)$$

Based on these results, we would conclude that there is a statistically significant difference between the GT and IPT means because 0 is not contained within the confidence interval. Table 5.1 reveals that the in-person group has higher mean math test scores than does the GT condition. There were no other significant mean differences, given that 0 is in the other two confidence intervals.

It is worth stating again that there are a large number of other post hoc comparison methods available to researchers. Of the three methods that we have considered here, Tukey's HSD may be the best option for most scenarios. It has been found to provide good control of the familywise error rate while at the same time yielding higher power than either Bonferroni or Holm (e.g., Jennions & Møller, 2003). These latter two approaches are worth knowing about, however, as they can be applied in a much wider array of contexts beyond just ANOVA, whereas HSD is limited to the comparison of means. Other post hoc techniques allow for treating one group as the focus (e.g., a control group) against which the others are compared. In addition, there is a set of approaches designed to control the false discovery rate (FDR), which is the probability that a rejection of the null hypothesis represents a false positive. The FDR is always less than or equal to the familywise error rate. This means that when we control the familywise error rate, we are also controlling the FDR, but the converse is not necessarily true.

Planned Comparisons

A final aspect of ANOVA that may be quite useful in some situations involves making planned comparisons among the means. For example, it may be interesting to know whether there is a population difference in the math test mean for the GT students versus the mean for those receiving one-on-one tutoring, whether it be in-person or online. We can make such a comparison using planned contrasts. These contrasts involve weighting each mean, summing these weighted means across the groups, and then dividing by a standard error. This equation is

$$t_c = \frac{\sum_{j=1}^{J} w_j \bar{x}_j}{\sqrt{\mathrm{MSW} \sum_{j=1}^{J} \frac{w_j^2}{n_j}}} \tag{5.9}$$

where $\bar{x}_j$ is the mean for group j, w_j is the weight for mean j, and n_j is the group j sample size.

The weights in equation (5.9) are selected so that they reflect the comparison we want to make with the test statistic. The w_j values must sum to 0 across the groups, but otherwise, there are no rules about what these values must take. We will use a common approach to setting these values in the example below.

Returning to the math tutoring example, we would like to compare group 3 (GT) with the combination of groups 1 and 2 (in-person and OLT). The null hypothesis for this test would be

$$H_0 : \mu_{GT} = \left(\frac{\mu_{OLT} + \mu_{IPT}}{2} \right)$$

We can use the weights –1, 0.5, and 0.5 to make this comparison. The numerator for t_c would then be

$$0.5 * 96.5 + 0.5 * 83.8 - 1 * 64.5 = 25.67$$

Using the MSW from the ANOVA table (Table 5.3) along with the sample sizes, the denominator would be

$$\sqrt{363.59 \left[\frac{0.5^2}{6} + \frac{0.5^2}{6} + \frac{-1^2}{6} \right]} = \sqrt{363.59 \left[0.04 + 0.04 + 0.17 \right]} = 9.53$$

The test statistic is then

$$t_c = \frac{25.67}{9.53} = 2.69$$

The p-value is obtained by referring this value to the t-distribution with $N - J$ degrees of freedom. For a value of 2.69, $p = 0.0167$, which is less than our α of 0.05. We reject the null hypothesis that the mean of the GT approach is equal to that of the combined one-on-one tutoring conditions. An examination of the means in Table 5.1 shows that those who participated in one-on-one tutoring (regardless of the mode) exhibited higher math performance than did those in the GT condition. In other words, being tutored one-on-one appears to yield better performance on the math test when compared with GT.

Welch ANOVA

We saw in Chapter 4 that when the assumption of equal group variances is violated, we have an alternative for comparing group means that does not rely on pooling the groups' standard errors. This Welch t-test has an analog in the context of ANOVA that was described by Welch (1951). Specifically, it is based on the heteroscedastic F statistic, which is calculated as

$$F_w = \frac{\sum_{j=1}^{J}\left(\frac{n_j}{s_j^2}\right)\left(\frac{\left(\bar{x}_j - x'\right)^2}{J-1}\right)}{\left[1+\frac{2}{3}\left((J-2)v\right)\right]}$$

(5.10)

where

$$x' = \frac{\sum_{j=1}^{J}\left(\frac{n_j}{s_j^2}\right)\bar{x}_j}{\sum_{j=1}^{J}\left(\frac{n_j}{s_j^2}\right)}$$

$$v = \frac{3\sum_{j=1}^{J}\left[\left(1-\frac{\left(\frac{n_j}{s_j^2}\right)}{\sum_{j=1}^{J}\left(\frac{n_j}{s_j^2}\right)}\right)^2 / (n_j - 1)\right]}{J^2 - 1}$$

This is certainly a nasty-looking equation, but in reality, it is fairly straight-forward. The numerator of F_w is simply a weighted version of the MSB, where the weights $\left(\frac{n_j}{s_j^2}\right)$ are based on both the group sample size and the group variance. The denominator replaces the standard MSW again using the weighted sum of the group variances in the form of $\left(\frac{n_j}{s_j^2}\right)$. Thus, although it doesn't look much like the standard F, F_w actually just weights the MSB and MSW by the individual group variances and sample sizes to yield an analog to Welch's t. The degrees of freedom for F_w are $J-1$ and $1/v$.

Although the assumption of homogeneous variances was met for the math tutoring data, we will demonstrate its use here for pedagogical purposes. It is important to note that when the variances are homogeneous, F_w will generally have lower statistical power than the standard F statistic (Welch, 1951). For our example, $F_w = 4.06$ with 2 numerator and 8.97 denominator degrees of freedom, yielding a p-value of 0.055. Because this p is not less than our α of 0.05, we would not reject the null hypothesis that the population means are equal across groups. Recall that the standard F was 4.29 with a p-value of 0.0337, which led to our rejecting the null hypothesis. In addition, remember that the denominator degrees of freedom in the standard case was 15, whereas for F_w, it is 8.97. It is this difference, as much (or more) than the value of the test statistic itself that leads to the difference in the p-values. If the Welch F-test had been statistically significant we would have followed it up with the Welch *t*-test controlling for the Type I error rate using the Holm or Bonferroni procedure. Because F_w was not significant in this case, we won't perform any follow-up tests.

Kruskal–Wallis Test

Another alternative to the parametric *F* statistic involves the use of ranks. As with the one- and two-sample cases, using ranks in place of the raw data can be quite helpful when we are faced with outliers and/or a highly skewed sample. In particular, we can extend the rank-based Wilcoxon test for comparing population medians from Chapter 4 to the case when we have more than two groups, using the Kruskal–Wallis (KW) test. This technique involves first ranking all members of the sample based on the values for the dependent variable, without regard to group membership. We can then calculate the *H* statistic testing the null hypothesis that the population distributions of the ranks are the same across groups. If we can assume that the groups have the same variability in ranks, then the null hypothesis of KW is that the median of the dependent variable is the same across groups: $H_0 : M_{GT} = M_{OLT} = M_{IPT}$. The test statistic is calculated as

$$H = \frac{12}{N(N+1)} \sum\nolimits_{j=1}^{J} n_j \left(\bar{R}_j - \frac{N+1}{2} \right)^2 \tag{5.11}$$

where $\bar{R}_j$ is the mean rank for group *j*, *N* is the total sample size, and n_j is the sample size for group *j*.

H is asymptotically distributed as a chi-square statistic with $J-1$ degrees of freedom. If the KW test is statistically significant, the Wilcoxon test is used as the follow-up to determine which groups' medians are different in the population, with the familywise Type I error rate controlled using an approach such as Holm or Bonferroni.

The math scores and their ranks appear in Table 5.4, as do the mean ranks for each group.

We can use these scores to calculate *H* as

$$H = \frac{12}{18(18+1)} \left(6*\left(5.17 - \frac{18+1}{2}\right)^2 + 6*\left(9.83 - \frac{18+1}{2}\right)^2 + 6*\left(12.83 - \frac{18+1}{2}\right)^2 \right)$$

$$= \frac{12}{18(18+1)} (112.67 + 0.67 + 66.67) = 6.32$$

We can then compare 6.32 to the chi-square distribution with $3-1=2$ degrees of freedom, which yields $p = 0.045$. Given that the *p*-value is less than 0.05, we will reject the null hypothesis that the groups' medians are equal in the population.

TABLE 5.4

Group Membership, Raw, and Ranks of Math
Test Scores

Treatment	Math Score	Rank
1	108	1
1	105	2
1	98	4
2	105	2
2	95	5
2	73	14
3	78	13
3	85	9
3	95	5
1	92	8
1	95	5
1	81	11
2	80	12
2	68	16
2	82	10
3	27	18
3	70	15
3	32	17
Group 1 Rank mean	5.166667	
Group 2 Rank mean	9.833333	
Group 3 Rank mean	12.83333	

As we discussed above, we can follow up a statistically significant KW test with the Wilcoxon test and control for the Type I error rate using the Bonferroni correction. We will not work through that example in detail here, but the interested reader is encouraged to do so if they want to practice with the Wilcoxon test. The follow-up comparisons revealed that the population median for the IPT group was significantly larger than the population median for those receiving GT ($p = 0.04$, Bonferroni corrected). No other pairwise difference was found to be statistically significant.

Bootstrap-based Analysis of Variance

The bootstrap can be extended in a straightforward manner to compare central tendency for more than two groups. The basic approach to the percentile bootstrap for this purpose is very similar to what we used for the two-sample case in Chapter 4. The steps are as follows:

1. Use ANOVA to obtain the F statistic for the observed data.
2. Create B bootstrap samples by sampling with replacement.
3. Replace the raw data with $C_{ij}^* = x_{ij} - \bar{x}_j$, which represents a sample from a distribution where there is not a difference in means among the groups. Notice that we are centering the data for each group, thereby forcing the means to be the same across groups but retaining the shape and spread of our original sample distribution.
4. Compute the F statistic using standard ANOVA for each bootstrap sample based on C_{ij}^*.
5. Compare the F based on the observed data (step 1) with the distribution of F from steps 2–4.
6. The p-value is the percentile of the observed F in this bootstrap distribution.

The basic idea behind using the percentile bootstrap as described above is that we create a distribution of the F statistic for our data assuming that there is no group mean difference to which we then compare the observed F. This null distribution is created by centering the individual data points around their group's mean.

For the math tutoring data, the observed F was 4.286. We then conducted the percentile bootstrap approach with 10,000 bootstrap samples. The resulting p-value for the test was 0.108, leading us not to reject the null hypothesis. Thus, we would conclude that there is not sufficient evidence to conclude that the population mean math test scores differ across the various tutoring groups. Had the bootstrap ANOVA results been statistically significant, we could have followed it up using a series of bootstrap t-tests for the various group pairs, employing the Bonferroni or Holm corrections to control the familywise Type I error rate.

Comparison of Trimmed Means for More Than Two Groups

In Chapters 2–4, we described the advantages of using trimming to deal with outliers when we are interested in making inferences about the population central tendency in the one- and two-group cases. As you might expect, it is fairly straightforward to extend the use of trimmed means to situations in which we want to compare central tendency for more than two groups. *In that case, the null hypothesis is about the population trimmed means, rather than the standard means for the full population.* In particular, for the math tutoring data $H_0 : \mu_{\text{Tr,GT}} = \mu_{\text{Tr,OLT}} = \mu_{\text{Tr,IPT}}$. The key statistic underlying the comparison of trimmed means is an adjusted F based on the Winsorized variances accounting for the amount of the data that is trimmed. The equation for this statistic is

$$F_T = \frac{\dfrac{1}{J-1} \sum\limits_{j=1}^{J} w_j \left(\bar{x}_{Tj} - \tilde{x} \right)^2}{\dfrac{2(J-2)}{J^2-1} \sum\limits_{j=1}^{J} \left(\dfrac{1-\dfrac{w_j}{u}}{h_j-1} \right)} \tag{5.12}$$

where $\bar{x}_{Tj}$ is the trimmed mean for group j,

$$w_j = \frac{1}{d_j}$$

$$d_j = \frac{(n_j-1)S_{wj}^2}{h_j(h_j-1)}$$

$$u = \sum_{j=1}^{J} w_j$$

$$\tilde{x} = \frac{1}{u} \sum_{j=1}^{J} w_j \bar{x}_{Tj}$$

h_j is the number of the sample not trimmed for group j, n_j is the number in group j prior to trimming, and S_{wj}^2 is the Winsorized variance for group j.

As with many of the other equations in the book, this one looks intimidating. However, in reality, it is fairly straightforward to understand. As with the standard F statistic, F_T compares differences in central tendency across groups in the numerator versus variability within the groups in the denominator. The numerator involves calculating the difference between the trimmed group means ($\bar{x}_{Tj}$) with a measure of central tendency for the entire sample ($\tilde{x}$). This latter value is a weighted (w_j) combination of the individual group means, where the weights reflect the inverse of the variability within each group as measured by its Winsorized variance (S_{wj}^2). In other words, groups that have more differences among individuals within them will receive a lower weight in the calculation of the overall measure of central tendency. Finally, groups that have had fewer individuals trimmed from their sample (h_j) will have their means upweighted by d_j. The resulting statistic, F_T, follows the F distribution with the numerator degrees of freedom being $J-1$ and the denominator degrees of freedom being

$$\left[\frac{3}{J^2-1} \sum_{j=1}^{J} \frac{\left(1-\dfrac{w_j}{\displaystyle\sum_{j=1}^{J} w_j}\right)^2}{h_j-1} \right]^{-1} \tag{5.13}$$

For the math tutoring data, $F_T = 2.96$ with 2 and 5.11 degrees of freedom. The p-value associated with this statistic is 0.1397, which is larger than our α of 0.05. Thus, we don't reject $H_0 : \mu_{TrGT} = \mu_{TrOLT} = \mu_{TrIPT}$ and conclude that there is no evidence of a population difference in the trimmed population math test means across the groups. The group means and medians for the original and trimmed data appear in Table 5.5.

Perhaps the most notable pattern apparent in Table 5.5 is that the trimmed mean and median for those receiving GT were much larger than those in the raw data. This outcome suggests that this group had multiple outlying observations that were removed when we trimmed the extreme values from our sample. Furthermore, the trimmed mean and median for the in-person individual group were somewhat lower than those for the raw data, and the raw and trimmed mean and median for the online individual group were quite similar to one another. Thus, it would seem that there were not outliers in the full sample for this group.

Had the trimmed test been statistically significant, we would have followed it up with a series of trimmed means comparisons for the various group pairs and controlled the Type I error rate using Rom's method (1990). This approach for controlling the familywise Type I error is similar in some respects to Holm's approach, which we described in Chapter 4, in that the p-values for the individual pairwise comparisons are ordered. In the case of Rom's correction with trimmed mean, the p-value for each pairwise comparison based on a standard trimmed means t-test is first calculated. These values are then ordered from largest to smallest. If the largest p is less than 0.05, all pairwise comparisons are deemed to be statistically significant; i.e.,

TABLE 5.5

Raw and Trimmed Means and Medians for Math Test Score by Treatment

Treatment	Raw Mean	Raw Median	Trimmed Mean	Trimmed Median
In-person individual	96.5	96.5	89.3	92
Online individual	83.8	81	82.5	81
Group	64.5	74	86	85

all groups' trimmed means are determined to differ. If the largest p is greater than 0.05, this pair of means is determined not to be significantly different, and the second largest p is compared to 0.025, and if smaller, then the pair of trimmed means is determined to be significantly different. Such sequential comparisons are made between the p-value for a trimmed means comparison and a predetermined α value.

Recall that a research question of interest for the math tutoring data example involved comparing the performance of individuals who received one-on-one tutoring versus those who received GT. In the parametric ANOVA framework, we used a planned contrast in which coefficients were applied to the individual means, and the sum of these weighted means was divided by a measure of variability. The same basic approach can be used in the context of trimmed means. Indeed, the rules around the coefficients are the same as for the parametric case. The test statistic for the contrast is

$$
T = \frac{\displaystyle\sum_{j=1}^{J} c_j \bar{x}_{tj}}{\displaystyle\sum_{j=1}^{J} \frac{c_j^2 (n_j - 1) s_{wj}^2}{h_j (h_j - 1)}}
\tag{5.14}
$$

where c_j is the coefficient for group j, $\bar{x}_{tj}$ is the trimmed mean for group j, n_j is the total sample size for group j, h_j is the retained sample size for group j after trimming, and s_{wj}^2 is the Winsorized variance for group j.

The comparison of trimmed means for groups 1 and 2 versus 3 yielded a test statistic value of 1.43, which is less than the critical value of 3.01, meaning that we do not reject the null hypothesis of equality between the combined means for the in-person and online one-on-one tutoring and the mean for the GT individuals. This result differs from the planned contrast for the full sample, which was statistically significant. Therefore, when potential outliers are removed from the sample, there does not appear to be a statistically significant difference in the math test performance for the one-on-one and GT methods.

Comparison of Medians for More Than Two Groups Using McKean–Schrader Estimate of the Standard Error

If we are interested in comparing the medians for more than two groups, we can use an approach based on McKean and Schrader's (1984) estimate of the standard error for the median. Given their resistance to outliers (high breakdown point), we might be particularly interested in comparing population medians when extreme values are an issue in our sample. McKean and Schrader proposed an estimator for the standard error of the median, which can in turn be used to calculate an F statistic to test the null hypothesis about the equality of the groups' medians, such as for our math tutoring problem: $H_0 : M_{GT} = M_{OLT} = M_{IPT}$. This test statistic takes the form:

$$F_M = \frac{\frac{1}{J-1} \sum_{j=1}^{J} w_j \left(M_j - \tilde{M}\right)^2}{1 + \frac{2(J-2)}{J^2-1} \sum_{j=1}^{J} \frac{\left(1 - \frac{w_j}{u}\right)^2}{n_j - 1}} \tag{5.15}$$

where M_j is the median for group j, n_j is the sample size for group j,

$$w_j = \frac{1}{SE_j^2}$$

SE_j is the standard error for median of group j,

$$u = \sum_{j=1}^{J} w_j$$

$$\tilde{M} = \frac{1}{u} \sum_{j=1}^{J} w_j M_j.$$

An examination of equation (5.15) reveals that F_M involves the ratio of a weighted comparison of the difference between group medians from an estimate of the overall median in the numerator to a measure of within-group variability in the form of the standard error for the medians. We compare F_M to the F distribution with DF1$=J-1$ and DF2$=\infty$.

For the tutoring problem, $F_M = 2.28$ with $p=0.117$. Therefore, we would not reject H_0, leading us to conclude that there is not sufficient evidence to assert that the population medians for the three groups differ from one another. The sample medians for the groups appear in column 2 of Table 5.5. The standard errors for the group medians are 5.24 (in-person), 7.18 (online), and 13.20 (group), respectively. We should make two final points regarding comparing medians across multiple groups. First, if the omnibus test is statistically significant, indicating statistically significant differences in medians across groups, we would follow up with pairwise median comparisons using the approach based on Rom's method that was outlined above for the trimmed means. The Type I error rate would then be controlled in the standard way, with the Holm or Bonferroni procedures. The second issue to note here is that the McKean–Schrader standard errors are not accurate when there are ties in the data, and the problem becomes more acute for samples with more ties. Therefore, it is recommended that F_M not be used in that case. This is a potentially major limitation in the application of the McKean and Schrader test that data analysts should be aware of. In the case of many ties in the sample, we recommend using the KW to compare medians across groups.

Permutation Test for Comparing Means from More Than Two Groups

It is possible to test $H_0 : \mu_{GT} = \mu_{OLT} = \mu_{IPT}$ using a permutation-based approach. The methodology is essentially the same as for the two-sample case outlined in Chapter 4. Basically, the algorithm will compare the observed F statistic with a distribution of F values based on randomly ordered data, which reflects what we would expect if H_0 is true in the population. Formally, the steps for the permutation-based approach appear below.

1. Compute the F statistic for the observed data using ANOVA.
2. Compute an F^* for each possible permutation of the data to create a distribution of F for the case when the null hypothesis of no group differences is true.
3. Compare the F for the observed data (step 1) with the distribution of F^*.
4. Reject the null hypothesis of no group differences if $F \geq F^*$ at the 95th percentile. The p-value is the percentile in the F^* distribution where F for the observed sample falls.

Pairwise comparisons can be made using a permutation-based approach akin to the two-sample technique described in Chapter 4. The error rate can be controlled using a variety of approaches, including the FDR, Holm, or Bonferroni.

The permutation test for the math tutoring data yielded an exact p-value of 0.021. Thus, based on this test, we would reject the null hypothesis and conclude that there were mean differences on the math test among the groups. Based on the pairwise comparisons using the FDR correction, there were no statistically significant differences found. Therefore, although the omnibus test was statistically significant, when we consider specific group pairs, there were not pairwise group differences. This result is not as surprising as it may seem. Remember that the omnibus test assesses a different hypothesis than the pairwise tests. Thus, although this type of result is not common, it is not unprecedented. Nonetheless, this combination of a statistically significant omnibus test with no significant pairwise comparisons does present a quandary in terms of how to report our findings. We will consider effect sizes later in this chapter, and these may be helpful in characterizing the overall ANOVA result. Given the results thus far, we would conclude that there is not sufficient evidence, based on the permutation approach, to make any definitive conclusions regarding differences in math test means across the various tutoring groups.

Bayesian Comparison of Means for More Than Two Groups

The Bayesian-based approaches to investigating group mean differences that we discussed in terms of two groups (Chapter 4) are also applicable when we have more than two groups. Given that we discussed the technical details for these techniques in Chapters 3 and 4, we will not review them here but instead will go straight to their application with ANOVA, using the math tutoring data as an example. First, we will obtain the BF for each of several hypotheses and compare them with one another. Our primary interest with the math data is to determine whether there are differences in the test performance for individuals who receive different types of tutoring. Recall that our null hypothesis is of no such difference in group means: $H_0 : \mu_{IPT} = \mu_{OLT} = \mu_{GT}$. Based on prior research, we have some expectations that in-person one-on-one tutoring will yield the best results, followed by one-on-one OLT, with GT being the least effective. We can express this hypothesis as: $H_1 : \mu_{IPT} > \mu_{OLT} > \mu_{GT}$. We were also interested in assessing whether receipt of one-on-one tutoring, regardless of how it is disseminated, will yield higher means than GT: $H_2 : \mu_{IPT} = \mu_{OLT} > \mu_{GT}$. Finally, we would also like to know whether in-person one-on-one tutoring yields higher mean math test scores than either of the other two tutoring approaches: $H_3 : \mu_{IPT} > \mu_{OLT} = \mu_{GT}$. The BF for each hypothesis, along with their likelihoods, is shown in Table 5.6.

Given the prior distributions and the data, the most likely hypothesis is that the average math test for IPT is greater than that of OLT, which is greater than GT (0.53). The second most likely hypothesis is that the two one-on-one

TABLE 5.6

Bayes Factors and Likelihoods Associated with H_0, H_1, H_2, and H_3

Hypothesis	BF	Probability
$H_0 : \mu_{IPT} = \mu_{OLT} = \mu_{GT}$	0.12	0.01
$H_1 : \mu_{IPT} > \mu_{OLT} > \mu_{GT}$	4.97	0.53
$H_2 : \mu_{IPT} = \mu_{OLT} > \mu_{GT}$	3.08	0.33
$H_3 : \mu_{IPT} > \mu_{OLT} = \mu_{GT}$	1.27	0.13

tutoring groups have a higher math test mean than that of the GT (0.33). Together, these hypotheses have a probability of 0.86, leading us to conclude that the GT approach likely yields the lowest mean score in the population and that it is somewhat more likely that IPT yields a higher mean than the online treatment. The ratio of BFs for the non-null hypotheses is:

$$H_1 \text{ vs } H_2 : \frac{4.97}{3.08} = 1.61$$

$$H_1 \text{ vs } H_3 : \frac{4.97}{1.27} = 3.98$$

$$H_2 \text{ vs } H_3 : \frac{3.08}{1.27} = 2.43.$$

We did not include the calculations for the BF ratios involving H_0 here because they are all larger than 10, indicating that it is very unlikely to hold true in the population. When considering the non-null hypotheses in light of the threshold of 3 or higher suggesting a difference in hypotheses (Kaplan, 2014), it appears that hypothesis 1 is clearly more probable than hypothesis 3. Otherwise, none of the ratios reach the level of "significance" in the context of the BFs. This result supports our conclusions from the hypothesis likelihoods; i.e., hypothesis 1 is somewhat more likely than hypothesis 2. However, the difference in BFs and likelihoods is not sufficiently large for us to definitively select it as the most likely to be true in the population, without a doubt. What we can be more certain about is that the GT method is likely to yield a lower mean math test performance than either of the one-on-one approaches.

As was the case for the *t*-test, we can examine the ROPE to the math tutoring data. The reviewer referred to Chapter 4 for a description of this method. Recall that the goal of this approach is to determine whether means are sufficiently different for us to conclude that they are unlikely to be equivalent in the population. The means of the posterior distributions for the group math test average, pairwise differences between these averages, and the standardized mean difference (δ), along with the 95% credibility intervals, appear in Table 5.7.

The results in Table 5.7 suggest that the mean score for those receiving in-person one-on-one tutoring is larger than those of either of the other two groups. We reach this conclusion because 0 did not lie in the 95% credibility intervals for the differences between the means and δ values involving either pairwise comparison of the in-person group. Conversely, the 95% credibility interval for the posterior mean difference of the online and GT methods did include 0.

The δ values and the ROPE, as well as the characterization of the δs in terms of magnitude, appear below in Figure 5.7. Of main interest here is that for comparisons involving the in-person group, the posterior distribution of δ lay above the ROPE, whereas for the comparison of online and GT δ, it did overlap somewhat with ROPE. For all three comparisons, the majority of δ values were in the large category.

TABLE 5.7

Posterior Means and 95% Credibility Intervals
for Group Math Test Means, Differences in
Group Means, and Standardized Mean
Differences

Parameter	Mean of Posterior Distribution	95% Credibility Interval
μ_{IP}	97.62	(95.81, 99.38)
μ_{OL}	82.63	(79.39, 85.89)
μ_G	67.48	(52.23, 82.62)
$\mu_{IP} - \mu_{OL}$	14.99	(11.27, 18.69)
$\mu_{IP} - \mu_G$	30.14	(14.95, 45.5)
$\mu_{OL} - \mu_G$	15.15	(-0.39, 30.9)
δ_{12}	4.49	(3.34, 5.59)
δ_{13}	7.02	(3.34, 10.51)
δ_{23}	3.37	(-0.09, 6.81)

Effect Size Estimation

As we have discussed in previous chapters, it is important to accompany null hypothesis testing results with effect size estimates. In the one- and two-sample cases (Chapters 3 and 4), we learned that these effect sizes included standardized comparisons of the difference in group means, both assuming equal variances and not. In addition, we discussed approaches that were based on trimmed means, ranks, and an approach yielding the proportion of variability in the dependent variable that is explained by the grouping variable (using Winsorized variances). We also discussed Cliff's Delta, which essentially compares the overlap in distributions for the two groups. Although different in form, we will see that when there are more than two groups, calculation and reporting of effect sizes will remain an integral aspect of statistical practice. We will now consider several of the effect size options available to the researcher working in the ANOVA context.

Perhaps the most common effect sizes used by researchers working with parametric ANOVA are eta squared (η^2) and omega squared (ω^2). Both of these statistics are designed to measure the proportion of variation in the dependent variable that can be accounted for by group membership. The value of η^2 is estimated as the ratio of the SSB (equation 5.3) to the sum of squares total (equation 5.5):

$$\eta^2 = \frac{\sum_{j=1}^{J} n_j \left(\bar{x}_j - \bar{x}_. \right)^2}{\sum_{i=1}^{N} \left(x_{ij} - \bar{x}_. \right)^2} = \frac{SSB}{SST} \qquad (5.16)$$

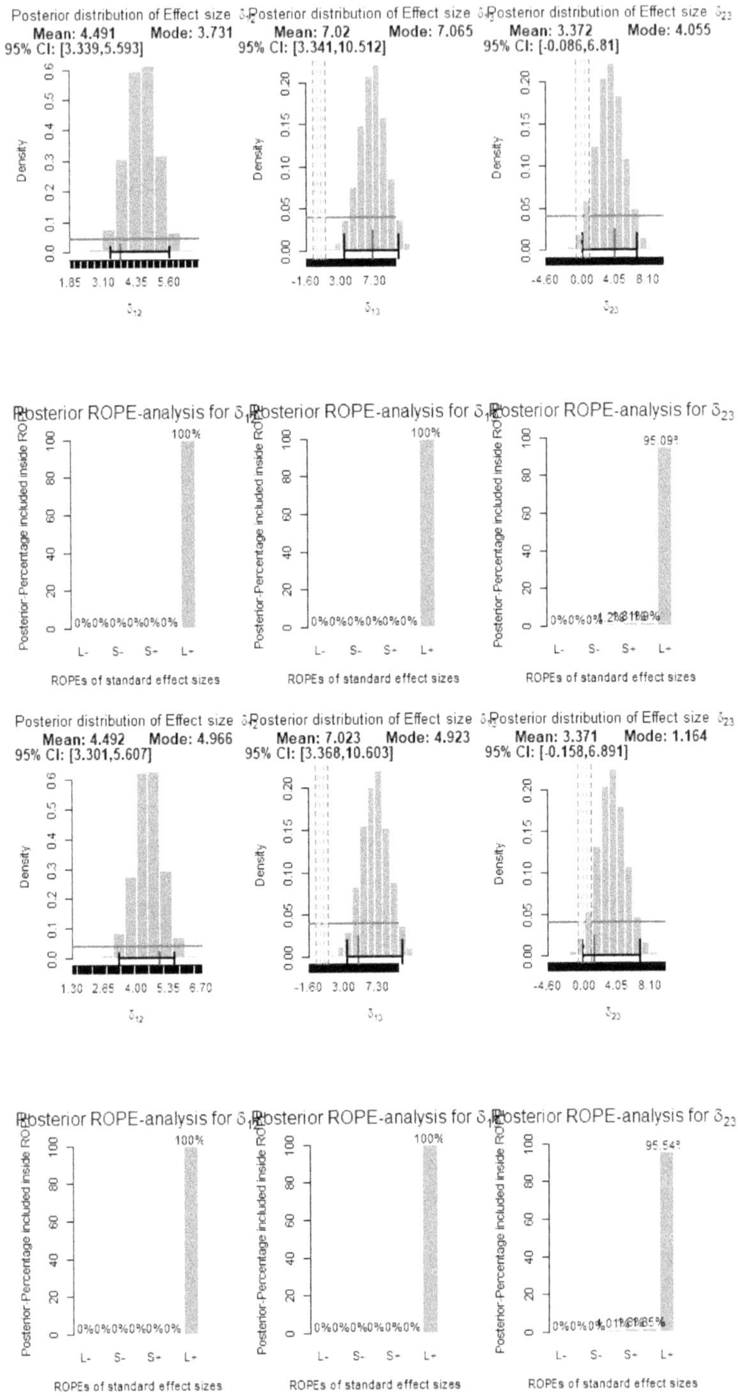

FIGURE 5.7
Posterior distribution of δ and classification of δ magnitudes.

This statistic ranges between 0 and 1 with higher values indicating that a greater proportion of the total variability in the data can be associated with the group to which an individual belongs. Cohen (1988) provided some general guidelines for interpreting the magnitude of η^2:

0.01–0.06:   Small

0.06–0.14:   Medium

0.14+:   Large

As for Cohen's guidelines for the two-sample case, which we discussed in Chapter 4, researchers should treat these as a very general rubric for interpreting effect size values. It is always preferable to place the effect size estimates for a given problem in the context of the literature in the field being studied.

Although η^2 is widely reported in the research literature and is available in many software packages, it has some limitations that make it less than optimal in most cases. Kroes and Finley (2023) provide an excellent discussion of these issues, and so we will simply summarize them briefly here. The problem is that η^2 is a positively biased estimate of the proportion of variability accounted for by group membership because the SSB includes variability due both to actual population differences in group means as well as differences due simply to random sampling variability. In other words, even if in the population the means are equal across groups, for a specific sample the means will differ somewhat from the overall mean simply because of sampling variability. As Kroes and Finley note, this random variability will add to any true systematic mean differences and thereby inflate the value of η^2 above what would be expected in the population.

To address this overestimation of the group effect, it is recommended that researchers instead use ω^2 (e.g., Myers & Well, 2003). The calculation of ω^2 removes the positive bias associated with η^2 by accounting for the random variability in group means in the numerator of the following calculation:

$$\omega^2 = \frac{\text{SSB} - \left((J-1)*\text{MSW}\right)}{\text{SST} + \text{MSW}} \tag{5.17}$$

The terms in equation (5.17) include the SSB and SST, as well as the MSW, and the number of groups J. It is generally recommended that in addition to the single estimate itself, researchers also report a confidence interval for the effect size estimate (Fritz et al., 2012), just as we did for the two-group case in Chapter 4. The details of this calculation are beyond our scope here, but the interested reader is referred to Steiger (2004) for a full explanation. Fortunately, software packages can provide these intervals to us.

Kelley (1935) described a third effect size measure for use with ANOVA, ϵ^2. As with ω^2, the calculation underlying ϵ^2 is designed to remove the bias inherent in η^2. It takes the form:

$$\epsilon^2 = 1 - \frac{(n-1)\text{SSW}}{(n-J)\text{SST}} \tag{5.18}$$

We can see that ϵ^2 and η^2 are similar in form, given that the numerator of the latter is SSB, which is SST – SSW. We can express 1 as $\frac{\text{SST}}{\text{SST}}$, yielding equation (5.19) $\frac{\text{SST}}{\text{SST}} - \frac{(n-1)\text{SSW}}{(n-J)\text{SST}}$, clearly showing the correspondence in the form of ϵ^2 and η^2. One final point that needs to be made here is that both ϵ^2 and ω^2 rest on an assumption of normality so that when the data are not normally distributed, these values can underestimate the actual proportion of variance explained in the dependent variable by group membership. We will discuss alternatives for such cases below.

For the math tutoring data, we can take the sums of squares from Table 5.3 and insert them into equations (5.16), (5.17), and (5.18).

$$\eta^2 = \frac{\text{SSB}}{\text{SST}} = \frac{3,116}{3,116 + 5,454} = 0.36$$

$$\omega^2 = \frac{\text{SSB} - \big((J-1)*\text{MSW}\big)}{\text{SST} + \text{MSW}} = \frac{3,116 - \big((3-1)*363.6\big)}{(3,116 + 5,454) + 363.6} = \frac{2,388.8}{8,933.6} = 0.27$$

$$\epsilon^2 = 1 - \frac{(n-1)\text{SSW}}{(n-J)\text{SST}} = 1 - \frac{(18-1)5,454}{(18-3)(3,116 + 5,454)} = 1 - \frac{92,718}{134,004} = 1 - 0.69 = 0.31$$

We can see evidence of the positive bias for η^2 versus ϵ^2 and ω^2. Given recommendations in the literature, we will focus our attention on the latter and conclude that approximately 27% of the variance in the math test scores is associated with the type of tutoring that a student received. If we used ϵ^2 as our measure of the tutoring effect, we would find that approximately 31% of the math test score variance is accounted for by tutoring treatment. Taking these values together, we see that the tutoring treatment accounts for somewhere around 30% of the variability in the math test scores. In addition, based on Cohen's guidelines, each of these effect size values falls in the large range, selecting a relatively strong relationship between tutoring method and math test score. As noted above, these values should be interpreted within the context of the literature within which this study sits.

Steiger (2004) described techniques for calculating the upper and lower bounds of confidence intervals for these three effect size estimates. The 95% confidence interval for ω^2 is (0, 1), indicating that we can be 95% confident that the actual proportion of test variability associated with group membership ranges from none (0%) to all (100%). The 95% confidence interval for ϵ^2 is also (0,1). Obviously, these results are not particularly informative and reflect the fact that we are working with a small sample, meaning that there is a great deal of uncertainty with respect to the population effect size values.

When we are working with nonparametric tests based on ranks, we can turn to an effect size estimate that is also based on ranks:

$$\eta_R^2 = \frac{H - J + 1}{n - J} \qquad (5.19)$$

where H is the value of H obtained from the KW test.

There is also a rank-based version of Epsilon-squared:

$$\epsilon_R^2 = \frac{H}{(n^2 - 1)(n + 1)}. \qquad (5.20)$$

The KW H value for the math tutoring data was 6.32, leading to effect size values of

$$\eta_R^2 = \frac{H - J + 1}{n - J} = \frac{6.32 - 3 + 1}{18 - 3} = \frac{4.32}{15} = 0.29$$

$$\epsilon_R^2 = \frac{H}{(n^2 - 1)/(n + 1)} = \frac{6.32}{(18^2 - 1)/(18 + 1)} = 0.37$$

Thus, we would conclude that the math tutoring method was associated with approximately 29%–37% of the variance in the rank of math test scores. As for the standard parametric effect size estimates, confidence intervals can also be calculated for these rank-based versions. For η_R^2, the 95% confidence interval is (0.04, 1.00), and for ϵ_R^2, the 95% confidence interval is (0.16, 1.00). As for the parametric versions, these intervals are really too wide to be very useful in practice.

In the context of robust statistical analysis, we can extend the effect size of Wilcox and Tian (2011), which we discussed in Chapter 4. Recall that this statistic is the square root of the ratio of the Winsorized variance for the dependent variable predicted by group membership to the Winsorized variance for the observed values of the dependent variable:

$$\xi = \sqrt{\frac{S_W^2(\hat{Y})}{S_W^2(Y)}} \qquad (5.21)$$

where $S_W^2(\hat{Y})$ is the Winsorized variance for the dependent variable predicted by group membership and $S_W^2(Y)$ is the Winsorized variance for the observed values of the dependent variable.

This value can be larger than 1 when there are more than two groups and Winsorization is used. For the math tutoring study, this statistic was 0.89 with a 95% bootstrap confidence interval of (0.44, 1.51). Based on the interpretation guidelines recommended by Wilcox and Tian, the effect could be in the moderate (0.3–0.5) or large (0.5+) range. This finding is similar to that of the parametric effect size estimates.

Summary

In this chapter, we have discussed a large number of statistical methods that can be used to compare means when we have more than two groups of interest. In many respects, these methods are quite similar to those for the two-group case. Indeed, it can be shown that the *t*-test is a special case of ANOVA. The basic approach to comparing two means that we learned in Chapter 4 can be directly transferred to the comparison of three or more means. We continue to use null and alternative hypotheses, test statistics, *p*-values, and effect size estimates. The inclusion of more than two means does necessitate some additional steps, however. In particular, when we reject the null hypothesis for ANOVA, then we must follow up with pairwise comparisons of the means to determine for which groups the dependent variable mean is likely to be different in the population. We saw that a primary goal in conducting these follow-up procedures is to control the familywise Type I error rate. There exists a large number of such techniques, with Tukey's test being a good choice to use in most cases. Finally, we saw that it is possible to test specific hypotheses of interest, beyond the standard ANOVA null, using contrasts.

The parametric ANOVA relies on the same assumptions that were required to use the *t*-statistic, namely, that the sample comes from a normally distributed population with respect to the dependent variable, that the group variances on the dependent variable be equal in the population, and that the individuals in the sample are independent of one another once group membership is accounted for. When these assumptions are violated, we will want to make use of the alternatives described in this chapter. The heuristics for selecting which alternative to use when for ANOVA are quite similar to those that we discussed at the end of Chapter 4 for the *t*-test. Therefore, we will not review them in depth here but rather point the reader to the Chapter 4 discussion. We will summarize those recommendations here, however. When the parametric assumptions are met, the standard *F*-test and associated parametric procedures will be optimal. In addition, if the sample size is sufficiently large, the researcher can rely on the central limit theorem to ensure that the use of *F* is fine. See the discussion in Chapter 3 regarding the necessary sample sizes for the central limit theorem to apply.

When the parametric assumptions are not met, the researcher should turn to one of the alternatives described in this chapter. The choice of which to use comes down to the nature of the assumption violation. As we discussed at the end of Chapter 4, if the data are skewed and/or there are outliers present, the rank-based methods may be preferable, as would the McKean and Schrader test for equality of medians (presuming that there are no ties in the data). If the primary problem is the presence of outliers, then a test comparing the trimmed means may be the best technique to use. It should be noted that when we use a test for the trimmed means, we are assuming that the population trimmed means excluding the extremes (and not the mean including all individuals in the population) are of most interest. If our interest is in testing hypotheses about the full population, including the extremes, then the trimmed means tests may not be the best. If our primary concern is a lack of equality in the groups' population variances but the normality assumption has not been violated, then the Welch F-test would be a good option.

Finally, we saw that the Bayesian framework can be brought to bear when comparing means for more than two groups. Bayes allows us to include prior information about the mean differences, which can be quite useful when we have strong theories and/or prior research that can help inform what is likely true about the population. In addition, the Bayesian paradigm also provides us with information about the relative likelihoods of various hypotheses in the form of BFs. This is not information that we can obtain from the standard frequentist ANOVA. Bayesian estimation can also give us detailed information about the full range of possible differences among the means of multiple groups in the form of ROPE. Indeed, using ROPE, we do not even need to conceive of the omnibus and follow-up procedures as being different tests. Rather, the pairwise comparisons and their associated credibility intervals provide us with a direct comparison of means without the multi-step process inherent in the frequentist ANOVA approach.

6

Factorial Designs

In Chapter 5, we devoted our attention to the comparison of central tendency across more than two groups, extending the two groups comparison work that we discussed in Chapter 4. We saw the correspondence between the t and F statistics and how each can be used to make inferences about differences in the population means. In this chapter, we will further extend our discussion of population mean comparison to the case where we have more than one independent grouping variable. The basic ANOVA approach that was described in the previous chapter can be very easily extended to incorporate multiple independent variables in what are known as factorial ANOVA designs. Indeed, the equations and approaches that we described in some detail in Chapter 5 are equally applicable in the scenarios described in Chapter 6. Factorial ANOVA models are typically indexed by the number of groups in each of the independent variables as being $K \times J$ models. Thus, a factorial ANOVA model where one independent variable has two groups and the other has three groups would be called a 2×3 ANOVA. So, with all of that in mind, let's get started.

Motivating Example

The motivating example data for this chapter are the same as we used in Chapter 5. Recall that the dataset includes math test scores for 18 high school students, each of whom was randomly assigned to one of three tutoring methods: 1=one-on-one tutoring in-person, 2=one-on-one tutoring online, and 3=group tutoring in-person. To explore approaches to factorial designs, we will consider a second independent variable, learning disability status (yes or no), giving us a 3×2 ANOVA model. This variable is crossed with the tutoring variable, meaning that within each of the tutoring groups, there are individuals in both learning disability groups. Indeed, for this example, there are equal numbers of students from each learning disability group within each of the tutoring groups.

DOI: 10.1201/9781003379324-6

Exploration of the Data

In Chapter 5, we explored the data with respect to tutoring treatment in some detail. We will not reiterate all of that work here but will do a similar exploration of the disability status variable and the combination of tutoring method and disability status. Table 6.1 includes descriptive statistics for the math test score by tutoring treatment and disability status.

For our sample, individuals identified with a math learning disability have a lower mean and median math test score than do those who were not so identified. And of course, we see the patterns in tutoring method means that we saw in Chapter 5, where in-person individual tutoring students had a higher mean math score than did the online individual tutoring students, who were followed by the group tutoring students.

The boxplot of math test scores by disability status appears in Figure 6.1. Note that a disability status of 1 is not identified and 2 is identified.

The median (solid horizontal line in the box) for the not identified group is clearly higher than for the identified individuals, *for this sample.* It is also interesting to note the two outliers at the bottom of the score distribution for the identified group. These are students with particularly low math test scores relative to the other identified students. The ridgeline plot in Figure 6.2 reveals that there is some overlap between the math scores for the two disability status groups. The outlying scores at the bottom of the distribution for the identified group are evident as well in this plot.

Finally, we can examine the mean math test scores for the six combinations of disability status and tutoring treatment using a line plot. Each line represents a different tutoring method, and the *x*-axis displays the disability status. The points represent the mean math test score, which appears on the *y*-axis (Figure 6.3).

TABLE 6.1

Descriptive Statistics for Math Test Score by Treatment

	Mean	Median	SD	MAD
Tutoring Treatment				
In-person individual	96.5	96.5	9.7	9.6
Online individual	83.8	81	13.8	15.6
Group	64.5	74	28.4	23.7
Disability Status				
Identified with disability	69.7	80	24.4	17.8
Not identified	93.6	95	12.4	14.8

SD, standard deviation; MAD, mean absolute deviation.

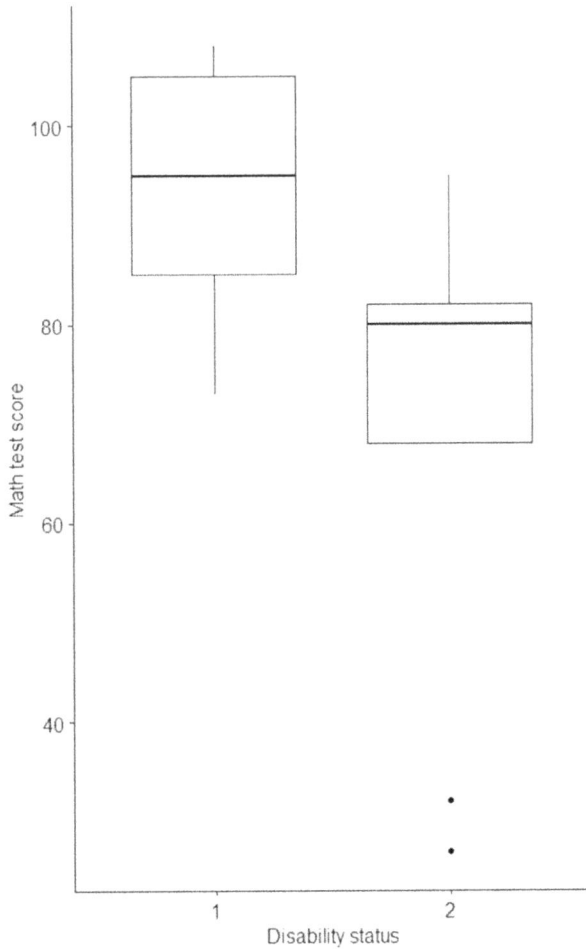

FIGURE 6.1
Boxplots of math test scores by disability status.

For this sample, mean math test scores were lower for those identified with a math learning disability regardless of the tutoring method condition to which they were assigned (denoted by the negative direction of each line). However, it also appears that the difference in means for those placed in the group tutoring condition was greater than for either of the one-on-one tutoring groups, given the more severe negative trend in the line for this group. We will investigate whether this apparent sample difference might be present in the population when we assess the interaction term in the ANOVA model, below.

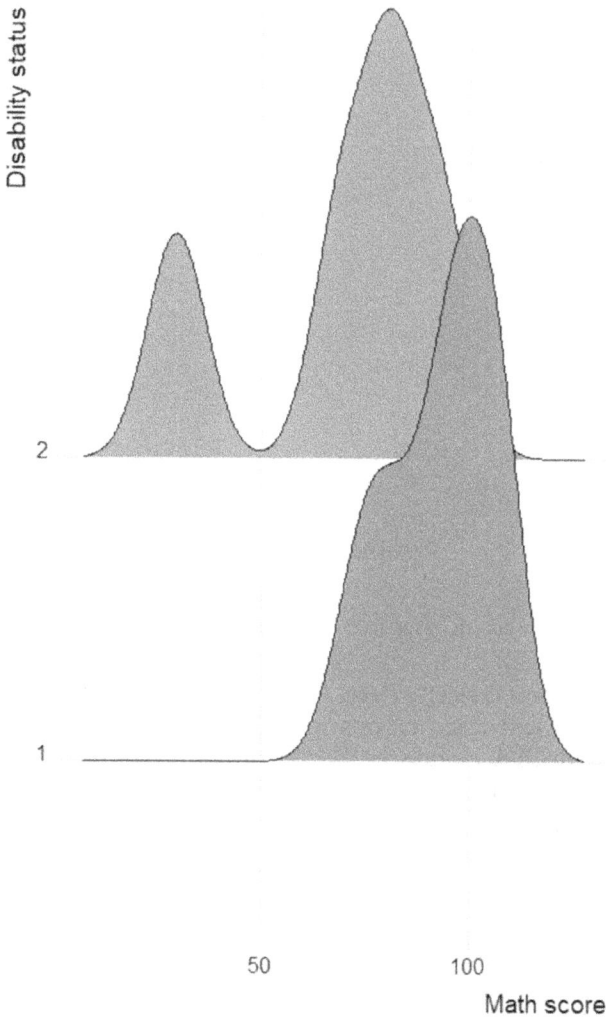

FIGURE 6.2
Ridgeline plot of math test scores by disability status.

Factorial Analysis of Variance

In the case of factorial ANOVA models, where we have multiple independent categorical variables, the basic statistical methodology that was outlined in Chapter 5 for the one-way ANOVA is also applied. Thus, the F statistic and its accompanying assumptions to which we devoted our attention in the previous chapter will all apply in the factorial ANOVA case as well. Therefore, because we described the F statistic in some detail in the previous chapter,

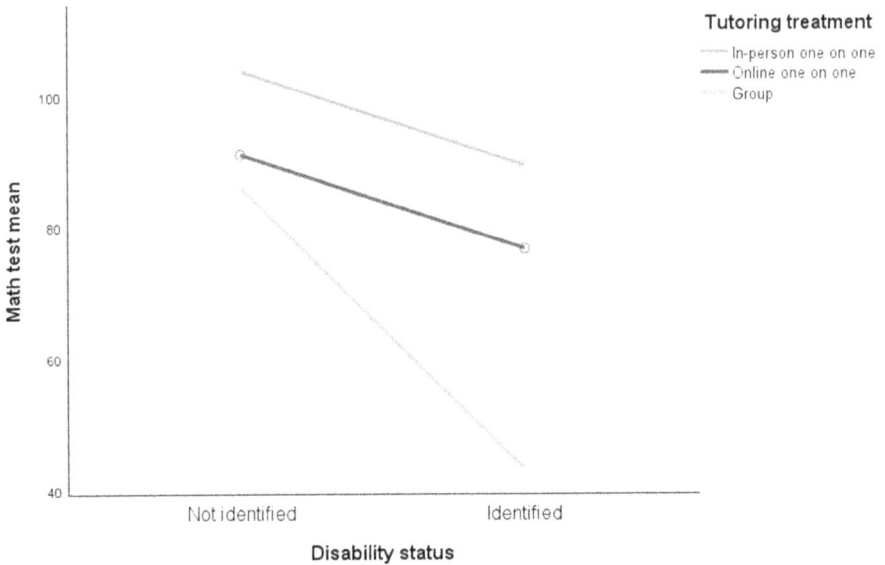

FIGURE 6.3
Mean of math test by disability status and tutoring treatment.

we will only review it briefly here. The equations, degrees of freedom, post hoc techniques, and assumptions are all the same in the two-way and higher-order ANOVA models as in the one-way ANOVA case. Specifically, remember that the sums of squares within (SSW), between (SSB), and total (SST) are expressed as:

$$SSW = \sum_{i=1}^{N} \left(x_{ij} - \bar{x}_j \right)^2 \tag{6.1}$$

$$SSB = \sum_{j=1}^{J} n_j \left(\bar{x}_j - \bar{x}. \right)^2 \tag{6.2}$$

$$SST = \sum_{i=1}^{N} \left(x_{ij} - \bar{x}. \right)^2 \tag{6.3}$$

When we have multiple independent variables, each will have a unique SSB. Thus, if our ANOVA model has two independent variables, then we will have two SSB terms (SSB1 and SSB2).

In addition to the individual independent variables, which are called main effects, the factorial ANOVA model will also include interactions between the main effects. An interaction term reflects the extent to which the relationship between an independent variable and the dependent variable is impacted by another independent variable. The sum of squares for the interaction can be calculated as

$$SSI = SST - SSB1 - SSB2 - SSW \tag{6.4}$$

We can think of the interaction as the variability remaining in the dependent variable after we've accounted for the variability associated with the independent variables and the within-group differences among individuals. The degrees of freedom for the interaction term are calculated as the product of the degrees of freedom for the independent variables that are a part of the interaction. For example, if the interaction involves two independent variables with 1 and 3 degrees of freedom, respectively, the interaction degrees of freedom would be $1 \times 3 = 3$. The mean square for each term in the factorial ANOVA model is all the sum of squares divided by the degrees of freedom. The F statistics for the main effects and interactions are their mean square divided by the mean square within groups, just as was the case for the one-way ANOVA. We demonstrate these calculations using our example, below.

Before going to the example, let's reframe our discussion of the factorial ANOVA model in terms of the sources of variation, as we did in Chapter 5. For our current example with two independent variables, the model takes the following form:

$$y_{ijk} = \mu_{...} + x_j + x_k + x_j x_k + \varepsilon_{ijk} \tag{6.5}$$

where y_{ijk} is the math test score for person i in tutoring method j with disability status k, $\mu_{...}$ is the overall mean for the dependent variable, x_j is the effect of being in tutoring method j, x_k is the effect of having disability status k, $x_j x_k$ is the interaction between tutoring method and disability status, and ε_{ijk} is the random error for person i in tutoring method j with disability status k.

This model framework allows us to see the potential sources of variability for the dependent variable (math test score). Some of these possible factors may ultimately not prove to be statistically meaningful, and we will test them using the F statistic, as demonstrated next.

The null hypotheses for our two-way ANOVA are

$H_0 : \mu_1 = \mu_2 = \mu_3$; No difference in math test means across three tutoring methods

$H_0 : \mu_D = \mu_{ND}$; No difference in math test means between disability status groups

H_0 : No interaction between disability status and tutoring treatment for math test means

The results of the ANOVA for the math test appear in Table 6.2. Note that the calculations for the elements in this table mirror the calculations from Chapter 5 and are thus not demonstrated here.

We can see that the ANOVA table for the higher-order model is quite similar to that for the one-way ANOVA, with added rows for the second independent variable and the interaction. This layout, including the degrees of freedom, sums of squares, means squares, F statistic, and p-value, is the standard format for presenting an ANOVA.

TABLE 6.2

Results of Two-Way ANOVA for Comparing Math Test Score Means

Variable	DF	SS	a	F	p
Tutoring method	2	3,116.4	1,558.2	9.06	0.004
Disability status	1	2,568.1	2,568.1	14.93	0.002
Interaction	2	821.8	410.9	2.39	0.13
Error	12	2,064.0	172.0		

ANOVA, analysis of variance.

TABLE 6.3

Results of One-Way ANOVA for Comparing Math Test Score Means

Variable	DF	SS	MS	F	p
Tutoring method	2	3,116.0	1,558.2	4.29	0.03
Error	15	5,454.0	363.6		

ANOVA, analysis of variance.

Before interpreting our results, let's examine how the elements in Table 6.2 are used to obtain F and p. The F values for each model term are calculated as

$$F_{Tutoring} = \frac{\frac{3,116.4}{2}}{\frac{2,064.0}{12}} = \frac{1,558.2}{172.0} = 9.06$$

$$F_{Disability} = \frac{\frac{2,568.1}{1}}{\frac{2,064.0}{12}} = \frac{2,568.1}{172.0} = 14.93$$

$$F_{Interaction} = \frac{\frac{821.8}{2}}{\frac{2,064.0}{12}} = \frac{410.9}{172.0} = 2.39$$

Given the F statistics and their associated degrees of freedom, we are able to obtain the p-values (the final column in the table) and reach conclusions regarding each null hypothesis. Using the standard null hypothesis testing approach, we would conclude that there are statistically significant mean differences for tutoring method and disability status, but not a statistically significant interaction.

As a point of comparison with the two-way ANOVA results, we can take a look at the ANOVA table for the one-way ANOVA from Chapter 5, for which only the tutoring treatment variable was included in the model (Table 6.3).

When we compare the results for the tutoring method from the one- and two-way ANOVA models, we can see that the sums of squares and mean squares for tutoring are quite similar. However, the values associated with the error are different between the two models. Of course, there are more degrees of freedom associated with error for the one-way ANOVA because those associated with disability status (1) and the interaction of disability and tutoring (2) are not included and thus subsumed into error. Likewise, the sum of squares for error in the one-way ANOVA includes the sums of squares for both disability status and the interaction term. Thus, when we include these terms in the two-way ANOVA model, their contributions to the error sum of squares are removed, causing it to decline in value.

If we think a bit more about the apparent link between model design and the error term, it becomes clear that the greater the differences in group means for a given variable, the greater the decline in the error sum of squares, leading to larger F values for each main effect and interaction in the model. And, larger F values lead to smaller p-values, assuming that the degrees of freedom are held constant. With these relationships in mind, we can see that including independent variables that are related to the dependent variable in the model will often result in greater statistical power (i.e., a greater likelihood of correctly rejecting the null hypothesis). It's important to note that such is not always the case, particularly when the newly included independent variable(s) is not related to the dependent variable. In that case, the within-group sum of squares may not be greatly reduced, but the degrees of freedom will be. Therefore, deciding on whether to include additional independent variables in a model should be based on theory and the hypotheses that we wish to test and not solely on a desire to reduce error variability. However, if we know from prior research and theory that a particular variable is likely to be associated with the dependent variable and we are able to measure it, then it may behoove us to include it in the model because doing so may reduce error variance and give us a clearer picture regarding the relationships between the independent and dependent variables. We certainly see evidence of the clarifying power of including important independent variables here. Finally, as we will discuss below, the inclusion of additional variables in the model will also impact the values of several effect size statistics.

Assessing Assumptions

The assumptions underlying factorial ANOVA are the same as those for one-way ANOVA: normality of the dependent variable, equality of variances across groups for each independent variable, and independence of observations after accounting for the terms in the model (main effects and interactions). We can assess the normality assumption using the QQ plot of the math test scores (Figure 6.4).

Recall that the QQ plot places the observed sample values on the y-axis and the value of the data, if it were normally distributed, on the x-axis. The solid

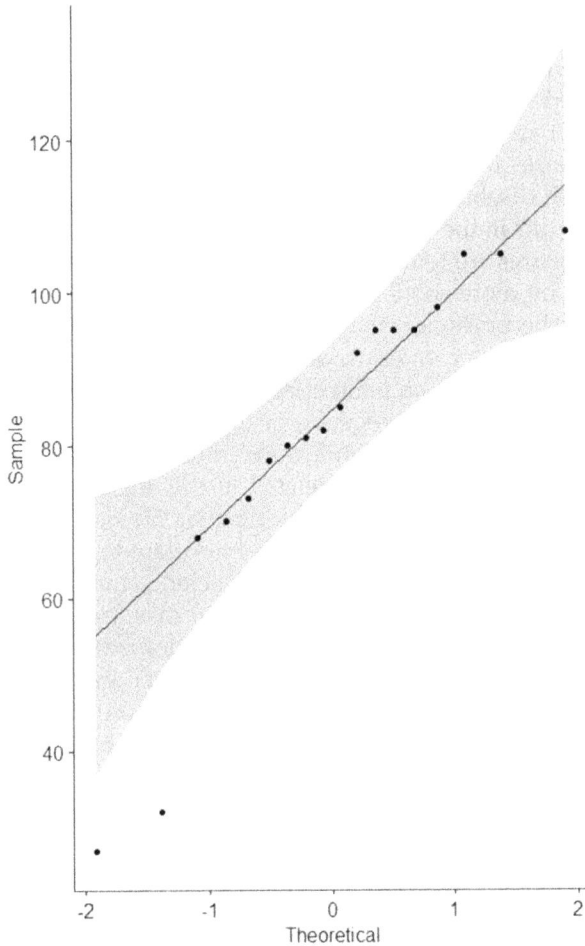

FIGURE 6.4
QQ plot of the math test scores.

line represents the normal distribution, and the individual data points are plotted as dots. The closer the dots are to the line of normality, the more confident we can be that the score follows the normal distribution. There is also a 95% confidence region around the line, reflecting estimation error. So, if all of the individual data values fall within the confidence region, we can conclude that the data are normally distributed. The more points that lie outside the confidence region and away from the normal line, the less confident we can be that the data are normally distributed. As we commented for these data in Chapter 5, most of the math test scores fall in the confidence region around the line of normality. However, at the lower end of the distribution, there are two points that lie outside of the normal region. The QQ plot suggests that it may be reasonable for us to assume normality for the math test scores, though there are two unusually low scores. We can also assess normality using the Shapiro–Wilk test, for which the null hypothesis is that the data are normally distributed. The p-value for the test was 0.01, leading us to reject the null of normality. In addition, the skewness estimate for the test is −1.36, which, along with the histogram below, suggests that the data are somewhat negatively skewed (Figure 6.5).

Histogram of Math Test Scores

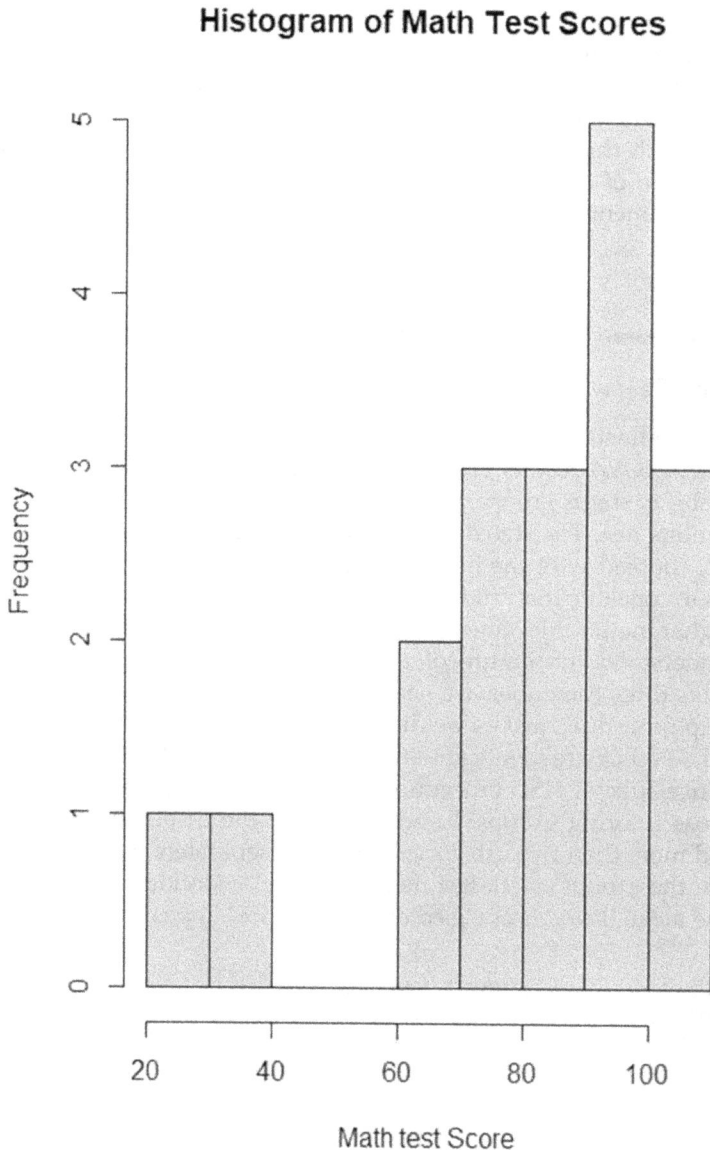

FIGURE 6.5
Histogram of math test scores.

We can assess the equality of variances assumption using Levene's test, which we described in some detail in Chapter 4. The null hypothesis for each independent variable is that the population variances are equal:

$$H_0 : \sigma_{GT}^2 = \sigma_{OLT}^2 = \sigma_{IPT}^2 \text{ Tutoring method}$$

$$H_0 : \sigma_{Disability}^2 = \sigma_{Non\text{-}disability}^2 \text{ Disability status}$$

For the tutoring method, Levene's test statistic is 2.20 with $p = 0.15$, whereas for disability status the test statistic is 1.15, yielding $p = 0.30$. Therefore, we don't reject either null hypothesis and conclude that homogeneity of variance holds for both the tutoring method and disability status. We will assume independence of observations, given the random sampling and treatment group assignment protocols that were used.

Post Hoc Pairwise Comparisons

Both main effects were found to be statistically significant based on the omnibus ANOVA results presented in Table 6.2. Given that there are only two disability status groups, we do not need to do any post hoc follow-up of the omnibus test. The statistically significant F-test for the disability status variable, coupled with the higher sample mean for the nondisabled group, leads us to conclude that students not identified with a disability are likely to have higher math achievement means than those identified with a disability.

To understand the nature of the differences for the tutoring method, which has three categories, we need to use a post hoc pairwise comparison follow-up procedure, just as we did in Chapter 5. The available methods for factorial ANOVA are identical to those for the one-way ANOVA. Therefore, we can use Tukey's HSD procedure to assess mean differences for the math test across tutoring groups. In addition, had the disability status variable included more than two categories, we could separately use Tukey's HSD to compare the groups' math test means. The 95% confidence intervals of the pairwise mean differences based on Tukey's HSD appear below:

$$\mu_{GT} - \mu_{OLT} : (-39.53, 0.87)$$

$$\mu_{GT} - \mu_{IPT} : (-52.20, -11.80)$$

$$\mu_{OLT} - \mu_{IPT} : (-32.87, 7.53)$$

We see that 0 is not in the confidence interval for the difference in means involving the group tutoring and in-person one-on-one tutoring conditions, but is included in the other intervals. Thus, based on the means in Table 6.1, we would conclude that students receiving in-person one-on-one tutoring have a higher population mean math test score than those receiving in-person group tutoring. However, there is likely not a population difference between the mean for the online one-on-one group when compared to the other two tutoring conditions.

Factorial ANOVA Using Welch's *F*-Test

When the equality of variances assumption is not met, we can use a variation of the *F* statistic based on work by Welch (1951). We outlined the calculations underlying this method in Chapter 5. This approach is the same in the factorial ANOVA context, and so we will not review the details again here. The assumption of variance heterogeneity held for both the tutoring method and disability status variables, as discussed above. Therefore, in actual practice, we would not need to use the Welch approach. However, for demonstrative purposes, we will report the results of our data (Table 6.4).

The results are qualitatively similar to those based on the standard *F* statistic. There are statistically significant differences among the tutoring method means and between the two disability status conditions. There was not a statistically significant interaction.

The pairwise comparisons involve Welch's *t*-test controlling for the Type I error rate using the Holm or Bonferroni correction. Of course, given that disability status has only two levels, no further testing is needed. Based on the means above, we would conclude that individuals who were identified with a math learning disability are likely to have lower math test means than for those who have not been so identified. The Welch pairwise results for the tutoring treatment using Holm's correction did not find any pairwise differences in mean math test scores among the various tutoring treatment groups. This result differs from Tukey's test results described above, likely due to different correction methods and/or the fact that the Welch approach does

TABLE 6.4

Results of the Welch ANOVA for Math Test by Tutoring Method, Disability Status, and the Interaction

Variable	DF	F_W	*p*
Tutoring method	3	360.21	<0.001
Disability status	1	5.09	0.04
Interaction	2	1.19	0.34

ANOVA, analysis of variance.

not use a pooled estimate of variability, whereas Tukey's technique does. As a reminder, given that the homogeneity of variance assumption was met, we would not use the Welch approach in this case.

Rank-Based Methods for Factorial ANOVA

As with the one-way ANOVA that we described in Chapter 5, we can use rank-based methods in the factorial context as well to test null hypotheses that there are no differences in medians among the groups for each of the independent variables and that there are no interactions between independent variables involving the median. There are two approaches for doing this, and we'll take a look at each of these here. One rank-based technique that we can employ in the context of ANOVA involves the use of what is known as relative effects, which measure the difference between the mean of ranks for a particular independent variable group (e.g., group tutoring) versus what we would expect the mean to be if there is no group effect. The larger this difference, the greater the effect of being in that particular group on the rank value of the dependent variable. This approach involves the following steps:

1. Rank the observed dependent variable scores for person i in group j (x_{ij}) to create the Rank scores (R_{ij}).
2. Calculate $\bar{R}_j$ for each group of the variable being tested (e.g., tutoring method, disability status).
3. Calculate the following elements for the relative effects:

$$Q_j = \frac{\bar{R}_j - 0.5}{N} \quad \text{The individual } Q_j \text{ values are placed together in a vector } \underline{Q}$$

(6.6)

where N is the total sample size.

$$S_j^2 = \frac{\sum_{i=1}^{N}\left(R_{ij} - \bar{R}_j\right)^2}{N^2\left(n_j - 1\right)}$$

(6.7)

where n_j is the sample size for group j and R_{ij} is the rank for individual i in group j.

$$F = \frac{N}{\text{Trace}\left(M_{11}\underline{V}\right)}\left(\underline{Q}M\underline{Q}'\right)$$

(6.8)

$$\underline{V} = N\text{diag}\left\{\frac{S_1^2}{n_1} \cdots \frac{S_j^2}{n_j}\right\}$$

(6.9)

$$\underline{M} = \underline{I} - \frac{1}{J}\underline{J}\qquad(6.10)$$

where $\underline{I}$ is the Identity matrix of size $J \times J$
$\underline{J} = J \times J$ matrix of 1s.

The F statistic from step 3 is distributed as an F with the following degrees of freedom:

$$DF_1 = \frac{M_{11}\left[\text{Trace}(\underline{V})\right]^2}{\text{Trace}(\underline{MVMV})}\qquad(6.11)$$

$$DF_2 = \frac{\left[\text{Trace}(\underline{V})\right]^2}{\text{Trace}(\underline{V}\Lambda)}\qquad(6.12)$$

$$\Lambda = \text{diag}\left\{(n_1 - 1)^{-1}, \dots, (n_j - 1)^{-1}\right\}\qquad(6.13)$$

The preceding formulas are indeed scary. But don't be intimidated by them because we can use the F and the degrees of freedom just as we would for a standard ANOVA to obtain the p-value. And fortunately, the computer will do all of the heavy lifting for us in terms of the calculations. In terms of understanding the methods, it's important for us to have a sense of what relative effects are (the differences between the observed mean ranks of the groups and the mean rank if the null hypothesis is true) but not necessarily how we calculate F. Using the relative effects, we obtain F and degrees of freedom for each main effect and interaction in the model. For the interaction term, we would have a mean rank for each combination of the variables (e.g., tutoring 1, disability status 1; tutoring 2, disability status 1; etc.).

We will not work through all of these calculations by hand, but it will be helpful to see the form of the relative effects for each group. These calculations for the tutoring method appear below:

$$Q_1 = \frac{5.17 - 0.5}{18} = 0.26$$

$$Q_2 = \frac{9.83 - 0.5}{18} = 0.52$$

$$Q_3 = \frac{12.83 - 0.5}{18} = 0.69.$$

We can see that the largest relative effect belongs to the group tutoring method. In a similar fashion, the calculations for the disability status are as follows:

$$Q_1 = \frac{6.11 - 0.5}{18} = 0.31$$

$$Q_2 = \frac{12.44 - 0.5}{18} = 0.66$$

Finally, the relative effects for the combinations of tutoring treatment and disability status are calculated as

$$Q_{11} = \frac{2.33 - 0.5}{18} = 0.10$$

$$Q_{21} = \frac{7 - 0.5}{18} = 0.36$$

$$Q_{31} = \frac{9 - 0.5}{18} = 0.47$$

$$Q_{12} = \frac{8 - 0.5}{18} = 0.42$$

$$Q_{22} = \frac{12.67 - 0.5}{18} = 0.68$$

$$Q_{32} = \frac{16.67 - 0.5}{18} = 0.90$$

The largest relative effects in the interaction belong to the combination of identified with a disability and group tutoring, followed by identified with a disability and online one-on-one tutoring. These values would then be used to calculate the F statistic as described above. We won't devote time to those calculations here but encourage the interested reader to do so on their own. We will present the results of this analysis below after we describe the second approach for using ranks in the factorial ANOVA context.

An alternative approach for using ranks in the conduct of ANOVA involves the aligned rank transformation (ART). In the two-factor case, we can visualize the data in a two-way table where the rows represent one of the variables (e.g., tutoring method) and the columns the other (e.g., disability status). We will then calculate the mean for each cell of the table (combination of tutoring and disability status), as well as the marginal means for each group across levels of the other. This example appears in Table 6.5.

We will now use the means in Table 6.5 to align the individual scores with respect to each independent variable. In other words, for each individual in the sample, we will have an aligned value, giving us two alignments for each person in this example. The alignments take the form:

$$x'_{k.i} = x_{ijk} - \overline{x}_{kj} + \overline{x}_{k.} - \overline{x}$$

$$x'_{.ji} = x_{ijk} - \overline{x}_{kj} + \overline{x}_{.j} - \overline{x} \tag{6.14}$$

$$x'_{kji} = x_{ijk} - \overline{x}_{kj} - \overline{x}_{k.} - \overline{x}_{.j} + \overline{x}$$

TABLE 6.5

Math Test Scores Represented in Row (Tutoring) and
Column (Disability)

Tutoring	Disability Status 1	2	Tutoring marginal
1	103.67	89.33	96.50
2	91.00	76.67	83.83
3	86.00	43.00	64.50
Disability marginal	93.56	69.67	81.61

where x_{ijk} is the score for individual i in group k for variable 1 (e.g., tutoring) and group j for variable 2 (e.g., disability), $\bar{x}_{ij}$ is the mean for combination of categories i and j for variables 1 and 2, $\bar{x}_{k.}$ is the mean for category k for variable 1, $\bar{x}_{.j}$ is the mean for category j for variable 2, and $\bar{x}$ is the overall mean.

The resulting aligned values are then ranked, and a separate ANOVA is conducted for each aligned ranked variable. The basic idea here is that alignment adjusts the scores in one independent variable for the presence of the other independent variable. Then by ranking these aligned scores, we can address issues of nonnormality and outliers as we would with any of the rank-based procedures discussed in this book. The aligned values for the tutoring data appear in Table 6.6. Given the space needed to show these by hand, the individual calculations are not included here. However, we encourage the reader to try a few, using the means in Table 6.5, to see how ART works.

The results of the two rank-based approaches appear in Table 6.7.

Qualitatively speaking, the ANOVA results for the two rank-based approaches were similar. The two main effects, the tutoring method and disability status, were statistically significant for the relative effects and aligned ranks method, whereas the interaction was not statistically significant for either approach. These results match one another as well as the parametric ANOVA, discussed above. The fact that the rank and parametric approaches lead us to the same conclusions regarding the main effects and interaction is not surprising given that our data do not depart markedly from the normal distribution.

The approach to following up statistically significant main effects for the relative effects rank-based approach is the same as the method that we used for the one-way ANOVA in Chapter 5, namely, pairwise comparisons using the Wilcoxon test. Given that we discussed this technique in detail there, we will not review it again here. In addition, we do not need to conduct a post hoc follow-up for the disability status variable given that it only has two levels. For the aligned rank ANOVA, we can conduct post hoc paired comparisons using the aligned ranks in conjunction with the Wilcoxon procedure and control for the familywise Type I error rate using approaches such as Tukey or Bonferroni. The post hoc comparisons for the tutoring method

TABLE 6.6

Individual Scores and Aligned Values by Disability Status and Tutoring Treatment

Disability	Tutoring	Math Score	Tutoring Alignment	Disability Alignment	Interaction Alignment
1	1	108	19.22	16.28	−0.45
1	1	105	16.22	13.28	−3.45
1	1	98	9.22	6.28	−10.45
1	2	105	16.22	25.95	9.22
1	2	95	6.22	15.95	−0.78
1	2	73	−15.78	−6.05	−22.78
1	3	78	−25.11	3.95	1.55
1	3	85	−18.11	10.95	8.55
1	3	95	−8.11	20.95	18.55
2	1	92	17.56	−9.27	7.44
2	1	95	20.56	−6.27	10.44
2	1	81	6.56	−20.27	−3.56
2	2	80	5.55	−8.61	8.11
2	2	68	−6.45	−20.61	−3.89
2	2	82	7.55	−6.61	10.11
2	3	27	−33.11	−27.94	−25.56
2	3	70	9.89	15.06	17.44
2	3	32	−28.11	−22.94	−20.56

TABLE 6.7

ANOVA Results for Rank-Based Methods

Variable	F	p
Relative Effects ANOVA		
Tutoring method	10.30	0.003
Disability status	19.09	0.001
Interaction	3.39	0.07
Aligned Ranks ANOVA		
Tutoring method	8.84	0.004
Disability status	16.58	0.001
Interaction	1.70	0.22

ANOVA, analysis of variance.

using the Wilcoxon paired comparisons with the Bonferroni comparison and the aligned ranks procedure appear in Table 6.8.

Both approaches revealed a statistically significant difference in the mean ranks between tutoring methods 1 and 3, with group 3 having the larger values.

TABLE 6.8

Post Hoc Paired Comparisons Results for Wilcoxon and Aligned Ranks Paired Comparisons

Tutoring Group	Mean Rank	Versus Group 1	Versus Group 2	Versus Group 3
Wilcoxon Paired Comparisons				
1	13.6	NA	4.67	7.58[a]
2	8.92	4.67	NA	2.92
3	6.00	7.58[a]	2.92	NA
Aligned Ranks				
1	14.58	NA	5.50	9.75[a]
2	9.08	5.50	NA	4.25
3	4.83	9.75[a]	4.25	NA

[a] Statistically significant at adjusted $\alpha=0.05$.

TABLE 6.9

Mean Ranks and Mean Aligned Ranks for Math Test Scores by Disability Status

Disability Status	Mean Rank	Mean Aligned Rank
Yes	6.44	5.56
No	12.60	13.44

The mean ranks and mean aligned ranks for disability status appear in Table 6.9.

Given the statistically significant result for disability status for both the relative effects and aligned ranks, we know that the mean ranks for the two groups are likely to differ in the population. The mean ranks in Table 6.9 show that those without a math learning disability had larger mean ranks than those who were identified as having a math learning disability. Finally, because the interaction was not statistically significant, no follow-up procedures were conducted for this term. Had the interaction been significant, we could have used the aligned ranks for comparing every pair of groups made from combining the tutoring group and disability status, using a Bonferroni correction.

Bootstrap-Based Methods for Factorial ANOVA

The bootstrap-based approach that we applied to the one-way ANOVA problem can also easily be extended to factorial ANOVA with two or more independent variables and the accompanying interactions. Recall that the basic idea with the bootstrap is to take a large number (e.g., $B=10,000$) of samples with replacement from the original dataset and then use the following steps:

1. Use ANOVA to obtain the F statistic for the observed data.
2. Create B bootstrap samples by sampling with replacement.
3. Replace the raw data with $C_{ij}^* = x_{ij} - \bar{x}_j$, which represents a sample from a distribution where there is not a difference in means among the groups. Notice that we are centering the data for each group, thereby forcing the means to be the same across groups but retaining the shape and spread of our original sample distribution.
4. Compute the F statistic using standard ANOVA for each bootstrap sample based on C_{ij}^*.
5. Compare the F based on the observed data (step 1) with the distribution of F from steps 2–4.
6. The p-value is the percentile of the observed F in this bootstrap distribution.

In this way, we create a distribution of the F statistic for our data assuming that the null hypotheses for the main effects and interactions are true. We then compare the observed F to this null distribution to determine whether there are statistically significant main effects or interaction(s).

The bootstrap ANOVA results for the math tutoring problem appear in Table 6.10.

The bootstrap ANOVA identified both the tutoring method and disability status as statistically significant, leading us to reject the null hypotheses of no group differences for each. The interaction was not statistically significant. The follow-up procedure for the bootstrap ANOVA involves a series of bootstrap t-tests (see Chapter 4), controlling the Type I error rate using the Bonferroni procedure. In this case, the pairwise bootstrap comparisons found a statistically significant difference between the in-person one-on-one and group tutoring math score means ($p=0.013$) but not for in-person versus online ($p=0.10$) or online versus group ($p=0.13$). We don't need to use a post hoc for disability status because it has only two groups. Taken together, these results reflect what we found with several of the other approaches:

TABLE 6.10

Bootstrap ANOVA Results for the Math Tutoring Data

Variable	F	p
Tutoring method	9.06	0.011
Disability status	14.93	0.005
Interaction	2.39	0.29

ANOVA, analysis of variance.

one-on-one in-person tutoring was associated with higher average math test scores than was group tutoring, and students who were identified with a math learning disability had lower average scores than did those not identified with a disability.

Comparing Trimmed Means for Factorial ANOVA

As in the one-way ANOVA case, we can compare trimmed means in the context of factorial ANOVA. The null hypotheses for our example would take the form:

$H_0 : \mu_{T1} = \mu_{T2} = \mu_{T3}$; No difference in trimmed means across three tutoring methods

$H_0 : \mu_{TD} = \mu_{TND}$; No difference in trimmed means between disability status

$H_0 :$ No interaction between disability status and tutoring treatment for trimmed means

The approach for comparing trimmed means involves a set of several calculations to obtain the F statistics needed to assess the hypotheses. Fortunately, the computer will do the work for us. We will take a look at how the calculations are done for the tutoring variable as a way to understand what the computer is doing. It is important to note that understanding these details is not necessary for us to be able to use the tool effectively. More important is that we remember that the hypotheses being tested involve the population trimmed means and not the full population means.

The F statistic for comparing trimmed means of tutoring conditions takes the form:

$$F_T = \frac{\sum_{j=1}^{J} r_j \left(T_j - \hat{T} \right)^2}{(J-1)\left[1 + \frac{2(J-2)B_a}{J^2-1} \right]} \tag{6.15}$$

where j is the tutoring group j, k is the disability status k; i.e., Yes or No, and $T_j = \sum_{k=1}^{K} \bar{x}_{Tjk}$ is the sum of trimmed means for tutoring group j across levels of disability status

$$\hat{T} = \frac{\sum_{j=1}^{J} r_j R_j}{\sum_{j=1}^{J} r_j}$$

$$r_j = \frac{1}{\displaystyle\sum_{k=1}^{K} d_{jk}}$$

$$d_{jk} = \frac{\left(n_{jk} - 1\right) S_{wjk}^2}{h_{jk}\left(h_{jk} - 1\right)}$$

where S_{wjk}^2 is the Winsorized variance for combination j and k for tutoring and disability status and h_{jk} is the number of cases for combination j and k for tutoring group and disability status that remain after trimming

$$B_a = \sum_{j=1}^{J} \frac{1}{\hat{V}_j}\left(1 + \frac{r_j}{\displaystyle\sum_{1=j}^{J} r_j}\right)^2$$

$$\hat{V}_j = \frac{\displaystyle\sum_{k=1}^{K} \left(d_{jk}\right)^2}{\displaystyle\sum_{k=1}^{K} d_{jk}^2 / \left(h_{jk} - 1\right)}$$

Again, the details are not so important as remembering that we are comparing the means for the trimmed population rather than the means for the full population. Similar types of calculations will be made to calculate the test statistics for the disability status and interaction null hypotheses.

To obtain the p-values for the F statistics, we can compare the values with the F distribution using the following degrees of freedom for each variable:

$$DF_1 = J - 1 \tag{6.16}$$

$$DF_2 = \frac{\left(J^2 - 1\right)}{3B_a}$$

where J is the number of groups for the term of interest; e.g., tutoring group, disability status, and interaction.

The ANOVA results for the trimmed means based on the methodology described above appear in Table 6.11.

TABLE 6.11

ANOVA Results for Trimmed Means

Variable	F	p
Tutoring method	19.85	0.011
Disability status	14.93	0.005
Interaction	3.57	0.29

ANOVA, analysis of variance.

These results would lead us to reject the null hypotheses for the tutoring method and disability status but not for the interaction. Thus, we would conclude that it is likely that there are differences in the population trimmed means among the tutoring treatment methods and disability status. However, there is not an interaction among the trimmed means for these two variables.

As we discussed with respect to the rank-based approaches described above, given that there are only two disability status groups, we do not need to use a post hoc test to assess for which groups the trimmed means differ. However, we will need to look at pairwise comparisons among the trimmed means for tutoring status. We could also make pairwise comparisons among the six interaction groups created by crossing tutoring treatment and disability status. However, given that we did not reject the null hypothesis associated with the interaction, we will not examine those here. The standard pairwise comparison that can be used to follow up the significant ANOVA controls the Type I error rate using the Bonferroni correction (Wilcox, 2012). The results of the pairwise tests for the trimmed means appear in Table 6.12.

These results indicate that there is a statistically significant difference between the trimmed means for groups 1 and 3 (one-on-one in-person and group in-person), but not for the other pairs. This result qualitatively matches the rank-based and bootstrap results described above. The trimmed means for the math test scores by tutoring method and disability status appear in Table 6.13.

TABLE 6.12

Paired Comparison Results for Trimmed
Means of Tutoring Groups

Tutoring Comparison	Adjusted p-Value
1 vs 2	0.09
1 vs 3	0.02
2 vs 3	0.09

TABLE 6.13

Trimmed Means for Math Achievement
Test for Disability Status and Tutoring
Method Groups

Disability Status	Trimmed Mean
Yes	72.1
No	94.4

Tutoring Group	Trimmed Mean
In-person one-on-one	97.5
Online one-on-one	82.5
Group in-person	66.2

Coupling the paired comparisons and trimmed means results, we see that the population trimmed math test mean scores are likely to be lower for those who have been identified with a learning disability than those who have not. With respect to the tutoring treatment, the population trimmed means are higher for those receiving in-person one-on-one tutoring than for those who were in the group tutoring condition.

Comparing Medians for Factorial ANOVA

When the researcher is interested in comparing population group medians, we can rely on an approach described in Wilcox (2012) that is based on a standard error developed by McKean and Schrader (1984) using the bootstrap. The test statistic involves a Welch-like approach where the standard errors are not pooled. And as with the trimmed means comparisons, the test statistic for assessing the hypotheses involves multiple steps. The null hypotheses in this case are:

$H_0 : M_{T1} = M_{T2} = M_{T3}$; No difference in medians across three tutoring
   methods

$H_0 : M_{TD} = M_{TND}$; No difference in medians between disability status

H_0 : No interaction between disability status and tutoring treatment for
   medians

The F statistic for comparing the population medians of the math achievement test across the tutoring groups takes the following form. Note that the same equation would be used for comparing disability status medians and the medians for the interaction.

$$F_M = \frac{\sum\nolimits_{j=1}^{J} r_j \left(M_j - \hat{M} \right)^2}{(J-1)\left(1 + \dfrac{2(J-2)B_a}{J^2 - 1} \right)} \tag{6.17}$$

$M_j = \sum\nolimits_{k=1}^{K} M_{jk} =$ Sum of medians for tutoring group j across disability status,

$$\hat{M} = \frac{\sum\nolimits_{j=1}^{J} r_j R_j}{\sum\nolimits_{j=1}^{J} r_s}$$

$$r_j = \frac{1}{\sum_{k=1}^{K} d_{jk}}$$

$$r_s = \sum_{j=1}^{J} r_j$$

where $d_{jk} = S_{Mjk}^2$ is the squared standard error of the median for combination j and k of tutoring and disability status

$$B_a = \sum_{j=1}^{J} \frac{1}{\hat{V}_j} \left(1 - \frac{r_j}{\sum_{1=j}^{J} r_j} \right)^2$$

$$\hat{V}_j = \frac{\left(\sum_{k=1}^{K} d_{jk} \right)^2}{\sum_{k=1}^{K} d_{jk}^2 / (h_{jk} - 1)}.$$

We would then compare F_M to the F distribution with $DF_1 = J - 1$ and $DF_2 = \infty$ to get the p-value.

For our math tutoring example, the ANOVA results appear in Table 6.14.

As we saw with the ranks and the trimmed means, there are statistically significant main effects for math test medians of the tutoring approach and disability status, but no interaction between the two. The post hoc median results appear in Table 6.15.

TABLE 6.14

ANOVA Results for Median

Variable	F	p
Tutoring method	31.66	<0.001
Disability status	47.65	<0.001
Interaction	3.31	0.20

ANOVA, analysis of variance.

TABLE 6.15

Paired Comparison Results Based on Math Test Medians of Tutoring Groups

Tutoring Comparison	Adjusted p-Value
1 vs 2	0.02
1 vs 3	<0.001
2 vs 3	0.06

TABLE 6.16

Medians for Math Achievement
Test for Disability Status and
Tutoring Method Groups

Disability Status	Median
Yes	69.7
No	93.6

Tutoring Group	Median
In-person one-on-one	96.5
Online one-on-one	83.8
Group in-person	64.5

The medians for the math test by tutoring method and disability are featured in Table 6.16.

Based on the results in Tables 6.15 and 6.16, we see that in the population, students identified with a math learning disability are likely to have lower median math test scores in the population than those not identified. With respect to the tutoring method, individuals who received in-person one-on-one tutoring have higher median math test scores than those in either of the two groups, and there appears to be no difference in the population medians between the group tutoring and online one-on-one tutoring groups.

Permutation Tests for Factorial ANOVA

The permutation technique for the one-way ANOVA can be extended to the case where there are multiple independent variables and interactions. The technique is essentially identical to that described in Chapter 5, in which ANOVA is conducted for the observed data, and the F statistics for the main effects and interactions(s) are retained. The data are then randomly mixed with respect to the dependent variable and levels of the independent variables for each possible permutation of the data. ANOVA is conducted for each of these permutations to create a distribution for the F statistics for when the null hypotheses for each independent variable and interaction are true; i.e., no differences in groups means across tutoring conditions, no difference in means between disability status groups, and no interaction of tutoring condition and disability status. The F value for the observed data is then compared to the permuted F distribution, and the null hypotheses are rejected if this value exceeds the 95th percentile of the permutation distribution. The pairwise comparisons for statistically significant main effects can be conducted in the same manner as for the one-way ANOVA (see Chapter 5).

The permutation ANOVA results for the math data appear in Table 6.17.

TABLE 6.17

Permutation-Based ANOVA
Results for Math Achievement Data

Variable	Permutation p
Tutoring method	0.004
Disability status	0.001
Interaction	0.133

ANOVA, analysis of variance.

TABLE 6.18

Paired Comparison Results for Tutoring
Group Means Based on Permutations

Comparison	Permutation-Adjusted p
1 vs 2	0.14
1 vs 3	0.10
2 vs 3	0.16

The permutation ANOVA results reveal that there are statistically significant differences among the math test means for the tutoring methods and disability status, but that the interaction of these variables is not statistically significant. The pairwise comparisons for the tutoring method means based on permutation methods appear below in Table 6.18.

Results from the permutation paired comparisons indicate that there are not statistically significant math test mean differences between any pair of tutoring methods after controlling for the familywise Type I error rate using Tukey's method. This result matches those from the standard ANOVA, discussed earlier in the chapter. The statistically significant result for disability status, coupled with the means in Table 6.1, leads us to conclude that those identified with a math disability are likely to have a lower math test mean in the population.

Bayesian Factorial ANOVA

The Bayesian techniques for one-way ANOVA that we discussed in Chapter 5 can be directly extended to the factorial case. We can examine the Bayes Factors (BF) and likelihoods of various hypotheses about each main effect, as well as for the interaction(s). In addition, we can also consider the region of practical equivalence (ROPE) for parameters of interest, such as the differences in group means. We have already explored the technical details of these methods in some detail in Chapters 4 and 5. Therefore, we will not

TABLE 6.19

Bayes Factors and Likelihoods Associated with Hypotheses for Mean Reading Test Score Associated with Tutoring Method

Hypothesis	BF	Probability
$H_0: \mu_{\text{IPT}} = \mu_{\text{OLT}} = \mu_{\text{GT}}$	0.12	0.01
$H_1: \mu_{\text{IPT}} > \mu_{\text{OLT}} > \mu_{\text{GT}}$	4.97	0.53
$H_2: \mu_{\text{IPT}} = \mu_{\text{OLT}} > \mu_{\text{GT}}$	3.08	0.33
$H_3: \mu_{\text{IPT}} > \mu_{\text{OLT}} = \mu_{\text{GT}}$	1.27	0.13

GT, group tutoring; OLT, online tutoring; IPT, in-person tutoring.

TABLE 6.20

Bayes Factors and Likelihoods Associated with Hypotheses for Mean Reading Test Score Associated with Disability Status

Hypothesis	BF	Probability
$H_0: \mu_{\text{disabled}} = \mu_{\text{nondisabled}}$	1.73	0.46
$H_1: \mu_{\text{disabled}} < \mu_{\text{nondisabled}}$	1.82	0.49
$H_2: \mu_{\text{disabled}} > \mu_{\text{nondisabled}}$	0.18	0.05

repeat that discussion here. Instead, we will focus our attention on examining how the approaches can be used for our two-way ANOVA problem, particularly with respect to mean differences for disability status and the interaction of tutoring method with disability status.

In Chapter 5, we considered four hypotheses regarding the population means for the tutoring methods. These hypotheses appear in Table 6.19, along with the BFs and the probability of each hypothesis based on the two-way ANOVA model.

We see that the most likely hypothesis is that in-person one-on-one tutoring has the highest population mean reading test, followed by online one-on-one tutoring, with in-person group tutoring having the lowest mean math test score. The next most likely hypothesis states that the two one-on-one tutoring methods have a higher mean reading test score than group tutoring.

For the disability status variable, there are three hypotheses to consider:

$$\mu_{\text{disabled}} = \mu_{\text{nondisabled}}$$

$$\mu_{\text{disabled}} < \mu_{\text{nondisabled}}$$

$$\mu_{\text{disabled}} > \mu_{\text{nondisabled}}$$

The BFs and probabilities for each hypothesis appear in Table 6.20.

The probabilities for the null hypothesis of no group mean difference and of H_1, which states that the mean math test score for nondisabled students is greater than that of disabled students. It is very unlikely that those identified with a disability exhibit higher average math test performance than those not identified with a disability (probability = 0.05). What is particularly interesting about the results in Table 6.20 is that the probability of the null hypothesis (no difference in average means across disability status) is almost as high as the probability for the hypothesis that students not identified with a disability have higher math test score means than those who have been identified. If we relied solely on the hypothesis test results and means, we would come to the conclusion that those in the disability status condition perform worse on the math test and think nothing more of it. However, also considering the posterior probabilities associated with each hypothesis provides greater nuance to our conclusions. Yes, it is true that students in the disability status condition are most likely to have lower mean math test performance. But it is also almost as likely that there is no difference in math test performance at all. This result should provide us with context as we move forward with our research in this area, keeping in mind that the math test performance for the two disability status groups identified in this sample may not be found in future studies, or indeed in the population as a whole.

Finally, we can consider hypotheses about the means associated with the interaction of disability status and tutoring method. For this study, we have three such hypotheses of interest based on prior work in the field. These hypotheses appear in Table 6.21.

Hypothesis H_0 states that there is not an interaction between tutoring method and disability status. H_1 states that math test means for those identified with a disability are lower than those not so identified, regardless of tutoring method, and that there are no differences in math test means across tutoring methods for those with a disability. It also states that for individuals who have not been identified with a disability, those receiving group tutoring have lower mean math test scores than do students receiving one-on-one tutoring. The third hypothesis is similar to H_1, except that for nondisabled students, individuals who receive in-person one-on-one tutoring have higher

TABLE 6.21

Bayes Factors and Likelihoods Associated with Hypotheses for Mean Reading Test Score Associated with Disability Status

Hypothesis*	BF	Probability
$H_0 : \mu_{T1Ld1} = \mu_{T2LD1} = \mu_{T3LD1} = \mu_{T1Ld2} = \mu_{T2LD2} = \mu_{T3LD2}$	0.83	0.79
$H_1 : \mu_{T1Ld1} = \mu_{T2LD1} = \mu_{T3LD1} \left\langle \mu_{T1Ld2} = \mu_{T2LD2} \right\rangle \mu_{T3LD2}$	0.09	0.08
$H_2 : \mu_{T1Ld1} = \mu_{T2LD1} = \mu_{T3LD1} \left\langle \mu_{T1Ld2} \right\rangle \mu_{T2LD2} = \mu_{T3LD2}$	0.14	0.13

T1, in-person one-on-one; T2, online one-on-one; T3, Group; LD1, identified with disability; LD2, not identified with disability.

mean scores than either of the other tutoring groups. The results indicate that H_0 (no interaction between tutoring and disability status) is by far the most likely hypothesis (probability = 0.83).

In addition to considering the probabilities associated with each hypothesis, we can also investigate the ROPE results for our factorial ANOVA. The approach used in this case is basically identical to that featured in Chapter 5 for the one-way ANOVA. We will not review the details of the method here, but readers should refer to Chapters 4 and 5 as needed. The posterior means and 95% credibility intervals for various parameters of interest for the factorial ANOVA appear in Table 6.22.

These means are quite similar to those obtained in the frequentist approach and displayed in Table 6.23. The credibility intervals are quite narrow for all of the means displayed here, meaning that we have a great deal of certainty about the mean of the posterior distribution.

The means of the posterior estimates for the difference in means for the tutoring method reveal that the in-person one-on-one approach yielded higher mean math test scores than either of the other methods. The 95% credibility interval for the reading score mean difference for the two disability status groups did not include 0, indicating that the not identified group had a higher mean score than did those who were identified (Table 6.24).

TABLE 6.22

Posterior Estimates of Means and 95% Credibility Intervals for Math Test

Parameter	Posterior Mean	95% Credibility Interval
Tutoring Method		
μ_{IP}	97.62	(95.81, 99.38)
μ_{OL}	82.63	(79.39, 85.89)
μ_G	67.48	(52.23, 82.62)
Disability Status		
μ_{Ld1}	94.10	(93.79, 94.40)
μ_{Ld2}	68.79	(67.61, 69.96)
Interaction		
μ_{T1Ld1}	87.81	(87.14, 88.47)
μ_{T2Ld1}	77.67	(75.96, 79.37)
μ_{T3Ld1}	43.79	(32.39, 55.41)
μ_{T1Ld2}	103.57	(103.19, 103.92)
μ_{T2Ld2}	92.00	(86.50, 97.39)
μ_{T3Ld2}	84.85	(83.34, 86.41)

TABLE 6.23

Posterior Estimates of Differences in Group Means and
Standardized Mean Differences by Tutoring Method,
Disability Status, and the Interaction

Parameter	Posterior Mean	95% Credibility Interval
Tutoring Method		
$\mu_{IP} - \mu_{OL}$	14.99	(11.27, 18.69)
$\mu_{IP} - \mu_{G}$	30.14	(14.95, 45.5)
$\mu_{OL} - \mu_{G}$	15.15	(−0.39, 30.9)
δ_{12}	4.49	(3.34, 5.59)
δ_{13}	7.02	(3.34, 10.51)
δ_{23}	3.37	(−0.09, 6.81)
Disability Status		
$\mu_{Ld1} - \mu_{Ld2}$	25.31	(24.08, 26.50)
δ_{12}	5.88	(5.58, 6.17)

TABLE 6.24

Posterior Estimates of Differences in Group Means and
Standardized Mean Differences by the Interaction of
Disability Status and Tutoring Treatment

Parameter	LQ	Mean	UQ	Std.Err
:------------	:------	:------	:------	:------
mu2-mu1	−17.12	−11.64	−6.1	2.81
mu3-mu1	−20.31	−18.73	−17.15	0.8
mu4-mu1	−16.46	−15.75	−15.02	0.37
mu5-mu1	−27.66	−25.92	−24.27	0.86
mu6-mu1	−71.39	−59.6	−48.12	6.01
mu3-mu2	−12.98	−7.09	−1.23	2.93
mu4-mu2	−9.62	−4.11	1.39	2.82
mu5-mu2	−19.91	−14.28	−8.51	2.93
mu6-mu2	−60.92	−47.96	−34.9	6.6
mu4-mu3	1.28	2.97	4.66	0.85
mu5-mu3	−9.48	−7.2	−4.94	1.16
mu6-mu3	−52.84	−40.87	−29.02	6.06
mu5-mu4	−11.96	−10.17	−8.43	0.9
mu6-mu4	−55.79	−43.84	−32.26	6.01
mu6-mu5	−45.57	−33.67	−22.1	6.07
delta12	1.87	3.61	5.28	0.87
delta13	6.58	7.34	8.02	0.37

(Continued)

TABLE 6.24 (*Continued*)

Posterior Estimates of Differences in Group Means and
Standardized Mean Differences by the Interaction of
Disability Status and Tutoring Treatment

Parameter	LQ	Mean	UQ	Std.Err
delta14	6.54	7.04	7.49	0.24
delta15	3.16	3.64	4.09	0.24
delta16	2.61	3.49	4.3	0.44
delta23	0.35	2	3.61	0.82
delta24	−0.41	1.24	2.87	0.84
delta25	0.61	1.07	1.49	0.23
delta26	1.64	2.34	2.99	0.35
delta34	−1.73	−1.11	−0.47	0.32
delta35	0.54	0.81	1.07	0.13
delta36	1.54	2.28	2.96	0.37
delta45	1.06	1.34	1.59	0.14
delta46	1.75	2.54	3.26	0.39
delta56	1.16	1.87	2.53	0.35

With respect to the interaction of disability status and tutoring treatment, we can see that the 95% credibility intervals for the differences in means do not include 0, with one exception (mu4-mu2). This means that, with this lone exception, it appears that math test means differ across the various combinations of the two variables of interest. The exception to this result was for the comparison of online one-on-one tutoring for individuals identified with a math learning disability and individuals in the in-person one-on-one tutoring conditions who were not identified with a math learning disability. In other words, those who were identified with a learning disability but received in-person one-on-one tutoring performed similarly on the math test to a nondisabled individual who received online one-on-one tutoring.

Effect Size Estimates

In this chapter, we have reviewed a variety of approaches for assessing inference about hypotheses focused on group mean differences. We will now consider the magnitude of the effects associated with these differences. Recall from Chapter 5 that in the context of ANOVA, we generally focus on measures that describe the proportion of variance in the dependent variable that is associated with the independent variable. In the case of one-way ANOVA, these included such statistics as η^2, ω^2, and ϵ^2. We learned that η^2 tends to overestimate the size of the effect, which led us to prefer ω^2, in particular. However, it is also true that in many cases these statistics yield fairly similar results. As a reminder, the equations for these effect size estimates appear below:

$$\eta^2 = \frac{SSB}{SST} \tag{6.18}$$

$$\omega^2 = \frac{SSB - \left((J-1)*MSW\right)}{SST + MSW} \tag{6.19}$$

$$\epsilon^2 = 1 - \frac{(n-1)SSW}{(n-J)SST} \tag{6.20}$$

When we have multiple independent variables in our model, these effect size estimates are adjusted to reflect the effect of a variable after accounting for the presence of the other variables in the model. These statistics, which are referred to as partial effect sizes, are each calculated as

$$\eta^2_{partial} = \frac{SSB}{SSB + SST} \tag{6.21}$$

$$\omega^2_{partial} = \frac{SSB - \left((J-1)*MSW\right)}{SST + \left(N - (J-1)\right)MSW} \tag{6.22}$$

$$\epsilon^2_{partial} = \frac{(F-1)*(J-1)}{\left(F*(J-1)+df_W\right)} \tag{6.23}$$

For each partial effect size estimate, we obtain information about the magnitude of the effect associated with a given variable (e.g., tutoring treatment) after accounting for the presence of the other terms in the model (e.g., disability status and interaction). Thus, in the context of factorial ANOVA, the partial effect sizes will almost always be preferable to the raw value. Cohen's (1988) guidelines for interpreting the magnitudes of these statistics can also be used for the factorial ANOVA problem.

The parametric partial effect size estimates and their 95% confidence intervals for the factorial ANOVA appear in Table 6.25.

There are several points to note with respect to the effect size estimates in Table 6.23. First, the confidence intervals are quite wide, suggesting that we have low precision regarding the effects for each model term. Second, tutoring treatment and disability status had comparable relationships to the math test scores with point estimates in the large range. However, the interaction accounted for a much smaller proportion of variance in the math test scores with a confidence interval including all possible values from 0 to 1. Thus, we would need to be careful in assigning much importance to the interaction.

TABLE 6.25

Parametric Effect Size Estimates (95% Confidence Intervals) for Factorial ANOVA

Variable	ω^2	ϵ^2	η^2
Tutoring treatment	0.47 (0.06, 1.00)	0.54 (0.13, 1.00)	0.60 (0.21, 1.00)
Disability status	0.44 (0.09, 1.00)	0.52 (0.16, 1.00)	0.55 (0.20, 1.00)
Interaction	0.13 (0.00, 1.00)	0.17 (0.00, 1.00)	0.28 (0.00, 1.00)

ANOVA, analysis of variance.

TABLE 6.26

Rank-Based and Robust Effect Size Estimates for Factorial ANOVA

Variable	ξ	η_R^2
Tutoring treatment	1.17 (−0.02, 3.46)	0.59
Disability status	0.79 (0.46, 1.69)	0.58
Interaction	1.17 (−0.02, 5.04)	0.22

ANOVA, analysis of variance.

In terms of the main effects, we can see that the effect for each variable is likely to be moderate to large based on the confidence intervals.

The rank-based and robust effect size estimates (see Chapter 5 for a description) appear in Table 6.26.

Currently, there is no available software to calculate the confidence intervals for η_R^2 in the factorial case, and thus only the point estimates can be reported. The rank-based results reveal a large effect for both tutoring treatment and disability status with a smaller effect for the interaction, mirroring the results for the parametric effect sizes. The point estimates for the robust effect size fall into the large range as suggested by Wilcox and Tian (2011). However, the confidence interval for both tutoring and the interaction includes 0, suggesting the possibility of no effect at all. Thus, we would want to interpret these results with some care.

Summary

In this chapter, we extended the work that we began in Chapter 5. We saw that it is possible to apply many of the robust and nonparametric techniques we have been discussing throughout this book to factorial ANOVA models. In addition, we saw that the parametric factorial ANOVA rests on the same set of assumptions as did one-way ANOVA (independence of observations, homogeneity of variance, and normality). When these assumptions are violated, the results obtained from the ANOVA may not be reliable, in which case one of the alternatives that we learned about may be preferable. We also saw in this chapter that the effect sizes and Bayesian techniques that proved so helpful in the one-way ANOVA can also be applied in the factorial case.

The choice of the nonparametric/robust alternative to use when the parametric assumptions don't hold can be made using the same logic that we applied to the one-way case. For example, when the data are skewed and/or there are outliers present, rank-based methods approaches such as relative effects and ART provide control over the Type I error rate and yield good power. Likewise, the ANOVA for comparing medians based on the McKean and Schrader standard error is also a reasonable option for the data analyst to use. If the primary problem is the presence of outliers, then a test comparing the trimmed means may be the best technique to use. If variance heteroscedasticity is the major issue but the data are normally distributed, then the Welch F-test would be a good option. The bootstrap and permutation test approaches are also viable alternatives, with the latter being particularly useful when we have a small sample size. And, as we mentioned above, Bayesian estimation provides us with an additional set of informative tools, particularly the BF and the ROPE estimates of mean differences. Indeed, the latter can be used to directly assess pairwise mean differences without needing the two-step approach (omnibus test followed by post hoc comparisons) that is inherent in most other ANOVA analyses.

7

Comparing Means for Repeated Measures Analysis of Variance and Split-Plot Designs

In Chapter 6, we learned about factorial designs where our interest was in comparing the central tendency of a dependent variable across groups associated with multiple variables. For example, we might be interested in comparing means on a measure of depression for individuals who have been assigned to a treatment or control group. We might also have information about their exposure to violence as preschool children, coded as yes or no, and want to compare mean depression scores between the genders. Furthermore, we saw that the ability to investigate the interaction of two variables (i.e., treatment condition by violence exposure) can be easily incorporated into factorial designs. Exploration of such interactions is a particularly useful feature of factorial designs, as we saw in the previous chapter.

In the factorial ANOVA framework that we discussed in Chapter 6, the data were assumed to come from a cross-sectional research design. This means that there is one measurement made for each individual in the sample. In the depression example, each subject will be randomly assigned to only one of the treatment conditions (treatment or control) and will only select one early childhood violence category (yes or no). However, in some studies, an individual may appear in multiple categories of one or more of the independent variables. A common example of such a situation occurs in repeated measures designs.

To illustrate the repeated measures design framework, let's consider a study where an exercise scientist is interested in the amount of time a subject can ride a stationary bicycle at 80% of maximum aerobic capacity before they ask to stop due to fatigue. In the first testing session, the subjects ingest 8 ounces of sugar water 30 minutes prior to the cycling session. In a second session 1 week later, the same subjects ingest 8 ounces of coffee with an equivalent amount of sugar as in session one. This represents a repeated measures design that includes a within-subjects variable, type of drink. We can expand the cycling study by including a variable reflecting whether an individual exercises 150 minutes per week or more, which we code as 1=yes and 0=no. Each person in the sample will only have one value for exercise minutes, making it a between-subjects variable. We have now extended the simple repeated measures design (including only within-subjects independent variables) to what is known as a split-plots design, with both within-subjects and between-subjects independent variables. We can answer the same types of

 DOI: 10.1201/9781003379324-7

questions about main effects and interactions with these repeated measures models as we could with the factorial designs from Chapter 6. However, the standard factorial ANOVA approach is not appropriate in this instance because we have multiple dependent variable values for each subject, leading to a violation of the independent subject's assumption. Repeated measures models can include multiple within-subjects and multiple between-subjects effects. In addition, the repeated measures variable can include two or more time points, as is the case for the between-subjects variables.

One of the primary advantages of the repeated measures framework is that the individual subjects serve as their own controls. This means that any qualities or traits specific to an individual that might impact the outcome variable will be associated with each of the repeated measures in the same manner. For example, in our cycling example described above, we would expect an individual's fitness level to have a direct influence on the time that they can cycle until fatigue forces them to stop. In the repeated measures case, the impact of a person's fitness will apply equally to the sugar water and coffee treatments. In contrast, if we were to randomly assign subjects to only one of the treatments, an individual's inherent aerobic capacity would only influence the outcome for the single treatment to which they were assigned. And while randomness should, in theory, balance such effects out, in practice, this might not be the case, especially for smaller samples. This fact represents a major advantage of repeated measures designs.

We can express the repeated measures model as

$$y_{ij} = \mu_{..} + t_j + p_i + \varepsilon_{ij} \qquad (7.1)$$

where y_{ij} is the dependent variable value for person i at time j, $\mu_{..}$ is the overall mean for the dependent variable, t_j is the time effect, p_i is the person effect, and ε_{ij} is the random error for person i at time j.

If we were to use the one-way ANOVA from Chapter 5 and ignore the repeated nature of the dependent variable, our model would be

$$y_{ij} = \mu_{..} + t_j + \varepsilon_{ij} \qquad (7.2)$$

Notice that the model in equation (7.2) does not account for p_i, meaning that it will be folded into the error term. Thus, if we apply the standard ANOVA model to the repeated measures data, we would be assuming that the error term is ε_{ij}. But it would actually be $p_i + \varepsilon_{ij}$. As we discussed in Chapters 4, 5, and 6, with standard cross-sectional means comparison models, ε_{ij} is random with a 0 correlation between any pair of values therein. However, when the data come from a repeated measures process, this assumption is very likely to be violated, given the effects of p_i. For this reason, we need to properly account for the repeated measurements on the same individuals using an approach other than the standard cross-sectional ANOVA models.

The fundamental equations for sums of squares, mean squares, degrees of freedom, and the F statistic in the repeated measures framework are similar to those presented in Chapters 5 and 6, with one major addition. In Chapter 5, we learned that the sums of squares reflect differences among group means (sum of squares between groups) and among individuals within the same group (sum of squares within groups). We can add these two values together to obtain the total sum of squares (sum of squares total), which reflects differences among individuals within the full sample. In equation form, these relationships are written as:

$$SST = SSB + SSW \tag{7.3}$$

where SST is the sum of squares total, SSB is the sum of squares between groups, and SSW is the sum of squares within groups.

In turn, the equations for these sums of squares for the standard ANOVA with a single between-subjects variable are:

$$SSB = \sum\nolimits_{j=1}^{j} n_j \left(\bar{x}_j - \bar{x}_. \right)^2 \tag{7.4}$$

$$SSW = \sum\nolimits_{i=1}^{N} \left(x_{ij} - \bar{x}_j \right)^2 \tag{7.5}$$

$$SST = \sum\nolimits_{i=1}^{N} \left(x_{ij} - \bar{x}_. \right)^2 \tag{7.6}$$

where x_{ij} is the score for individual i in group j, $\bar{x}_j$ is the mean for group j, and $\bar{x}_.$ is the overall sample mean.

Equation (7.1) implies that in the context of repeated measures data, the SST consists not only of between and within groups effects (SSB and SSW) but also of a person effect. This person effect can be captured by the sum of squares associated with the person (SSP), which reflects differences among individuals within the sample. It is calculated as:

$$SSP = k \sum\nolimits_{p=1}^{P} \left(\bar{x}_P - \bar{x}_. \right)^2 \tag{7.7}$$

where k is the number of measurements made on each person and $\bar{x}_P$ is the mean of measurements made on person p.

Notice that when we are working in the repeated measures framework, x_{ij} becomes x_{ijp} and is now the value of the dependent variable for measurement i (e.g., pre and posttest) for person p in between-subjects variable group j (e.g., treatment or control). The total sum of squares for the repeated measures design, which represents the total variability contained in the sample data, is then written as

$$SST = SSB + SSP + SSW \qquad (7.8)$$

We can see the importance of accounting for a person when we analyze data obtained in a repeated measures framework. If we do not do so appropriately and instead incorrectly use a model based on SST = SSB+ SSW, we will be ignoring an important source of variation (SSP) in the dependent variable. We will illustrate the calculation of the mean squares and F statistic for this design below.

As mentioned above, we can extend the repeated measures model to include one or more between-subjects effects, as well as multiple within-subjects effects. Given their importance in many areas of research, we will focus on a model with one between-subjects and one within-subjects effect, often referred to as a split-plot design. This model combines the between-subjects factorial model from Chapter 6 with the repeated measures model described above and can be written as:

$$y_{ijk} = \mu_{..} + t_j + \gamma_k + t_j\gamma_k + p_i + \varepsilon_{ijk} \qquad (7.9)$$

where γ_k is the effect of being in group k of the between-subjects variable and $t_j\gamma_k$ is the interaction of within and between-subjects variables.

We can see that the dependent variable is a function of the overall mean, the time effect, the group effect, the interaction of group and time, the person effect, and random error. In many respects, the model in equation (8.4) is quite similar to the two-way ANOVA model in Chapter 7. The one additional factor in the split-plot model is the person effect, which is not present in purely cross-sectional models, as those in Chapter 7.

As an example of the model in equation (7.9), we can consider a study in which individuals were assigned to one of two treatment groups (treatment or control) and measured on an outcome variable prior to the treatment and then again after the treatment. The within-subjects term, t_j, has two levels (pretest and posttest) as does the between-subjects term, γ_k, treatment, and control. The interaction term, $t_j\gamma_k$, assesses whether the change in mean scores from the pretest to the posttest differs for the treatment and control groups. Next, we will examine the assumptions underlying the parametric repeated measures model, followed by a full example of how this model can be fit to data from a split-plot design.

Assumptions Underlying Repeated Measures ANOVA Models

The parametric repeated measures ANOVA model rests on three fundamental assumptions about the data. First, we must assume that the data are normally distributed and will assess it in the same manner that we did in Chapters 4–6, using the QQ plot and tests of normality. The second assumption is that the individual data points are independent of one another once we've accounted for the time and person effects. The final assumption underlying the repeated measures ANOVA is sphericity. Sphericity plays a role in repeated measures designs similar to the homogeneity of variance assumption in the cross-sectional ANOVA context. Sphericity occurs when the variances of the differences among *all pairs* of the dependent variables are equal. As an example, imagine that we have three measurements made over time: $T1$, $T2$, and $T3$. We can calculate the following difference scores using these time variables:

$D12 = T1–T2$

$D13 = T1–T3$

$D23 = T2–T3$

The sphericity assumption is met when the variances of $D12$, $D13$, and $D23$ are all equal. When the sphericity assumption is violated, the Type I error rate is inflated (the F value is positively biased). We can use Mauchly's test to assess the sphericity assumption for our data. The null and alternative hypotheses for this test are:

H_0: sphericity is present.

H_A: sphericity is not present

Thus, if Mauchly's test returns a statistically significant result, we would conclude that sphericity does not hold for our sample.

Huynh and Feldt (1976), as well as Greenhouse and Geisser (1959), developed a measure for the degree of departure from sphericity for a set of sample data, ε. When sphericity is present, $\varepsilon = 1$. If sphericity doesn't hold, we can use the value of ε to adjust the df for our test statistic, which is necessary because a violation of sphericity results in fewer df for the F. These are the methods laid out by Huynh and Feldt, and Greenhouse and Geisser. One final point to make here is that when we have only two repeated measurements, sphericity is not an issue, given that there is only one difference (e.g., $T1–T2$) and thus only one variance present.

Example of Parametric Repeated Measures Analysis

We will demonstrate the application of the repeated measures ANOVA model with an example dataset featuring a sample of 15 adults who have been diagnosed with autism (ASD). Each individual was administered a scale to measure their interpersonal relationship skills, where higher scores indicate that the individual is relatively more skilled in interacting with other people.

After the initial measurement, participants were randomly assigned to either a treatment or control group. Each individual in the treatment group participated in a 4-week training program in which they were taught strategies for successfully interacting with others. Individuals who were assigned to the control group received no special instruction. After 4 weeks, study participants from both groups were again given the interpersonal skills assessment. The program coordinators believed that the training session would result in improved interpersonal skills on the part of those assigned to the treatment condition, with no change anticipated for the control group.

The means and standard deviations of the interpersonal assessment scores by time and treatment condition appear in Table 7.1.

For this sample, there was an increase in mean interpersonal skills scores of approximately six points from time 1 to time 2 for the treatment group, whereas for the control group, the change over time in mean scores was under one point.

The distributions of scores by group and time appear in the boxplots in Figure 7.1.

TABLE 7.1

Group Size (*N*) and Means (Standard Deviations) of Interpersonal Skills Assessment Scores by Time and Treatment Condition

Condition	N	Time 1	Time 2
Treatment	8	6.75 (4.20)	12.63 (2.97)
Control	7	9.43 (4.16)	10.14 (4.95)
Total	15	8.00 (4.26)	11.47 (4.07)

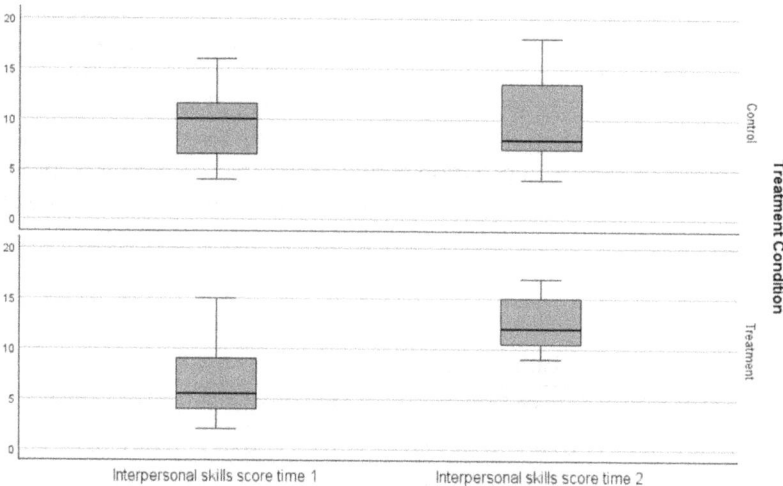

FIGURE 7.1
Distribution of interpersonal skills scores by time and treatment group.

The boxplots show that there appears to have been an increase in scores for the treatment condition but not for the control group, reflecting the patterns evident in Table 7.1. In addition, it appears that there was greater variability in the time 2 scores for the control group than was the case for those in the treatment condition. We also see something similar in the standard deviations in Table 8.1 as well.

We will begin with a simple repeated measures model considering only the time effect and ignoring treatment conditions. This model corresponds to equation (7.2). The null and alternative hypotheses of interest are:

$H_0 : \mu_{T1} = \mu_{T2}$; i.e., the interpersonal score means are the same at time 1 and time 2.

$H_A : \mu_{T1} \neq \mu_{T2}$; i.e., the interpersonal score means are different at time 1 and time 2.

The results of the parametric repeated measures ANOVA appear in Table 7.2. Notice that we refer to the within-group variability as an error.

Using the terms from Table 7.2, we can calculate the sums of squares, mean squares, and F statistic.

$$MSW = \frac{SSW}{N-J} = \frac{205.87}{15-2} = 14.71$$

$$MSB = \frac{SSB}{J-1} = \frac{90.13}{2-1} = 90.13$$

$$F = \frac{MSB}{MSW} = \frac{90.13}{14.71} = 6.13$$

The p-value for $F_{1,14}$ is 0.03. If we set $\alpha = 0.05$, we would reject the null hypothesis and conclude that the mean interpersonal skills score is unlikely to be equal at times 1 and 2 in the population. The overall time mean at time 2 (Table 8.1) was higher than that of time 1. Taking the results together leads to the conclusion that in the population of those identified as ASD, interpersonal skills scores are likely to be higher at time 2 than at time 1. Finally, the effect size (η^2) for this model is 0.31, meaning that time accounts for approximately

TABLE 7.2

Results of Parametric Repeated Measures ANOVA

Term	SS	df	MS	F	p
Time	90.13	1	90.13	6.13	0.03
Error	205.87	14	14.71		

ANOVA, analysis of variance.

TABLE 7.3

Results of ANOVA *Incorrectly* Ignoring the Repeated Measures Nature of the Data

Term	SS	df	MS	F	p
Time	90.13	1	90.13	5.20	0.03
Error	485.73	28	17.35		

ANOVA, analysis of variance.

31% of the variability in the interpersonal skills scores. The calculation for this statistic is the same as for the one-way ANOVA in Chapter 5.

$$\eta^2 = \frac{90.13}{90.13 + 205.87} = \frac{90.13}{296} = 0.305$$

If we *incorrectly* ignored the repeated nature of our dependent variable, the ANOVA results would be

There are some interesting results to notice here. First, the SSB is the same for time, regardless of how we model the data. This makes sense because the calculation of this term only involves the means of time and the overall mean. However, the SSW and its associated degrees of freedom are quite different for the two approaches. The reason for this difference is that for SSW in Table 7.2, the person effect has been separated from the error, whereas for the results in Table 7.3, the person effects have been folded into SSW. It is important to state again that this is not the correct way to fit this model and that the repeated measures approach should be used for problems such as this.

Now that we have worked with the repeated measures model including a single within-subjects variable, let's extend it to the split-plot model by including treatment condition as well as time. This model corresponds to equation (7.9). The null and alternative hypotheses associated with this model are:

Time

$H_0 : \mu_{T1} = \mu_{T2}$; i.e., the interpersonal score means are the same at time 1 and time 2.

$H_A : \mu_{T1} \neq \mu_{T2}$; i.e., the interpersonal score means are different at time 1 and time 2.

Treatment

$H_0 : \mu_{Trt1} = \mu_{Trt2}$; i.e., the interpersonal score means are the same for the two treatments.

$H_A : \mu_{Trt1} \neq \mu_{Trt2}$; i.e., the interpersonal score means are different for the two treatments.

Interaction

H_0 : There is not an interaction between time and treatment

H_A : There is an interaction between time and treatment

The results of the ANOVA model appear in Table 7.4.

Using the terms from Table 7.4, we can calculate the sums of squares, mean squares, and F statistics.

TABLE 7.4

Results of Parametric Repeated Measures ANOVA

Term	SS	df	MS	F	p
Time	81.05	1	81.05	6.75	0.02
Time × treatment	49.72	1	49.72	4.14	0.06
Error for within-subjects effects	156.15	13	12.01		
Treatment	0.072	1	0.072	0.003	0.96
Error for between-subjects effect	279.80	13	21.52		

ANOVA, analysis of variance.

$$MSW_{Within} = \frac{SSW_{Within}}{N - df_{Time} - df_{Time \times Trt}} = \frac{156.15}{15 - 1 - 1} = 12.01$$

$$MSB_{Time} = \frac{SSB_{Time}}{J - 1} = \frac{81.05}{2 - 1} = 81.05$$

$$F_{Time} = \frac{MSB_{Time}}{MSW_{Within}} = \frac{81.05}{12.01} = 6.75$$

$$MSW_{Between} = \frac{SSW_{Between}}{N - df_{Trt} - df_{Time \times Trt}} = \frac{279.80}{15 - 1 - 1} = 21.52$$

$$MSB_{Time} = \frac{SSB_{Time}}{K - 1} = \frac{0.072}{2 - 1} = 0.072$$

$$F_{Time} = \frac{MSB_{Time}}{MSW_{Within}} = \frac{0.072}{21.52} = 0.003$$

There are several points of interest in the results from the split-plots analysis. First, notice that the denominators for the within (time) and between (treatment) effects are different. This difference reflects the fact that the between-subjects error term includes both random error (ε_{ijk}) and person effects (p_i), whereas the within-subjects error term includes only random error (ε_{ijk}). Therefore, the error term for the between-subjects effect, estimated by $MSW_{Between}$, is larger than the error term for the within-subjects effect, MSW_{Within}. In addition, the MSW_{Within} is smaller for the split-plot model (12.01) than for the simple repeated measures model (14.71). This difference in magnitude reflects the fact that by including treatment and the interaction of treatment by time, we are accounting for some of the variability in the interpersonal skills score that was not accounted for in the original model, leading to a decrease in the error term. Similarly, we see that MSB_{Time} also changes value when we include treatment and the interaction because some of the time variability in the original model is associated with either the treatment condition, the interaction, or both. Finally, note that the interaction of a between- and a within-subjects effect is itself a within-subjects effect.

Now that we have reviewed the mathematics underlying the model terms, let's consider what these results actually tell us about our research questions. Given the *p*-values in Table 8.4 and using $\alpha=0.05$, we would conclude that the time effect ($p=0.02$) is statistically significant but not treatment ($p=0.96$) nor the interaction of time by treatment ($p=0.06$). Given these results, we can reject the null hypothesis for time and conclude that the interpersonal score means do differ between time 1 and time 2. However, there was not sufficient statistical evidence for us to reject the null hypotheses for either the treatment effect or the interaction. Putting these results together with the means in Table 8.1, we conclude that interpersonal skills scores increased over time, and these changes were not different between the two groups.

The η^2 effect size is calculated in the same fashion for the split-plot design as it was for the repeated measures with only a within-subjects variable. The values for this example appear below.

$$\eta^2_{Time} = \frac{81.05}{81.05+156.15} = \frac{81.05}{237.20} = 0.34$$

$$\eta^2_{Trt} = \frac{0.072}{0.072+279.80} = \frac{0.072}{279.87} = 0.00$$

$$\eta^2_{Time \times Trt} = \frac{49.72}{49.72+156.15} = \frac{49.72}{205.87} = 0.24$$

Time was associated with 34% of the variability in interpersonal skills scores and the interaction of time by treatment was associated with 24% of the score variation. Treatment was associated with essentially none of the variability in interpersonal skills scores.

Assumption Check for Repeated Measures and Split-Plot ANOVA

Now that we've fit the split-plot ANOVA, we will assess whether the assumptions underlying it have been met. As we discussed above, these assumptions are quite analogous to those underlying the between-subjects ANOVA models from Chapters 5 and 6. Likewise, assessment of these assumptions can also be carried out much as we did for those models. For example, the QQ plot and Shapiro–Wilk test for normality can be used in the repeated

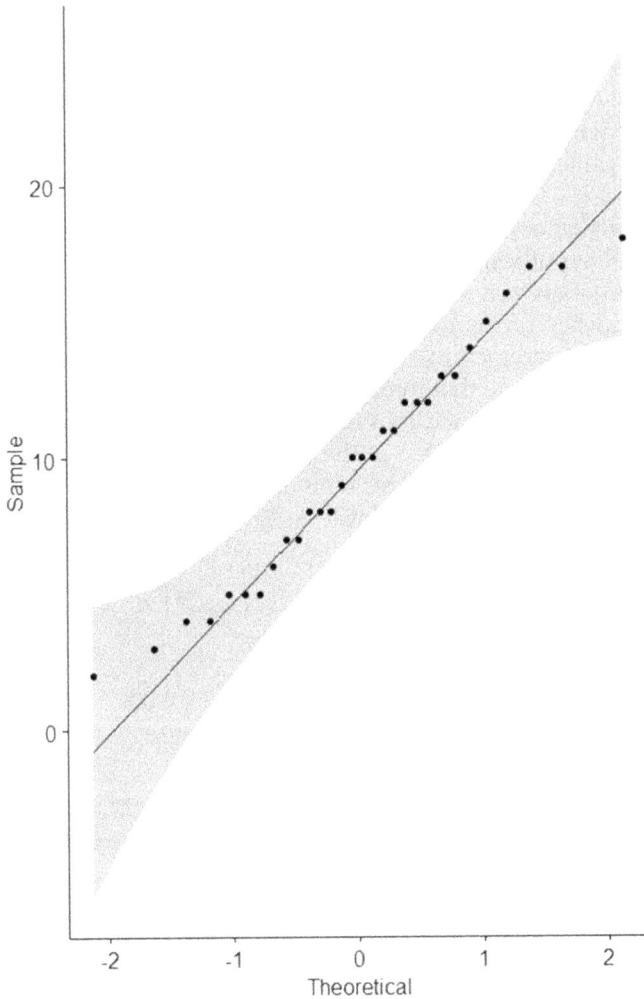

FIGURE 7.2
QQ plot for the interpersonal skills score.

measures context. Figure 7.2 contains the QQ plot for the combined repeated measures data from times 1 and 2 for our interpersonal relations data.

The individual data points all lie close to the line representing normality and fall within the 95% confidence band. Therefore, we can conclude that the data do appear to be normally distributed. The histogram of interpersonal skills scores across time appears in Figure 7.3.

The Shapiro–Wilk $W=0.97$ with a p-value of 0.5398. Thus, we would not reject the null hypothesis that the data come from the normal distribution. Finally, the skewness for our data is 0.13, and the kurtosis is 1.93,

Histogram of Interpersonal Scores

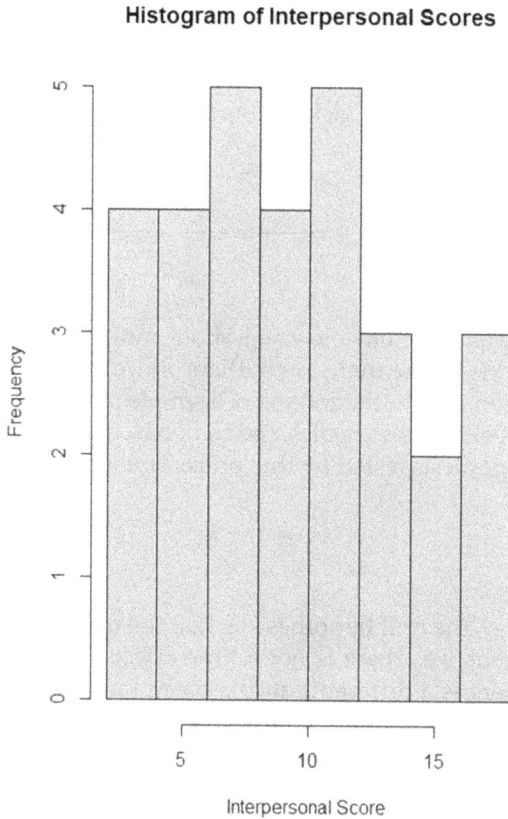

FIGURE 7.3
Histogram for the interpersonal skills score.

both of which fall within reasonable bounds for the normal distribution
(George & Mallery, 2010).

The sphericity assumption is not an issue for this dataset because we only
have two time points. Recall that sphericity involves comparing the variances of difference scores for the repeated measures data. However, since
we only have a single such difference (Time 2–Time 1), there is only a single
variance, and thus no comparison to be made. Finally, we will assume that
the data are independent given that our sample has been randomly selected
and randomly assigned to the treatment condition. One final assumption
that we need to assess in the split-plot model is homogeneity of the dependent variable variance across the between-subjects groups. This is the same
assumption about equal group variances that we make in the standard factorial ANOVA context, and the method that we use (i.e., Box's Test) is the same
as well. The null hypothesis of the test for equality of variances is that they
are, in fact, equal across groups. The result of Box's test for the interpersonal
skills data and treatment condition is $F_{3,121191.448} = 3.578$, $p = 0.013$. If we set

$a = 0.05$, we would reject the null hypothesis and conclude that the interpersonal skills score variances are not equal across the treatment groups, thus violating the homoscedasticity assumption. Below, we will consider methods that properly account for this lack of homoscedasticity.

Friedman's Test

In previous chapters, we have learned about rank-based methods for data analysis in contexts where there are outliers and/or the parametric assumptions have not been met. In the context of a simple repeated measures design, with no between-subjects variables, the rank-based procedure is Friedman's test. The null hypothesis tested by this procedure is

$$H_0 : \mu_{RT1} = \mu_{RT2}$$

$H_A : H_0$ is not true

Put another way, the null hypothesis is that the mean of the ranks is equal across time periods; i.e., there is not a time effect. If we reject H_0, then we conclude that there is a difference in the mean rank of the dependent variable across time. The assumptions underlying Friedman's test are that the observations are independent and that the residuals (differences between the observed and model-predicted interpersonal skills score) come from the same distribution across times.

Friedman's test is calculated as

1. Rank the scores for the times within each subject.
2. Calculate Friedman's test statistic as:

$$\text{FM} = \frac{12n}{k(k+1)} \sum_{k=1}^{K} \left(\bar{R}_k - \frac{K+1}{2} \right)^2 \tag{7.10}$$

   where K is the number of measurement occasions, n is the number of individuals in the sample, and $\bar{R}_k$ is the mean rank for time k.
3. To obtain the p-value, compare the value of FM with the χ^2 distribution with $K - 1$ degrees of freedom.

   When there are ties in the data, the test statistic is calculated as:

$$\text{FM}_{\text{tie}} = \frac{n^2}{s_j^2} \sum_{k=1}^{K} \left(\bar{R}_k - \frac{K+1}{2} \right)^2 \tag{7.11}$$

s_j^2 is the variance of ranks within individuals.

For a simple repeated measures analysis of the interpersonal skills data, the results of Friedman's test were FM = 4.571, which yields a p-value of 0.033. Therefore, we would reject the null hypothesis and conclude that there is a difference in the mean ranks of interpersonal skills scores across time. The mean ranks for the sample are $\bar{x}_{RT1} = 1.23$ and $\bar{x}_{RT2} = 1.77$. Therefore, we conclude that the mean ranks were higher at time 2 than at time 1, leading us to conclude that interpersonal skills scores were higher at time 2, which is quite similar to what we found with the parametric repeated measures approach. This congruence of results between parametric ANOVA and Friedman's test is not surprising given that the parametric assumptions were met.

A common effect size for use with Friedman's test is Kendall's W (Tomczak & Tomczak, 2014), which ranges between 0 and 1, with larger values indicating a stronger effect associated with time. This statistic is calculated as

$$w = \frac{FM}{N(K-1)} \qquad (7.12)$$

where FM is the Friedman's test statistic value, N is the sample size, and K is the number of time points.

Cohen's (1988) guidelines for interpreting the magnitude of w are the same as for the correlation coefficient:

0.1–0.3: Small
0.3–0.5: Medium
0.5–1.0: Large

We can also obtain confidence intervals for w using the bootstrap (see Chapter 2 to review the bootstrap). For the interpersonal skills data, $w = 0.31$ with a 95% confidence interval of (0.00, 0.75). Based on the point estimate, we would conclude that there is a moderate relationship between time and the interpersonal skills score. Given the very wide confidence interval (which is likely due to the small sample size), there is a great deal of uncertainty as to the actual magnitude of the population time effect.

It is very straightforward to include a between-subjects independent variable in the analysis of ranks with an approach that is not dissimilar from those used for factorial ANOVA (Chapter 6). As an example, let's see the steps to construct this statistic for the test of a treatment effect in our example.

1. Rank the raw data, x_{ijk} into R_{ijk} for subject i in group j at time k.
2. Calculate the mean ranks:

$$\bar{R}_{.jk} = \frac{\sum_{i=1}^{N} R_{ijk}}{n_j} \qquad \text{mean rank for group } j \text{ at time } k \text{ across subjects} \qquad (7.13)$$

$$\bar{R}_{ij.} = \frac{\sum_{i=1}^{N} R_{ijk}}{n_j} \quad \text{mean rank for subject } i \text{ in group } j \text{ across times} \quad (7.14)$$

$$\bar{R}_{.j.} = \frac{\sum_{i=1}^{N} \bar{R}_{.jk}}{n_j} \quad \text{mean rank for group } j \text{ across persons and times} \quad (7.15)$$

3. Calculate the mean squares within for ranks:

$$s_j^2 = \frac{\sum_{i=1}^{n_j} \left(\bar{R}_{ij.} - \bar{R}_{.j.} \right)^2}{n_j - 1}, \; s^2 = \sum_{j=1}^{J} \frac{s_j^2}{n_j} \quad (7.16)$$

4. Calculate the F statistic for ranks:

$$F_R = \frac{J}{\sqrt{s^2 (J-1)}} \sum_{j=1}^{J} \left(\bar{R}_{.j.} - \bar{R}_{...} \right)^2, \text{ where } \bar{R}_{...} = \text{overall mean rank} \quad (7.17)$$

5. Calculate degrees of freedom for the F statistic.

$$DF_1 = \frac{(J-1)^2}{\dfrac{J(J-2)U}{s^2}} \quad (7.18)$$

$$DF_2 = \frac{s^2}{D} \quad (7.19)$$

$$D = \sum_{j=1}^{J} \frac{1}{n_j} \left(\frac{s_j^2}{n_j} \right)^2 \quad (7.20)$$

$$U = \sum_{j=1}^{J} \left(\frac{s_j^2}{n_j} \right)^2 \quad (7.21)$$

To obtain the p-value for F_R, we refer to the F distribution with the degrees of freedom calculated above. When testing for the time effect and the interaction, we will use the same basic approach as outlined above.

The results of the rank-based analysis for the interpersonal skills example, including both time and treatment groups, appear in Table 7.5.

Given the results in Table 7.5, we would reject the null hypotheses for time and the interaction of time by treatment. This latter result stands in contrast to the conclusion we reached for the parametric approach, where the p-value

TABLE 7.5

ANOVA Results for Rank-Based
Split-Plot Model of Interpersonal
Skills Data

Variable	F_R	p
Treatment	0.0001	0.98
Time	6.14	0.01
Time × treatment	3.83	0.05

ANOVA, analysis of variance.

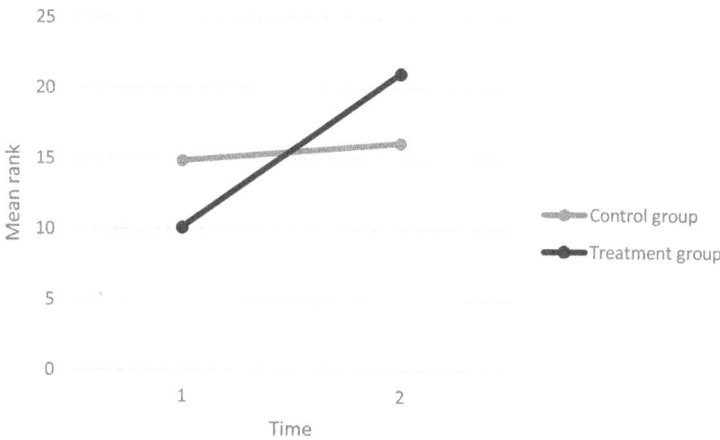

FIGURE 7.4
Mean rank by treatment condition and time.

ANOVA, analysis of variance.

for the interaction was 0.06. Of course, the actual difference in the *p*-values is quite small, but if we use α=0.05, the decisions we reach using the two approaches would indeed differ. *This divergence in conclusions based on a difference in p-values of 0.01 highlights the problem with making a simple dichotomous decision based solely on p-values.* The mean ranks for the treatment groups by time appear in Figure 7.4.

The change in the mean rank of interpersonal skills scores increased in value more from time 1 to time 2 for the treatment group than for the control, leading to the statistically significant interaction term.

In Chapter 6, we learned how to fit a factorial ANOVA model using the aligned ranks transformation (ART). This technique involves transforming the observed scores for an individual to reflect their divergence from group means on the variables in question. Aligned values are calculated for each independent variable (e.g., time and treatment condition) in turn, and then ranking occurs. These alignment calculations take the following form:

TABLE 7.6

ANOVA Results for Aligned Rank
Split-Plot Model of Interpersonal
Skills Data

Variable	F_{ART}	p
Treatment	0.02	0.89
Time	6.30	0.03
Time × treatment	4.30	0.06

ANOVA, analysis of variance.

$$x'_{i.k} = x_{ijk} - \bar{x}_{ij} + \bar{x}_{i.} - \bar{x} \tag{7.22}$$

$$x'_{.jk} = x_{ijk} - \bar{x}_{ij} + \bar{x}_{.j} - \bar{x}$$

$$x'_{ijk} = x_{ijk} - \bar{x}_{ij} - \bar{x}_{i.} - \bar{x}_{.j} + \bar{x}$$

where x_{ijk} is the score for individual k in group i for variable 1 (e.g., time) and group j for variable 2 (e.g., treatment), $\bar{x}_{ij}$ is the mean for combination of categories i and j for variables 1 and 2, $\bar{x}_{i.}$ is the mean for category i for variable 1, $\bar{x}_{.j}$ is the mean for category j for variable 2, and $\bar{x}$ is the overall mean.

The resulting aligned values are then ranked, and a separate ANOVA is conducted for each aligned ranked variable.

The ART results for the interpersonal skills data appear in Table 7.6.

The results are quite similar to those from other methods that we have seen for this example, with a conclusion of no difference in central tendency between the two treatment conditions, but a significant difference across time. The interaction was not statistically significant in this case, mirroring the results for the parametric repeated measures ANOVA.

Trimmed Means

Another nonparametric alternative for comparing central tendency in the repeated measures and split-plot contexts involves the use of trimmed means, which we spent some time discussing in earlier chapters. Remember from our earlier discussions (Chapters 2–6) that the core feature underlying analyses based on trimmed means is the removal of potential outliers from the data by removing the extreme dependent variable values from either end of the sample. For our interpersonal skills scores example, the null and alternative hypotheses for the repeated measures ANOVA based on trimmed means are:

$$H_0 : \mu_{\text{TrimmedTime1}} = \mu_{\text{TrimmedTime2}}$$

$$H_A : \mu_{\text{TrimmedTime1}} \neq \mu_{\text{TrimmedTime2}}$$

It is important to remember when using analyses with trimmed means that we are not testing null hypotheses about the overall population mean. Rather, we are testing null hypotheses about trimmed means only; i.e., we are examining a subset of the population in which the extreme values have been removed. We need to keep this nuance in mind when reporting the results of our analyses.

The statistic for the test associated with trimmed means in the repeated measures context takes the form:

$$F_T = \frac{\dfrac{(n-2g)\sum_{k=1}^{K}\left(\bar{x}_{Tk}-\bar{x}_{T}\right)^2}{K-1}}{\dfrac{\sum_{k=1}^{K}\sum_{i=1}^{N}\left(x_{wik}-\bar{x}_{wk}-\bar{x}_{wi}+\bar{x}_{w}\right)^2}{(h-1)(k-1)}} \tag{7.23}$$

where $g =$ is the number trimmed at each end of the sample distribution, n is the total sample size, h is the number retained from the original sample; i.e., untrimmed, K is the number of measurement occasions, $\bar{x}_{Tk}$ is the trimmed mean at time k, $\bar{x}_T$ is the overall trimmed mean, x_{wik} is the Winsorized value for person i at time k, $\bar{x}_{wk}$ is the Winsorized mean at time k, $\bar{x}_{wi}$ is the Winsorized mean for person I, and $\bar{x}_w$ is the overall Winsorized mean.

The degrees of freedom for F_T involve a relatively complex calculation that we won't discuss here. Fortunately, the software packages that allow for this test statistic take care of these calculations for us. We compare the F_T with the F distribution for the aforementioned degrees of freedom.

The trimmed means approach (20% trimming) to the interpersonal skills repeated measures comparison yielded $F_T = 6.13$ with $p = 0.038$, leading us to reject H_0. The 20% trimmed means were 7.56 and 11.33 for times 1 and 2, respectively. Thus, we would conclude that the population-trimmed mean interpersonal skills score was higher at time 2 than at time 1. In practice, for the 20% trimmed population, the interpersonal skills score increased over time, without consideration of the treatment group.

The trimmed means (20% trimming) split-plot model results appear in Table 7.7.

The trimmed means technique yielded quite similar results to those for the full sample. Namely, there was a statistically significant time effect and non-significant results for both treatment and the time by treatment interaction.

TABLE 7.7

ANOVA Results for Trimmed
Means Split-Plot Model of
Interpersonal Skills Data

Variable	F_T	p
Treatment	0.01	0.93
Time	7.14	0.03
Time × treatment	4.88	0.06

ANOVA, analysis of variance.

The Bootstrap

The bootstrap offers another method for comparing means in both the repeated measures and split-plot ANOVA designs. The basic structure of this approach, involving repeated sampling with replacement, is quite similar to its use in other contexts that we've reviewed in earlier chapters. For the repeated measures model with no between-subjects effect, the steps are quite similar to those for the one-way ANOVA in Chapter 5:

1. Calculate the standard F statistic for the observed data.
2. Center the data by subtracting the mean for each time point from the individual observations for that time point.
3. Draw B (e.g., 10,000) bootstrap samples from the centered data and calculate the F^* statistic to create a distribution of F^* when the null hypothesis of no mean differences is true.
4. The p-value is the proportion of F^* values that exceed F.

For the split-plot model in which there are both between- and within-subjects variables, the bootstrap takes the following steps (very similar to those for factorial ANOVA in Chapter 6):

1. Calculate the standard F statistic for the term of interest (e.g., group, time) using the original data.
2. Center the data for each combination of group (j) and time (k): $C_{ijk} = x_{ijk} - \bar{x}_{.jk}$
3. Generate B bootstrap samples of C_{ijk}
4. For each bootstrap sample, compute the F^* statistics for time, group, and interaction using standard split-plot ANOVA.
5. The p-value is the proportion of F^* values that exceed F.

With the bootstrap, we can test for differences using ranks, trimmed means, the median, M estimates, and MOM estimates.

For the repeated measures model including only time as the predictor of interpersonal skills score, we will use the bootstrap with 1,000 draws. The p-value of the observed F (6.13) based on the $F*$ distribution is 0.01, leading us to reject the null hypothesis $H_0 : \mu_{T1} = \mu_{T2}$; i.e., the interpersonal score means are the same at time 1 and time 2. Given the observed means (Table 7.1), we can further conclude that the interpersonal skills scores mean is higher at time 2 vis-à-vis time 1. We can also use the bootstrap ($B = 1,000$) to compare the medians at times 1 and 2 to test the following null hypothesis:

$H_0 : M_{T1} = M_{T2}$; i.e., the interpersonal score medians are the same at time 1 and time 2.

The p-value is 0.02, again leading us to reject the null hypothesis of equal medians for times 1 and 2. The medians are 8 and 12 for times 1 and 2, respectively, leading us to conclude that interpersonal skills scores increased from time 1 to time 2.

Next, we can expand our analysis to include the treatment condition along with time and use the bootstrap to test the following null hypotheses:

Time
> $H_0 : \mu_{T1} = \mu_{T2}$; i.e., the interpersonal score means are the same at time 1 and time 2.

Treatment
> $H_0 : \mu_{Trt1} = \mu_{Trt2}$; i.e., the interpersonal score means are the same for the two treatments.

Interaction
> H_0 : There is not an interaction between time and treatment.

The critical values (95th percentiles for the $F*$ distribution) for treatment, time, and interaction were 5.56, 5.57, and 5.32. The observed F statistics from Table 8.2 were 0.003, 6.75, and 4.14 for treatment, time, and interaction. Thus, we would reject the null hypothesis for time because $6.75 > 5.57$. However, we would not reject the null for either treatment or the interaction because the observed F statistics are not greater than the 95th percentile from the bootstrap.

The Permutation Test

Given what we have learned together in this book, it should not be a surprise that the permutation approach to hypothesis testing is an option for use in the context of repeated measures and split-plot ANOVA models. The basic idea underlying this technique involves the conduct of repeated measures

ANOVA on every possible permutation of the observed, thereby creating a reference distribution corresponding to the null hypothesis of no mean differences. The F value from ANOVA applied to the observed data is then compared with this permutation reference distribution to obtain the p-value. Formally, the permutation test steps for simple repeated measures ANOVA appear below.

1. Compute the F statistic for the observed data.
2. Permute the scores *within* individuals for the within-subjects variable.
3. Calculate F^* for each permuted sample.
4. Compare the value of F with the distribution of F^* values.
5. The p-value is the proportion of values in the F^* distribution greater than F.

Similarly, the permutation approach for the split-plot involves the following steps:

1. Compute the F statistic for the observed data.
2. Permute the scores across time *within* individuals, not across individuals.
3. Calculate F^* for each permuted sample for each model term (e.g., time, treatment, and interaction).
4. Compare the value of F with the distribution of F^* values for each model term.
5. The p-value is the proportion of values in the F^* distribution greater than F.

The permutation repeated measures model for interpersonal skills scores yielded a p-value of 0.024, meaning that we reject the null hypothesis of no mean difference over time. The mean scores by time (cited above) indicate that mean interpersonal skills improved over time. The permutation p-values for the split-plot ANOVA appear in Table 7.8.

TABLE 7.8

ANOVA Results for Trimmed Means Split-Plot Model of Interpersonal Skills Data

Variable	p
Treatment	0.95
Time	0.02
Time × treatment	0.05

ANOVA, analysis of variance.

Based on these results, we would reject the null hypotheses for time and for the interaction of time by treatment. We have discussed what this means in the real world above. Essentially, mean interpersonal skills scores increased over time, overall, and this increase was greater for those in the treatment group (see Figure 7.4). As we have discussed in this and other chapters, when the interaction term is statistically significant, we focus on that rather than on the main effects.

Summary

In Chapter 7, we have extended the ANOVA model to situations in which we have multiple values for our dependent variable. Most often, these values come from measurements made on the same individuals at multiple points in time. In many respects, repeated measures ANOVA models are similar to models for cross-sectional data. We continue to rely on the F statistic to test the null hypothesis, and we use common effect sizes such as η^2. The basic framework of interpretation is also the same as for ANOVA. However, there are some fundamental differences that we need to be aware of when comparing the repeated measures and cross-sectional ANOVA models. Chief among these differences is the way in which error is calculated. With cross-sectional models, we have a single data point per individual in the sample. Thus, there is no within-subject variability. However, when we make multiple measurements on an individual, the random error within subjects becomes an issue. The standard ANOVA models from Chapters 5 and 6 will not account for this error appropriately, as we demonstrated at the beginning of this chapter. It is for this reason that the repeated measures framework is necessary.

When considering the various options available to us in the repeated measures context, we have seen in this chapter that the rank-based, bootstrap, trimmed, median, M, and permutation techniques are all possibilities. It is also the case that the heuristics we applied for considering which approach to use when for the cross-sectional ANOVA models also apply in the repeated measures and split-plot contexts. When the assumptions of normality, sphericity, and independence (once we account for time) hold, then the parametric model is going to be optimal. When the data are not normally distributed, and in particular if there are outliers present in the data, then the rank-based methods may be our best choice. If sphericity doesn't hold, Friedman's test and the other rank techniques may not be the best option. In that case, we might be best off using the bootstrap or permutation approaches, which create null sampling distributions rather than relying on a mathematical sampling distribution. Finally, if there are some severe outliers, we might be best served comparing medians, M estimators, or trimmed means.

8

Correlation

Throughout this book, we have examined statistical methods that can be used to address questions and hypotheses posed by researchers in a variety of contexts. One commonality among most of these techniques is that they involved the comparison of some measure of central tendency among two or more groups. This was certainly not the case in Chapter 3, where our focus was on problems involving inference for a single variable. However, Chapters 4–7 were all concerned with the comparison of some measure of central tendency (e.g., mean, median, trimmed mean) among groups and/or across measurement occasions. In this chapter, we will turn our attention to the task of estimating relationships among variables. Thus, rather than asking questions centered around differences, we will instead consider the extent to which two variables are related to one another and if so, in what way. The characterization of relationships will take center stage and remain there from this chapter through the end of the book. Indeed, we can think of the topic of this chapter, correlation, as the entry to subsequent topics including regression, logistic regression, and multilevel models, which are all topics in subsequent chapters. One final point to make here is that, although these models will appear to be quite different from the mean comparison methods that dominated our discussion in the first half of the text, in fact, they are all part of the same broad family of statistical techniques that are known as the general linear model (GLM). We will have more to say about the GLM in the next chapter.

Pearson's Correlation Coefficient

When we think about examining relationships among variables, we might consider using a graphical tool such as a scatterplot. This graph will certainly provide us with a nice visual description of the relationship between variable pairs. In addition to visualizing the relationship between variables, it would also be useful to obtain a numeric characterization of this relationship. Such a number would be more precise and efficient than a graph in describing how two variables are related. This is not to diminish the importance and utility of graphs as a part of the data analysis process. Rather, we should see graphical and numerical investigations of relationships among variables as

DOI: 10.1201/9781003379324-8

complementary tools that together can yield a more complete picture than either approach could separately.

One way in which relationships between variables can be quantified is with the covariance. The covariance between two variables is a measure of how they change or covary together. The sample covariance for two variables, x and y, is calculated as

$$\text{COV}(x,y) = \frac{\sum_{i=1}^{N}(x_i - \bar{x})(y_i - \bar{y})}{N-1} \tag{8.1}$$

The numerator in equation (8.1) expresses the relationship between variables x and y, with positive values indicating that individuals with larger values of x also have larger values of y. In contrast, a negative covariance means that those with larger values of x have smaller values of y. Larger values of the covariance indicate that the relationship between the two variables is stronger.

A major problem with using the covariance in practice is that it is not scale-free; i.e., its value is directly tied to the scales of the two variables used to calculate it. The fact that the covariance is scale-dependent in this way makes it difficult to interpret. What value of the covariance connotes a strong relationship between a pair of variables? It depends on the scale of the variables used to calculate it. Therefore, it would be helpful if we could standardize the covariance and use that value to characterize the relationship between two variables. The Pearson correlation coefficient, or Pearson's r, is a statistical tool for just this purpose. It is simply a standardized version of the covariance involving division by the standard deviations of each variable as below:

$$r = \frac{\text{COV}(x,y)}{S_x S_y} \tag{8.2}$$

Pearson's r ranges between –1 and 1. A value of 0 means that the variables are completely unrelated to one another, whereas values near positive or negative 1.0 indicate strong relationships between the two. Indeed, a value of 1.0 means that the two variables covary perfectly with each other. As with the covariance, a positive correlation means that as x increases in value, so does y, with a negative correlation demonstrating that as x increases, y decreases.

The fact that r sits on a known scale means makes it easier to interpret than the covariance. In addition, these interpretations are universal across applications. For example, a Pearson's r value of 0.6 has the same statistical meaning regardless of the variables being correlated with one another. Having made this last point, it is very important to state that what might be considered a large correlation, or a strong relationship between variables, will differ across research contexts. For example, in disciplines where measurements tend to have more inherent variation (e.g., psychology, education, biology), a Pearson's *r* value of 0.6 might be considered quite large. However, when relatively less variability is present in the measurements (e.g., materials science, chemistry, physics), a correlation coefficient of 0.6 may be considered fairly small, with researchers expecting to see correlations more in the range of 0.8 or larger.

Cohen (1988) developed a heuristic for interpreting the magnitude of correlation coefficients. These guidelines are much like Cohen's guidelines for interpreting effect size values that we discussed in Chapters 3–7. They are meant to serve only as very loose suggestions for how we might interpret *r*. However, Cohen (and subsequent authors) was very clear that this heuristic should not replace informed professional judgment regarding the magnitude of *r* in a given setting. As we noted above, in the final analysis, the relative magnitude of *r* based on sample data must be made in light of the field under study and prior work in the area. With this caveat in mind, Cohen's guidelines for interpreting *r* appear below:

Small: 0.1–0.3

Medium: 0.3–0.5

Large: 0.5+

Again, we would always recommend that researchers consider the magnitude of a correlation coefficient in light of the field under study and not rely as much on these very general guidelines.

In addition to describing the relationships among variables in our sample based on the value of *r*, it is also possible to make inferences about the population correlation coefficient using the sample correlation estimate. We can, for example, test the null hypothesis that the population correlation (ρ) is equal to 0. More formally, this null hypothesis would be

$$H_0 : \rho = 0$$

where ρ is the population parameter, which we estimate with the sample statistic *r*.

The statistic for testing this hypothesis is

$$t = r\sqrt{\frac{n-2}{1-r^2}} \tag{8.3}$$

where *r* is the value of the correlation between a variable pair and *n* is the sample size.

The statistic in equation (8.3) is distributed as a *t* with $n-2$ degrees of freedom. If the *p*-value associated with the *t* in equation (8.3) is less than a (e.g., 0.05), we would reject H_0 and conclude that scores on the two variables are related to one another.

In addition, it is possible to construct confidence intervals for ρ such that we can be confident to some known degree (e.g., 95% confident) that the value of ρ lies within the bounds of the confidence interval. As is hopefully clear, we interpret this confidence interval for the correlation in much the same way that we would for the confidence interval of the population mean. To calculate the confidence interval, we first need to transform the correlation coefficient to a z value using Fisher's z transformation technique, which appears in equation (8.4):

$$z_r = 0.5 \ln\left(\frac{(1+r)}{(1-r)}\right) \tag{8.4}$$

Once we have converted r to z, we can then construct the lower and upper bounds of the confidence interval for Fisher's z as

$$z_L = z_r - z_{cv}\sqrt{\frac{1}{n-3}}$$

$$z_U = z_r + z_{cv}\sqrt{\frac{1}{n-3}} \tag{8.5}$$

where z_{cv} is the value of the normal distribution associated with $1-\frac{\alpha}{2}$ (e.g., 0.95) level of confidence.

The upper and lower bounds in equation (8.5) are in the standard normal (z) scale and need to be transformed back to the correlation scale to be interpretable. These conversions are

$$r_L = \frac{e^{2z_L} - 1}{e^{2z_L} + 1}$$

$$r_U = \frac{e^{2z_U} - 1}{e^{2z_U} + 1} \tag{8.6}$$

We would then conclude that we are 95% confident that ρ lies between r_L and r_U.

While Pearson's r is a useful tool for characterizing relationships between variables, it does have some limitations that we need to be aware of when using it. First, r can be greatly influenced by the presence of outliers, particularly when the sample size is relatively small. Second, it is limited to characterizing linear relationships and does not provide us with any information about nonlinear associations between variables. Third, if one or both of the variables involved in calculating r have a restricted range of values (i.e., small variance), it will tend to be negatively biased. Conversely, when the variances of one or more of the variables are large, r will also be diminished in value. In other words, r is quite sensitive to the variability in the variables. Finally, the statistic in equation (8.3) that is used to test the null hypothesis is based on an assumption of bivariate normality of the two variables involved in the calculation. This assumption also underlies the confidence interval calculation in equation (8.5). When the data are not bivariate normal, the hypothesis test and confidence interval may not provide the researcher with accurate information. Given the limitations associated with r, it is worthwhile to consider some alternative statistics for estimating relationships between variables, which we will do in the remainder of this chapter. It is also worth keeping in mind, however, that Pearson's r remains a very useful statistic and one that you will almost certainly make liberal use of in conducting your own research.

Spearman's Rho

We have seen in previous chapters that the use of ranks can be a powerful tool for dealing with data that are skewed and/or contain outliers. Ranks can be applied to the problem of correlation estimation as well. The rank-based alternative to Pearson's r that is particularly useful when there are outliers in the data is Spearman's rho. To calculate Spearman's rho, we simply replace the observed values of the two variables with their ranks and then calculate Pearson's r between the sets of ranks. In other respects, the discussion above pertaining to the interpretation of and inference for the correlation coefficient applies to Spearman's rho as well.

Kendall's Tau

Another robust alternative to r that is especially resistant to outliers is Kendall's tau. Tau is based upon identifying concordant and discordant pairs of observations in the data with respect to the ordering of variable values. A concordant pair is defined as follows. Assume that we have two variables of interest, x and y. Individuals i and j are concurrent when $x_i > y_i$ and $x_j > y_j$. In other words, individuals i and j are concordant when the same variable takes a larger value for both. Obviously, the pairs would also be concurrent in our example if y is larger than x for both individuals. These pairs would be discordant if $x_i > y_i$ and $x_j < y_j$, or vice versa.

Once we have identified concordance and discordance for every pair of observations in our sample, tau is calculated using the following steps.

1. Determine the number of concordant pairs and assign them the value 1, while discordant pairs are assigned the value 0.
2. Sum up the number of concordant pairs to get the value v.
3. Divide v by the total number of pairs (T) present in the dataset.
4. Tau is then calculated as $2\left(\dfrac{v}{T}\right) - 1$. The subtraction of 1 ensures that tau will range between -1 and 1, as the standard measures of correlation do.
5. To test the significance of tau, calculate the standard error, which takes the form: $\sqrt{\dfrac{4n+10}{9\left(n^2-n\right)}}$, and divide it into the value obtained in step 4.

When x and y are equal for both individuals in the pair, they are not counted as concordant or discordant and do not appear in the number of pairs, T.

M Correlations

There are several measures of association for pairs of variables based on the M estimation technique that we described in Chapter 2. We have seen how useful the M estimators have been in the context of means comparison, particularly when outliers are present. M estimators will prove useful for correlation estimation in the presence of outliers as well. Perhaps the most important of these techniques is the percentage bend (PB) Correlation. As with other M estimators, PB works by applying a weight to each member of the sample. Individuals who are further from the center mass of the data (as measured by the median) receive a smaller weight than those who are near the center. The maximum weight is 1.

The PB correlation is calculated by applying the following steps to each variable to be used in the correlation calculation. While it is worthwhile for readers to see the steps used to obtain the PB correlation coefficient, our main goal is that the reader have an understanding of the general principle underlying this approach (downweighting of outliers) and that it is particularly useful when outliers are present.

1. $w_i = |x_i - M|$ for each variable where M is the sample median.
2. Place w_i in ascending order.

3. $m = \left[(1-\beta)n\right]$ where β is a tuning parameter that we select reflecting the breakdown point, ranging between 0 and 0.5. A common value for β is 0.2.

4. $\hat{\omega}_x = w_m$.

5. $s_x = \sum_{i=i_1+1}^{n-i_2} x_i$ where i_1 is the number of x_i values such that $\dfrac{(x_i - M)}{\hat{\omega}_x} < -1$, and i_2 is the number of x_i values where $\dfrac{(x_i - M)}{\hat{\omega}_x} > 1$.
Note that i_1 and i_2 essentially reflect the boundary below and above which an observation is considered an outlier.

6. $\hat{\phi}_x = \dfrac{\hat{\omega}_x (i_2 - i_1) + s_x}{n - i_1 - i_2}$.

7. $u_i = \dfrac{\left(x_i - \hat{\phi}_x\right)}{\hat{\omega}_x}$.

8. $\psi_x = \text{Max}\left[-1, \text{Min}(1, x)\right]$.

Once the previous steps have been applied to each variable in the pair to be correlated, the PB coefficient is then calculated as

$$r_{Pb} = \frac{\sum_{i=1}^{N} \psi(u_i T_1) \psi(u_i T_2)}{\sqrt{\sum_{i=1}^{N} \psi(u_i T_1)^2 \sum_{i=1}^{N} \psi(u_i T_2)^2}} \tag{8.7}$$

Interpretation of r_{Pb} is the same as for Pearson's r and the other correlation coefficients described in this chapter. In addition, we can test the significance of this correlation using

$$T_{r_{Pb}} = r_{Pb} \sqrt{\frac{n-2}{1 - r_{pb}^2}} \tag{8.8}$$

Biweight Midcovariance

The biweight midcovariance is closely related to the PB correlation. It is calculated using the following steps:

1. For each variable in the pair to be correlated, calculate:

- $u_i = \dfrac{x_i - M}{9(\text{MAD})}$

- $\text{MAD} = |\text{Median}| \text{ of } (x_i - M)$

2. Set $a_i = 1$ if $-1 \le u_i \le 1$ and 0 otherwise. This gives outliers a weight of 0.

3. $S_{bT_1 T_2} = \dfrac{n \sum_{i=1}^{N} a_{iT_1} \left(x_{iT_1} - M_{T_1} \right) \left(1 - u_{iT_1}^2 \right)^2 a_{iT_2} \left(x_{iT_2} - M_{T_2} \right) \left(1 - u_{iT_2}^2 \right)^2}{\left(\sum_{i=1}^{N} a_{iT_1} \left(1 - u_{iT_1}^2 \right) \left(1 - 5u_{iT_1}^2 \right) \right) \left(\sum_{i=1}^{N} a_{iT_2} \left(1 - u_{iT_2}^2 \right) \left(1 - 5u_{iT_2}^2 \right) \right)}$

4. $r_{bw} = \dfrac{S_{bT_1 T_2}}{\sqrt{S_{bT_1} S_{bT_2}}}$

As with the r_{pb}, interpretation of r_{bw} is much the same as Pearson's r.

Winsorized Correlations

Another approach to dealing with the presence of outliers when estimating the degree of association between two variables involves Winsorizing each variable separately and then calculating Pearson's r between the two. Recall that Winsorization involves replacing a proportion (γ) of individuals at each extreme of the sample by the γ proportion with the next most extreme scores. As an example, if we set $\gamma = 0.2$, then the individuals in the sample with the lowest 20% of values on x will be replaced with members of the sample having the next lowest 20% of scores. Similarly, individuals with the highest 20% of scores are replaced by the next highest 20% of scores. Pearson's r is then calculated using this Winsorized data to obtain r_w. Everything that applied to the use of the r from the original data (i.e., interpretation and inference) is also applicable for r_w.

The previous three correlation coefficients (r_{bw}, r_{pb}, and r_w) are all part of the M estimator family. These correlation estimates are sensitive to heteroscedastic variances, meaning that the test of the null hypothesis of no relationship might not be accurate in such cases. Thus, when the variances of one or more of the variables differ at different levels, an alternative approach to inference is necessary. One approach that has been shown to be effective for identifying accurate confidence intervals for these M correlation estimators is the percentile bootstrap. To obtain the confidence interval, we simply draw B (e.g., 1,000) bootstrap samples (see Chapter 2 for a reminder of how the bootstrap works), and for each, calculate the correlation coefficient of interest. This set of correlation values creates a proxy for the sampling distribution of the coefficient. The 95% confidence interval then corresponds to the 2.5th and 97.5th percentile values of the bootstrap distribution.

Bayesian Correlation Estimation

Throughout this book, we have seen that the Bayesian paradigm provides us with a useful alternative to frequentist techniques. Bayes allows us to incorporate prior information into the parameter estimation/inference process and quite often yields more accurate and efficient estimates than frequentist methods with small samples. As we might expect, these advantages associated with Bayesian estimation are also available in the context of correlation. The basic ideas that we introduced in Chapter 2 are present with correlation estimation as well, including specifics of the MCMC algorithm such as the burn-in period, prior distributions, the length of the chains, the number of chains, thinning, and the use of the median and credibility intervals from the posterior. Given that we devoted time to these ideas in Chapter 2, we will not review them again here.

Example of Correlation Coefficients

To demonstrate the application of the correlation estimates that we discussed above, we will work with a common set of data that involves a set of cognitive ability and academic achievement tests that were given to a sample of 50 children who were between the ages of 14 and 16 years. The small battery of assessments included measures of listening skills, spatial acumen, verbal communication, and mathematical reasoning. Our primary interest is to gain insights into how scores on these scales are related to one another. We will explore these relationships using a variety of statistical tools.

To demonstrate the correlation estimates, we will focus our attention on the measures of spatial acumen and mathematics achievement. We will always want to start our investigation of the relationships between variables by examining a scatterplot. The plot for math and spatial appears in Figure 8.1.

There appears to be a positive relationship between the two variables. Individuals who are better at spatial tasks also tended to have higher math achievement scores.

We can assess the assumption of multivariate (which includes bivariate) normality using Mardia's test, which in practice is essentially an extension of the Shapiro–Wilk test of normality for individual variables that we have used throughout this book. The null hypothesis is that the data are multivariate normal. The p-value for our data is 0.68, which is larger than our $\alpha=0.05$. Thus, we do not reject the null hypothesis and conclude that our data are bivariate normal.

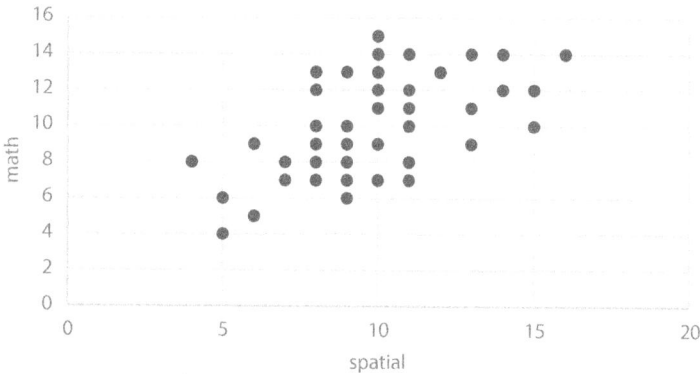

FIGURE 8.1
Scatterplot of spatial acumen and mathematics achievement.

TABLE 8.1

Correlation Coefficient Estimates for Relationship between Spatial Acumen and Mathematics Achievement

Coefficient	Value	95% Confidence Interval
Pearson's r	0.59[a]	0.38, 0.75
Spearman's rho	0.58[a]	0.28, 0.59
Kendall's tau	0.45[a]	0.35, 0.74
Percentage bend	0.57[a]	
Biweight midcovariance	0.59[a]	
Winsorized	0.55[a]	
Bayesian (uninformative prior)	0.57[a]	0.38, 0.74

[a] Statistically significant for $\alpha = 0.05$.

We have calculated each of the correlation coefficients discussed above for the relationship between math achievement and spatial acumen. The resulting values appear in Table 8.1.

There is a great deal of concordance among the various correlation estimates in Table 8.1. This result is not particularly surprising given the relative absence of outliers in the data, as evident in Figure 8.1. In addition, the fact that math achievement and spatial acumen are bivariate normal means that the hypothesis test for Pearson's r is reliable. In all cases, the correlation estimates were positive, statistically significant, and fell in Cohen's large range. Thus, we can conclude that individuals with higher spatial acumen scores also tended to have higher math achievement test scores. Finally, a variety of prior distributions were used with this example, and the correlation estimates were all within 0.02 of the value presented in Table 8.1.

Biserial and Point Biserial Correlations for Dichotomous Data

In some research scenarios, we may be interested in estimating the relationship between a continuous and a dichotomous variable. For example, we might want to know whether responses to a dichotomous survey item (e.g., agreement with a statement indicating the respondent uses TikTok) and a continuous variable (e.g., age). The correlation coefficients described in this chapter would not be appropriate given that they were designed for continuous (or at least ordinal) variable. One viable alternative for researchers interested in assessing whether there are differences in age for those who use TikTok versus those who do not would be the t-test from Chapter 4. This approach would allow us to test the null hypothesis that the means ages for the two groups are equal in the population. However, the t-test would not provide us with an estimate of the strength of the relationship between age and TikTok use. A statistic that can be used for this purpose is the biserial correlation coefficient. The biserial correlation coefficient is calculated under the assumption that underlying the observed dichotomous variable (e.g., TikTok use) there exists a normally distributed latent variable. The value of the observed variable (Yes, I use TikTok or No, I don't use TikTok) is a function of an underlying continuous variable (propensity to use TikTok). For this example, when the level of the continuous latent propensity to use the TikTok variable for an individual exceeds a threshold value, the observed response would be, Yes, I use TikTok. However, when the latent variable value falls below the threshold, the response would be, No, I don't use TikTok.

The biserial correlation is calculated as

$$\rho_{BIS} = \frac{(\bar{x}_+ - \bar{x}_.)}{s_.} \frac{P^*}{Y} \tag{8.9}$$

where $\bar{x}_+$ is the mean test score for examinees answering item correctly, $\bar{x}_.$ is the mean test score for all examinees, $s_.$ is the standard deviation for test scores of all examinees, and Y is the Y value for a z score (from the standard normal distribution) associated with P^*.

Interpretation of the biserial correlation is the same as for the other correlation coefficients described in this chapter.

We will apply the biserial correlation coefficient to the question of whether TikTok use (1=Yes, 0=No) is related to age in years. Our sample consists of 147 individuals participating in a study of social media use. The biserial correlation estimator of the relationship between our target variables is −0.36. This result suggests that there is a positive relationship between the item response and age. We can conclude that the older an individual is, the less likely they are to use TikTok.

TABLE 8.2

Cross-Tabulation of TikTok Use by Age

TikTok Use	Age ≤ 30	Age > 30
No	6	22
Yes	17	29

Phi and Cramer's *V* Coefficients

When both variables for which we would like to estimate a correlation coefficient are dichotomous, we can use the phi coefficient (ϕ). When one or more of these variables are nominal with more than two categories, the phi coefficient can be extended to Cramer's *V*. Both of these statistics are based upon the chi-square statistic assessing the association between the variables. As an example of how this statistic is calculated, let's consider an example in which members of the sample are asked whether they use TikTok (yes or no) and whether they are younger than 30 or older than 30. We can create a cross-tabulation of the responses to these items as below (Table 8.2).

The correlation between these two variables can be calculated as

$$\phi = \sqrt{\frac{\chi^2}{n(df)}} \tag{8.10}$$

where

$$\chi^2 = \sum_{k=1}^{K} \sum_{j=1}^{J} \frac{\left(O_{kj} - E_{kj}\right)^2}{E_{kj}}$$

E_{kj} is the expected frequency in cell kj if the two variables are unrelated and O_{kj} is the observed frequency in cell kj if the two variables are unrelated.

The closer ϕ is to 1, the stronger the relationship between the two variables. The guidelines for interpreting ϕ are the same as those used for r and outlined by Cohen (1988):

Small: 0.1–0.3

Medium: 0.3–0.5

Large: 0.5+

The χ^2 statistic can be used to test the null hypothesis
  H_0 : There is not a relationship between the variables.

The alternative hypothesis is that the two variables are related. The χ^2 is compared to the chi-square distribution with $(R-1)(C-1)$ degrees of freedom, where R is the number of rows in the contingency table and C is the number of columns in the table.

For our example, $\phi = 0.16$, which falls in Cohen's small range. The χ^2 test is 1.96 with $(2-1)(2-1)=1$ degree of freedom, yielding a p-value of 0.162. Because this value is greater than 0.05, we would not reject the null hypothesis and conclude that there is not sufficient evidence for us to conclude that TikTok use and being 30 years old or younger are related in the population.

Comparing Correlation Coefficients from Independent Samples

We will conclude this chapter by considering how we can compare the magnitudes of two correlation coefficients. There are three instances in which this question might be of interest to a researcher. First, we may want to compare the correlation between two variables, x and y, for individuals belonging to one of two groups. For example, we may be interested in whether students who participated in a mindfulness course will exhibit a different Pearson's r value for the relationship between listening skills and verbal reasoning than students who did not participate in the mindfulness program. A second scenario in which we may wish to compare correlation coefficients involves a set of three overlapping variables measured on the same sample. In this context, overlapping variables mean that one of the variables in each pair is the same. For example, we might be interested in comparing the correlation of listening skills and verbal reasoning with the correlation between listening skills and mathematics achievement. The third correlation comparison scenario that may be of interest is the non-overlapping case for the same sample. An example of this situation would be comparing the correlation between listening skills and verbal reasoning with the correlation between spatial skills and mathematics achievement. We will consider each scenario below.

The null hypothesis for the comparison of correlations between the same variables between two groups is

$$H_0 : \rho_{xy.1} = \rho_{xy.2}$$

In this null, $\rho_{xy.1}$ is the correlation between the two variables for group 1 and $\rho_{xy.2}$ is the correlation for group 2. To test this null hypothesis, we must first convert the correlation coefficients to Z scores using Fisher's r to Z transformation, as in equation (8.4). We would then use these transformed variables to construct the Z test for independent samples:

$$Z = \frac{z_1 - z_2}{\sqrt{\left[\left(\dfrac{1}{N_1}\right) + \left(\dfrac{1}{N_2}\right)\right]}} \qquad (8.11)$$

where z_1 is the Fisher z value for group 1, z_2 is the Fisher z value for group 2, N_1 is the sample size for group 1, and N_2 is the sample size for group 2

We would then compare the Z from equation (8.11) to the standard normal distribution to obtain the p-value.

Zou (2007) proposed a confidence interval for the difference in group correlations. If this interval includes 0, we would conclude that there is no difference between the correlation coefficients for the groups. In this way, the interpretation of Zou's confidence intervals is much the same as interpreting confidence intervals in the other contexts that we encountered earlier in the book. The calculation details of the confidence interval are not given here, but the reader is referred to Zou for those.

When we are interested in comparing correlations for overlapping pairs, the null hypothesis would be

$$H_0 : \rho_{xy.} = \rho_{xk}$$

This null states that in the population, the correlation between variables x and y is equal to the correlation between variables x and k. A test statistic for this null hypothesis is

$$t = \left(r_{xy.} - r_{xk}\right) \sqrt{\frac{(n-1)\left(1 + r_{yk}\right)}{2\left(\dfrac{n-1}{n-3}\right)|S| + \dfrac{\left(r_{xy.} + r_{xk}\right)^2}{4}\left(1 - r_{yk}\right)^3}} \qquad (8.12)$$

where $|S|$ is the determinant (multivariate variance) of a function of the correlation coefficients

The t from equation (8.12) is compared to the t distribution with $n-3$ degrees of freedom to obtain the p-value for the test. Zou (2007) also developed confidence intervals for this correlation difference comparison that are similar to those for comparing independent sample correlations that are discussed above.

Finally, the null hypothesis for the comparison of correlations for non-overlapping variable pairs is

$$H_0 : \rho_{xy.} = \rho_{lk}$$

This null hypothesis states that the correlation between variables x and y is equal to the correlation between variables l and k. A test statistic for this null hypothesis is

$$z = \frac{\sqrt{n}\left(r_{xy.} - r_{lk}\right)}{\sqrt{\left(1 - r_{xy.}^{2}\right)^{2} + \left(1 - r_{xy.}^{2}\right)^{2} - k}} \tag{8.13}$$

where k is a complex function of the correlation coefficients.

The test statistic from equation (8.13) is referenced to the standard normal distribution to obtain a p-value. There is also a Zou (2007) confidence interval for this comparison.

We will conclude our discussion of the correlation comparison methods by referring back to the example involving the measures of spatial acumen, verbal reasoning, listening skills, and math achievement. First, we will compare the correlation between listening skills and verbal reasoning for individuals who received mindfulness training and those who did not. We will not include the full set of calculations here but encourage the interested reader to do them. The correlation between the variables for the two groups was

$$r_{\text{listening, verbal . mindfulness}} = 0.67$$

$$r_{\text{listening, verbal . not mindfulness}} = 0.51$$

The Z value is 1.20, yielding a p-value of 0.23. Thus, we would not reject the null hypothesis and conclude that there is not sufficient evidence to conclude that the correlation between listening skills and verbal reasoning differs in the population between those who received mindfulness training and those who did not. Zou's 95% confidence interval for the difference was $(-0.10, 0.43)$. Given that 0 falls within this interval, we would conclude that there is no difference between the correlation of listening and verbal scores for the mindfulness training and control groups.

Next, we will compare the correlation between the listening skills and verbal reasoning scores with the correlation between the listening skills and math achievement scores for our sample. The relevant Pearson's r estimates appear below:

$$r_{\text{listening, verbal}} = 0.45$$

$$r_{\text{listening, math}} = 0.11$$

$$r_{\text{math, verbal}} = 0.32$$

The t-statistic value for this test is 3.20 with a p-value of 0.002. Given the total sample size of 100, the degrees of freedom is 97. The 95% Zou confidence interval for the difference is $(0.12, 0.55)$. Because 0 is not in the interval, we conclude that it is likely that the correlations differ in the population. In addition, we are fairly confident that in the population, the difference between the correlation coefficients lies between 0.12 and 0.55. Thus, the differences might be quite different (0.55) or not very different (0.12). This wide interval reflects our uncertainty about this difference.

We will conclude the chapter with a comparison of correlations for the non-overlapping variable pairs of listening skills with verbal reasoning versus spatial acumen with mathematics achievement. The Pearson's r values are

$$r_{\text{listening, verbal}} = 0.451$$

$$r_{\text{spatial, math}} = 0.447$$

The Z test statistic is 0.09 with a p-value of 0.93. Given that the p-value is not less than 0.05, we don't reject the null hypothesis and conclude that there is not sufficient evidence to say that the correlations between listening and verbal differ from spatial and math in the population. The Zou 95% confidence interval is (−0.21, 0.23). Because 0 is within the interval, we would conclude that we don't have sufficient evidence to conclude that the correlations are likely to differ in the population.

Summary

The focus of this chapter was on characterizing relationships between pairs of variables using correlation. We learned that there are a number of correlation coefficients available to researchers, with by far the most popular of these being Pearson's r. Correlation coefficients typically fall between −1 and 1, with values near 0 indicating that two variables are not related to one another and values near the extremes suggesting a strong relationship. In addition, positive correlation values tell us that as one variable increases in value, so does the other, whereas a negative correlation coefficient means that as one variable increases in value, the other decreases. Cohen (1988) provided researchers with a very general set of rules for interpreting the correlation magnitude, but we noted in the chapter that it is much better for researchers to consider correlation coefficient values in light of the area under study. In addition to characterizing relationships descriptively based on the coefficient value, we can also use statistical inference to test null hypotheses about the correlation between two variables in the population. Most often, this null hypothesis would be that the population correlation, ρ, is equal to 0 in the population. However, it is certainly possible to test other null hypotheses using the methods outlined in this chapter.

In addition to Pearson's r, we also learned about several alternative correlation coefficients that might be preferable in some circumstances. Specifically, we introduced a set of correlation coefficients designed to reduce the impact of outliers, which are known to have a deleterious impact on Pearson's *r*. These robust alternatives should be used when an examination of the data reveals the presence of potential outliers. As we noted in the example, scatterplots are invaluable tools for exploring the data and will quite frequently clue us in on potentially outlying values. In that case, we have a number of worthwhile alternatives to consider. We also saw in this chapter that Bayesian estimation can be used to obtain correlation coefficients. We also discussed the fact that the typical advantages associated with it in other contexts (the ability to incorporate prior information into our analyses and its robustness to small sample sizes) are also present for estimating correlation coefficients. Next, we concluded the chapter with a correlation measure for assessing the relationship between a continuous and a categorical variable (biserial correlation) and two categorical variables (ϕ and Cramer's *V*). An important point to make here is that there exist other statistics for estimating the correlation between categorical variables, the tetrachoric and polychoric correlations. We selected ϕ and Cramer's *V* because they are applicable for situations in which we have either nominal or ordinal variables, whereas the polychoric family of correlations is applicable in the ordinal case only. However, we do recognize the utility of these other statistics and encourage the interested reader to investigate them. Finally, we concluded the chapter with a discussion of techniques for comparing correlation coefficient values with one another in three distinct situations.

Having discussed the concept of relationships among variables in the context of correlation, we are now ready to extend these ideas to a more complex paradigm, regression analysis. Many of the ideas that we learned with respect to correlation coefficients will appear again in regression, such as the ideas of direction and strength of relationships between variables. Indeed, the correlation coefficient and regression slope that we will study in the next chapter are deeply connected to one another. One final point that we need to return to here is that the correlation coefficients we have studied in Chapter 8 are all used to assess the degree of *linear* relationship between two variables. If the relationship is nonlinear in nature, these statistics will not be particularly effective at characterizing it. However, as we will see in Chapter 12, there are tools available to us for estimating and describing such nonlinearities in our data. But before we get there, we need to now consider linear regression models.

9

Ordinary Least Squares Linear Regression

As we have seen throughout this book, statistical models of various types provide researchers with powerful tools to investigate research questions and test hypotheses in a wide array of contexts. We have seen that such tools allow for the comparison of means among multiple groups both when data are collected cross-sectionally and when repeated measurements are made on the same individuals at multiple points in time. In the previous chapter, we turned our attention to correlation coefficients, which express the degree of relationship between two variables. We saw that the correlation ranges in value between –1 and 1, with values further from 0 indicating a stronger relationship between the variables. There are correlation coefficients for a wide variety of cases, including various combinations of continuous, ordinal, and nominal variables. In all cases, the interpretation of these statistics is similar, with more extreme values indicating stronger relationships between the variables.

In this chapter, we will build upon the fundamental purpose underlying the correlation coefficient with linear regression. Regression amplifies the ideas underlying the correlation, namely, estimating the strength and direction of the relationship between two variables. This amplification comes through a more nuanced, and in some ways, more informative estimate of the relationship between variables, as we will see below. In addition, the linear regression model allows for the estimation of the relationship between a dependent variable and multiple independent variables simultaneously. This is not something that is easily done using correlation coefficients. Such models allow for the examination of relationships among multiple variables, which in turn can lead to a better understanding of the world. For example, sociologists use linear regression to gain insights into how factors such as ethnicity, gender, and level of education are related to an individual's income. Biologists can use the same type of model to understand the interplay between sunlight, rainfall, industrial runoff, and biodiversity in a rainforest. And using linear regression, educational researchers can develop powerful tools for understanding the role that different instructional strategies have on student achievement. In addition to providing a path by which various phenomena can be better understood, statistical models can also be used as predictive tools. For example, econometricians might develop models to predict labor market participation given a set of economic inputs, whereas higher education administrators may use similar types of models to predict the grade point average for prospective incoming freshmen to identify those who might need academic assistance during their first year of college.

DOI: 10.1201/9781003379324-9

An important point to make here is that the regression framework that we discuss in Chapter 9 is part of a larger family of models known collectively as the general linear model (GLM). The GLM links a dependent, or outcome, variable to one or more independent variables and encompasses virtually all of the inferential methods that we have discussed in this book, including the t-test, analysis of variance (ANOVA), correlation, and regression, which is our focus in this chapter. Indeed, we will see direct correspondence between the linear regression model and the ANOVA model that was presented in Chapters 5 and 6. Our focus in the following pages will be on linear regression as estimated using the ordinary least squares (OLS) technique. We will devote a great deal of time to this model, as it is the foundation for so much of the work that is done in statistical practice. Chapter 9 will also serve as the springboard for our discussion of robust regression techniques in Chapter 10, regression for categorical dependent variables in Chapter 11, and advanced modeling techniques in Chapter 12. It is no exaggeration to say that the GLM is one of the major pillars of statistical modeling and that regression is a key component of the GLM.

Motivating Example

The running example featured in this chapter involves measures of perfectionism and personality that were given to a group of college students. The primary dependent variable of interest is a score reflecting the degree of self-oriented perfectionism (SOP) for each study participant. Higher scores on this measure indicate that the individual has higher levels of perfectionism that is derived from their own internal standards. Sample items include "When I am working on something, I cannot relax until it is perfect" and "One of my goals is to be perfect in everything I do". Other measures of interest in this example include effortful control (EFFC) (ability to make plans and follow through with them), negative affect (feelings of fear, frustration, etc.), and extraversion, as well as whether the individual identifies as male (yes or no). We will examine statistical models that relate these independent variables with the SOP outcome. First, we'll consider the relationship between SOP and EFFC, after which we will expand the model to include the rest of the independent variables of interest.

Exploration of the Data

As with the other analyses that we have learned about together in this book, we will start by exploring the data. Such explorations will provide us with insights into the distributions of the variables that will be used in the

TABLE 9.1

Descriptive Statistics for Personality Variables

Variable	N	Mean	Median	Standard Deviation	Skewness	Kurtosis
N_AFF	353	4.15	4.12	0.73	−0.06	−0.23
EFFC	353	4.33	4.32	0.72	−0.03	0.07
EXT	353	4.38	4.41	0.78	−0.26	0.10
SOP	353	73.50	73.37	15.40	−0.06	−0.03

SOP, self-oriented perfectionism; EFFC, effortful control; EXT, extraversion; N_AFF, negative affect.

regression models, as well as the nature of the relationships among them. We will start by examining descriptive statistics for each variable in our dataset.

The statistics in Table 9.1 provide information about our sample that we can use both for descriptive purposes and to place it within the broader literature with respect to the variable scores. In other words, we can check to see whether our sample appears to be typical with respect to central tendency and variation vis-à-vis samples used in other studies. Such is the case here, given that the means and standard deviations for our sample are similar to those reported in other studies using the same measures. In addition, the skewness and kurtosis values for each variable are close to 0, meaning that the variables are neither highly skewed nor kurtotic.

We can also gain a great deal of insight into the relationships among the variables as well as their distributions through a scatterplot matrix (Figure 9.1).

The upper right-hand section of the graph displays the Pearson correlation coefficients among the variables. We can see that the correlations between SOP and negative affect (N_AFF), EFFC, and extraversion (EXT) are 0.299, 0.204, and 0.036, respectively, suggesting the presence of positive relationships. The correlations of SOP with N_AFF and EFFC fell in Cohen's (1988) small range, whereas that with EXT was negligible in magnitude, using these guidelines. In addition, we see that the first two of these are statistically significantly different than 0 (denoted with stars), whereas the third is not. In terms of the independent variables, the correlations ranged between −0.09 and −0.145, all of which fall into the negligible to small range as suggested by Cohen (1988).

The bottom left portion of the matrix displays scatterplots for variable pairs. From these, we can see that the relationships among the variables are not particularly strong, given that they show little direction, either positive or negative. These relatively weak relationships were also reflected in the aforementioned Pearson correlation coefficients. The diagonal of the scatterplot matrix displays density plots for each variable. Density plots display the distributions for the variables, and we interpret them in much the same way we would histograms. Though not perfectly symmetric, these plots do suggest that the variables generally follow a normal distribution. We will discuss this issue later when we consider the assumptions underlying linear regression.

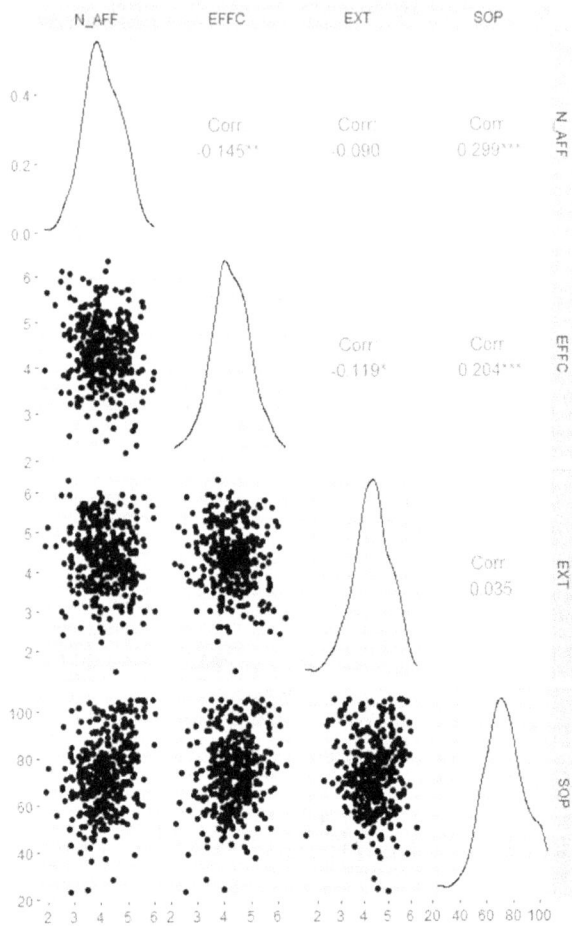

FIGURE 9.1
Scatterplot matrix for variables of interest.

Ordinary Least Squares Regression

We will begin our discussion of the regression modeling framework for the case with one dependent and one independent variable, known as simple linear regression. The simple linear regression model in population form is

$$y_i = \beta_0 + \beta_1 x_i + \varepsilon_i \tag{9.1}$$

where y_i is the value of the dependent variable for individual i, x_i is the value of the independent variable for individual i, β_0 is the intercept, and β_1 is the slope relating the independent and dependent variables to one another.

The intercept is the point where the line in equation (9.1) crosses the y-axis at $x = 0$. It is also the mean of y for individuals with a value of 0 on x, and it is this latter definition that will be most useful in actual practice. The slope expresses the relationship between y and x. Positive slopes indicate that larger values of x are associated with larger values of y, while negative slopes mean that larger x values are associated with smaller y. In addition, larger values of β_1 (positive or negative) indicate a stronger linear relationship between y and x. Finally, ε_i represents the random error in the model and encompasses all of the factors that might influence the dependent variable (y) other than the independent variable (x).

The regression model in equation (9.1) can be used to obtain a predicted value of y_i (known as $\hat{y}_i$) given a value of x_i. In the population, this prediction is expressed as

$$\hat{y}_i = \beta_0 + \beta_1 x_i \tag{9.2}$$

We will see that $\hat{y}$ plays a key role in obtaining sample estimates for the population parameters β_0 and β_1. Also, notice that equation (9.2) doesn't include the random error term. Given that the error includes all of the factors that influence y other than our measured independent variable x, this is not a surprise. However, this fact also means that our prediction of y will very likely not be perfectly accurate for any one individual. Furthermore, the larger the error term, the less accurate our prediction is likely to be. Now that we have considered the population simple linear regression model, let's turn our attention to obtaining sample estimates for the population parameters.

Estimating Regression Models with Ordinary Least Squares

In virtually all real-world contexts, the population is unavailable to the researcher. Therefore, β_0 and β_1 must be estimated using sample data taken from the population. There exist in the statistical literature several methods for obtaining estimated values of the regression model parameters (b_0 and b_1, respectively) given x and y. By far the most popular and widely used of these methods is OLS. The goal of OLS is to minimize the sum of the squared differences between the observed values of y and the model-predicted values of y across the sample. This difference, known as the residual, is written as

$$e_i = y_i - \hat{y}_i \tag{9.3}$$

Therefore, the method of OLS seeks to minimize

$$\sum_{i=1}^{n} e_i^2 = \sum_{i=1}^{n} \left(y_i - \hat{y}_i \right)^2 \tag{9.4}$$

The actual mechanism for finding the linear equation that minimizes the sum of squared residuals involves the partial derivatives of the sum of squared function with respect to the model coefficients. We will leave these mathematical details to excellent references, such as Fox (2016). It should be noted that in the context of simple linear regression, the OLS criteria reduce to the following equations, which can be used to obtain b_0 and b_1 as

$$b_1 = r \left(\frac{s_y}{s_x} \right) \tag{9.5}$$

and

$$b_0 = \bar{y} - b_1 \bar{x} \tag{9.6}$$

where r is the Pearson correlation coefficient between x and y, s_y is the sample standard deviation for y, s_x is the sample standard deviation for x, $\bar{y}$ is the sample mean for y, and $\bar{x}$ is the sample mean for x

Distributional Assumptions Underlying OLS Estimation

Estimation of the linear regression model coefficients using OLS rests upon several assumptions about the distribution of the residuals in the population. These assumptions will be familiar to us, as they are quite similar to assumptions that we made when using ANOVA and the t-test. Other assumptions underlying OLS are unique to regression. The first assumption that must hold true for linear models to function optimally is that the relationship between y and x is linear. If the relationship is not linear, then clearly an equation for a line will not provide an adequate fit, and the model is thus misspecified. A second assumption is that the variance in the residuals is constant regardless of the value of x. This assumption is typically referred to as homoscedasticity and is a generalization of the homogeneity of error variance assumption in ANOVA (Chapter 5). The third assumption is that the residuals are normally distributed in the population. Fourth, it is assumed that the independent variable x is measured without error and that it is unrelated to the model error term, ε. It should be noted that the assumption of x measured without error is not as strenuous as one might first assume. In fact, for most real-world problems, the model will work well even when the independent variable is not error-free (Fox, 2016). Fifth and finally, the residuals for any two individuals in the population are assumed to be independent of one another. This independence assumption implies that the unmeasured factors influencing y are not related from one individual to another. This is identical to the independence assumption that we discussed in Chapters 4–6. In the example below, we will describe how to assess these assumptions with our sample data.

Coefficient of Determination

When the linear regression model has been estimated, researchers generally want to measure the relative magnitude of the relationship between the variables. One useful tool for ascertaining the strength of the relationship between x and y is the coefficient of determination, which is the squared multiple correlation coefficient between the observed and model-predicted dependent variables. It is denoted as R^2 and reflects the proportion of the variation in the dependent variable that is explained by the independent variable. Mathematically, R^2 is calculated as

$$R^2 = \frac{SS_R}{SS_T} = \frac{\sum_{i=1}^{n}(\hat{y}_i - \bar{y})^2}{\sum_{i=1}^{n}(y_i - \bar{y})^2} = 1 - \frac{\sum_{i=1}^{n}(y_i - \hat{y})^2}{\sum_{i=1}^{n}(y_i - \bar{y})^2} = 1 - \frac{SS_E}{SS_T} \tag{9.7}$$

The terms in equation (9.7) are as defined previously.

The value of R^2 always lies between 0 and 1, with larger numbers indicating a stronger linear relationship between x and y, implying that the independent variable accounts for more variance in the dependent variable. R^2 is a very commonly used measure of the overall fit of the regression model and, along with the parameter inference discussed below, serves as the primary mechanism by which the relationship between the two variables is quantified.

Inference for Regression Parameters

In addition to R^2, a second method for understanding the nature of the relationship between x and y involves making inferences about the relationship in the population given the sample regression equation. Because b_0 and b_1 are sample estimates of the population parameters β_0 and β_1, respectively, they are subject to sampling error as is any sample estimate. This means that, although the estimates are unbiased given that the aforementioned assumptions hold, they are not precisely equal to the population parameter values. Furthermore, if we were to draw multiple samples from the population and estimate the intercept and slope for each, the values of b_0 and b_1 would differ across samples, even though they would be estimating the same population parameter values for β_0 and β_1. The magnitude of this variation in parameter estimates across samples can be estimated from our single sample using a statistic known as the standard error. The standard error of the slope, denoted as σ_{b_1} in the population, can be thought of as the standard deviation of slope values obtained from all possible samples of size n taken from the population. Similarly, the standard error of the intercept, σ_{b_0}, is the standard deviation of the intercept values obtained from all such samples. Clearly, it is not possible to obtain census data from a population in an applied research context. Therefore, we will need to estimate the standard errors of both the slope (s_{b_1}) and intercept (s_{b_0}) using data from a single sample, much as we did with b_0 and b_1.

To obtain s_{b_1}, we must first calculate the variance of the residuals,

$$S_e^2 = \frac{\sum_{i=1}^{n} e_i^2}{N - p - 1} \tag{9.8}$$

where e_i^2 is the residual value for individual i, N is the sample size, and p is the number of independent variables (1 in the case of simple regression).

Using S_e^2, we can calculate the standard errors for the regression model parameter estimates.

$$S_{b_1} = \frac{1}{\sqrt{1-R^2}}\left[\frac{S_e}{\sqrt{\sum_{i=1}^{n}(x_i - \bar{x})^2}}\right] \tag{9.9}$$

$$S_{b_0} = S_{b_1}\sqrt{\frac{\sum_{i=1}^{n}x_i^2}{n}} \tag{9.10}$$

Given that the sample intercept and slope are only estimates of the population parameters, researchers are quite often interested in testing hypotheses to infer whether the data represent a departure from what would be expected in the null case. Most frequently (though not always), the inference of interest concerns testing that the population parameter is 0. In particular, a non-0 value of β_1 means that x is linearly related to y. Therefore, researchers are typically interested in using the sample to make inferences about whether the population slope (β_1) is 0 or not. Inference can also be made regarding the intercept, and again the typical focus is on whether this value is 0 in the population.

Inferences about regression parameters can be made using both confidence intervals and hypothesis tests, just as we can use either of these approaches to make inferences about means (see Chapters 4–6). Much as with the confidence interval of the mean, the confidence interval of the regression coefficient yields a range of values within which we have a set level of confidence (e.g., 95%) that the population parameter value resides. If we are interested in whether x is linearly related to y, then we would simply determine whether 0 is in the sample confidence interval for β_1. If 0 does lie in this interval, we would not be able to conclude that the population value differs from 0. Such a result does not imply that the null hypothesis is true, but rather it means that there is not sufficient evidence available in the sample data to reject the null. Similarly, we can construct a confidence interval for the intercept, and if 0 is within the interval, we would conclude that the value of y for an individual with $x = 0$ could plausibly be 0. The confidence intervals for the slope and intercept take the following forms:

$$b_1 \pm t_{CV}S_{b_1} \tag{9.11}$$

and

$$b_0 \pm t_{CV}S_{b_0} \tag{9.12}$$

Here the parameter estimates and their standard errors are as described previously, while t_{CV} is the critical value of the t distribution for $1-\alpha/2$ (e.g., the 0.975 quantile if $\alpha=0.05$) with $n-p-1$ degrees of freedom. The value of α is equal to 1 minus the desired level of confidence. Thus, for a 95% confidence interval (0.95 level of confidence), α would be 0.05.

In addition to confidence intervals, inferences about the regression parameters can also be made using hypothesis tests. In general, the forms of this test for the slope and intercept, respectively, are

$$t_{b_1} = \frac{b_1 - \beta_1}{S_{b_1}} \tag{9.13}$$

$$t_{b_0} = \frac{b_0 - \beta_0}{S_{b_0}} \tag{9.14}$$

The terms β_1 and β_0 are the parameter values under the null hypothesis. Again, most often the null hypothesis posits that there is no linear relationship between x and y (i.e., $\beta_1 = 0$) and that the value of $y = 0$ when $x = 0$ (i.e., $\beta_0 = 0$). For simple regression, each of these tests is conducted with $n - 2$ degrees of freedom. We will demonstrate how these calculations can be made by hand later in the chapter.

Identification of Influential Cases

An important aspect of conducting regression analysis is the identification of unusual or outlying cases. In other words, we need to determine whether any observations in our dataset are unusual in some way from the other observations, and if so, which ones they are. Being unusual in the context of regression can be defined in multiple ways, and we will consider each of these in this chapter. More specifically, an individual can be unusual with respect to the independent variables (high leverage), in terms of the predicted value of the outcome (high distance), or with regard to its impact on the prediction equation (high influence). There are different statistical tools available to identify each type.

An observation has high leverage if its scores on the set of independent variables are unusual when compared with the other observations. It's important to note that an individual with high leverage may not have unusual values on any single independent variable but may have an unusual pattern on the full set of predictors. Perhaps the most popular statistic for identifying high leverage points is the hat value. This statistic is so named because it plays a role in obtaining the predicted value of the dependent variable ($\hat{y}$) in terms of the observed dependent variable values (y). The hat value for individual i is denoted as h_i and reflects the distance between the set of independent variable values for individual i and the multivariate mean (centroid) of the independent variables for the full sample. Larger values indicate a greater divergence between the individual and the sample as a whole with respect to the independent variable values. Hat values range between $1/n$ and 1 with a mean of

$$\bar{h} = \frac{(p+1)}{n} \tag{9.15}$$

where p is the number of independent variables.

One common guideline (Fox, 2005) for identifying cases with high leverage is $2*\bar{h}$ or $3*\bar{h}$.

In addition to being unusual in terms of the set of independent variables, an observation can also be unusual with respect to the difference between the predicted ($\hat{y}_i$) and observed dependent (y_i) variable values, known as the residual (e_i), as defined in equation (9.3). Remember that in OLS regression, our goal is to minimize the squared residuals. The value of e_i is in the scale of the dependent variable and therefore not particularly easy to interpret in terms of its magnitude. Therefore, we will standardize this value in some way so as to make it easier to interpret. The studentized residual is calculated as

$$e_i^* = \frac{e_i}{S_{e(-i)}\sqrt{1-h_i}} \tag{9.16}$$

where

$$S_{e(-i)} = \sqrt{\frac{\sum_{i=1}^{n} e_{-i}^2}{(n-k-1)}}$$

$S_{e(-i)}$ is the standard error of regression with individual i removed from the sample. The calculation of this standard error, excluding each individual in turn, ensures that the presence of a large residual will not lead to inflation of the standard error and thus make identification of outliers difficult. In practice, e_i^* for each individual is calculated in turn, and those with $\left| e_i^* \right| \geq 2$ are by convention viewed as outliers in terms of distance. However, as noted by Fox (2016), this cut-off for identifying outliers is likely to lead to inflation of the Type I error rate. In other words, if we assess each $\left| e_i^* \right|$ in the sample, we are essentially conducting n separate hypothesis tests. Thus, it is recommended that researchers use the Bonferroni adjustment to assess whether the largest e_i^* in the sample is significantly different from 0. As an example, if we have ten individuals and two independent variables and we want our overall Type I error rate to be 0.05, then the Bonferroni-corrected threshold for the largest e_i^* to be statistically significant would be $\alpha_{\text{Bonferroni}} = \frac{\alpha_{\text{original}}}{n} = \frac{0.05}{10} = 0.005.$

The third aspect of outlier detection in regression analysis is in terms of individual influence on the regression model itself. In other words, we want to identify individuals whose presence in the sample has an unusually large impact on the regression model parameter estimates. There are multiple statistics designed for this purpose. Cook (1977) proposed a distance measure that combines information about the extent to which an individual is an outlier with respect to both the residuals and the independent variables. Cook's distance is calculated as

$$D_i = \frac{e_i'^2}{k+1} * \frac{h_i}{1-h_i} \tag{9.17}$$

where

$$e_i' = \frac{e_i}{\sqrt{\dfrac{\sum_{i=1}^{n} e_i^2}{(n-k-1)}}\sqrt{1-h_i}}$$

There are no well-defined cut-values for what constitutes a large value of D_i, and it is generally recommended that researchers simply examine their data for unusually large values to identify potentially influential observations (e.g., Fox, 2016).

A second statistic that can be used to identify influential cases is DFBETA, which reflects the change in regression coefficients when individuals are removed from the sample. DFBETA begins with the following calculation:

$$D_{ij} = b_j - b_{j(-i)} \tag{9.18}$$

where b_j is the regression slope estimate for independent variable j and $b_{j(-i)}$ is the regression slope estimate for independent variable j with individual i removed from the sample.

D_{ij} is then scaled through dividing it by the standard error for the coefficient with individual i removed from the sample ($SE_{(-i)}(b_j)$).

$$D_{ij}^* = \frac{D_{ij}}{SE_{(-i)}(b_j)} \tag{9.19}$$

As with Cook's distance measure, there are no agreed-upon standards for what connotes a large value of D_{ij}^*. Therefore, we look for cases with unusually large values as potential outlying observations.

Finally, Belsley et al. (1980) provided an additional measure of an individual's distance from the sample based on the magnitude of the residuals and the hat value, much in the spirit of Cook's statistic. The DFFITS for individual i is calculated as

$$DFFITS_i = e_i^* \sqrt{\frac{h_i}{1-h_i}} \tag{9.20}$$

As with DFBETA, there are no cut-values for identifying extreme cases and thus we will look for unusually large or small values as potential outliers in regards to the regression model.

Simple Linear Regression Example

Now that we have discussed the basic simple linear regression model, parameter estimation, inference, assumptions, model fit, and outlier detection, let's apply these ideas to our example. We will start with simple linear regression where SOP is the dependent variable and EFFC is the independent variable. Typically, regression results are reported in the form of a table (Table 9.2), similar to the ANOVA table from Chapter 5.

These results show that EFFC has a statistically significant positive relationship with SOP, meaning that individuals who have higher EFFC also have higher levels of socially oriented perfectionism. We know that this relationship is likely to differ from 0 in the population because the 95% confidence interval (2.27, 6.66) does not include 0. In addition, the p-value associated with $H_0 : \beta_1 = 0$ is less than our α of 0.05. The confidence interval for the coefficient reveals that in the population, the value of β_1 is likely to lie between 2.27 and 6.66. The intercept of 54.18 is the mean SOP score if EFFC is equal to 0. In practice EFFC cannot be 0, meaning that the intercept is not particularly useful from a practical perspective in this particular instance. Finally, we can write the estimated regression equation as

$$\widehat{\text{SOP}}_i = b_0 + b_1 x_i = 54.18 + 4.47\,\text{EFFC}_i$$

We can see how the coefficient and intercept can be calculated by hand using equations (9.5) and (9.6). Recall that the coefficient is a function of the correlation between the dependent and independent variables and the ratio of their standard deviations.

$$b_1 = r\left(\frac{s_y}{s_x}\right) = 0.21\left(\frac{15.40}{0.72}\right) = 0.21(21.39) = 4.49$$

This value is within the rounding error of the coefficient estimate that we obtained using computer software. The intercept can then be calculated as

$$b_0 = \bar{y} - b_1\bar{x} = 73.5 - 4.49 * 4.33 = 73.5 - 19.44 = 54.06$$

TABLE 9.2

Simple Linear Regression Table with SOP as the Outcome and EFFC as the Independent Variable

Variable	Coefficient	Standard Error	t	p	95% Confidence Interval
Intercept	54.18	4.89	11.07	<0.001	44.55, 63.80
EFFC	4.47	1.12	4.00	<0.001	2.27, 6.66

SOP, self-oriented perfectionism; EFFC, effortful control.

Again, our hand-calculated intercept is within the rounding error of the one obtained using the software. We can visualize the relationship using an effects plot, as in Figure 9.2.

The y-axis reflects the predicted value of SOP, and the x-axis reflects values for EFFC. The solid line displays the predicted relationship between the two variables, and the shaded region is the 95% confidence bound for the line.

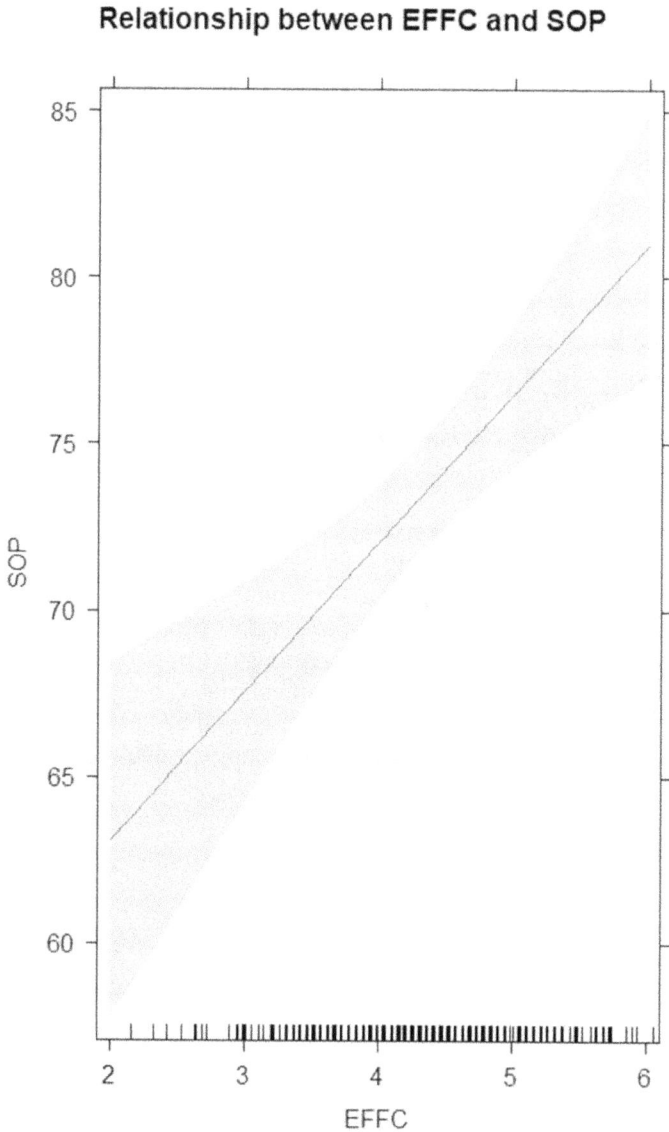

FIGURE 9.2
Effects plot for simple linear regression.

The R^2 value for our model is 0.04, meaning that approximately 4% of the variation in SOP is associated with EFFC scores. Interpretation of R^2 is largely a matter of context, meaning that what is a large value for one discipline may be quite small for another and vice versa. With that in mind, we would want to interpret the 4% for this study in the context of the broader research literature using these constructs. For example, in other studies examining relationships between perfectionism and individual personality traits, R^2 values generally ranged between 0.03 and 0.14. Thus, our value falls at the lower end of what is typically seen in research linking personality and perfectionism. Cohen (1988) provided the following general guidelines for interpreting R^2:

Negligible: $R^2 < 0.02$
Weak: $0.02 \leq R^2 < 0.13$
Moderate: $0.13 \leq R^2 < 0.26$
Large: $R^2 \geq 0.26$

These guidelines are frequently used in the social sciences. However, as with all such recommendations (see our discussion of Cohen's guidelines in Chapters 3–6), they should serve only as a starting point for researchers when interpreting R^2. Ideally, the value should be considered within the broader context of the literature pertinent to the problem being addressed.

To be sure that the results reported above are reliable, we need to ensure that the assumptions underlying OLS estimation have been met. Such an assessment is particularly important for the inferences associated with the coefficient and intercept. Pena and Slate (2006) proposed an omnibus test for simultaneously assessing the assumptions of linearity, homoscedasticity, and normality based on the results of an estimated regression model. This test combines multiple statistics that can be used to assess various aspects of data shape and variation. These statistics appear below, followed by that used in the omnibus test. Note that we are using the notation from Pena and Slate here, so that some of the terms that have appeared in the text above are denoted slightly differently in the following equations.

$$\hat{S}_1^2 = \left\{ \frac{1}{\sqrt{6n}} \sum_{i=1}^{n} R_i^3 \right\}^2 \qquad (9.21)$$

$$\hat{S}_2^2 = \left\{ \frac{1}{\sqrt{24n}} \sum_{i=1}^{n} R_i^4 - 3 \right\}^2 \qquad (9.22)$$

$$\hat{S}_3^2 = \frac{\left\{ \frac{1}{\sqrt{n}} \sum_{i=1}^{n} (Y_i - \bar{Y})^2 R_i \right\}^2}{\left(\hat{\Omega} - \left(\mathbf{b}' \hat{\Sigma}_x \mathbf{b} \right)^2 - \hat{\Gamma} \hat{\Sigma}_x^{-1} \hat{\Gamma}' \right)} \qquad (9.23)$$

$$\hat{S}_4^2 = \left\{ \frac{1}{\sqrt{2\hat{\sigma}_v^2 n}} \sum\nolimits_{i=1}^{n} (v_i - \bar{v})(R_i^2 - 1) \right\}^2 \tag{9.24}$$

where $R_i = Y_i - \hat{Y}_i$ is the residual for individual i, Y_i is the observed dependent variable value for individual i, $\hat{Y}_i$ is the model-predicted dependent variable value for individual i, $\bar{Y}$ is the sample mean of the dependent variable, n is the sample size, $\hat{\Sigma}_x$ is the estimated covariance matrix for the independent variables, and b is the set of regression model coefficient estimates

$$\hat{\Gamma} = \frac{\sum\nolimits_{i=1}^{n} (Y_i - \bar{Y})^2 (X_i - \bar{X})}{n}$$

$$\hat{\Omega} = \frac{\sum\nolimits_{i=1}^{n} (\hat{Y}_i - \bar{Y})^4}{n}$$

X_i is the independent variable value for individual i and $\bar{X}$ is the mean of independent variable.

The omnibus test for the null hypothesis that all assumptions have been met is then calculated as

$$\hat{G}_4^2 = \hat{S}_1^2 + \hat{S}_2^2 + \hat{S}_3^2 + \hat{S}_4^2 \tag{9.25}$$

More specifically, the statistic $\hat{G}_4^2$, which is distributed as a chi-square statistic with four degrees of freedom, can be used to test the following null hypothesis:

H_0 : Assumptions of linearity, homoscedasticity, independence, and normality hold

H_A : At least one of the assumptions in H_0 does not hold.

If H_0 is rejected based on $\hat{G}_4^2$, its individual component statistics can be used to learn which assumption(s) may have been violated. These individual statistics are distributed as a chi-square with one degree of freedom. If $\hat{S}_1^2$ is statistically significant, we would conclude that the residuals are skewed. A statistically significant $\hat{S}_2^2$ indicates kurtotic residuals, whereas a statistically significant $\hat{S}_3^2$ indicates a lack of linearity. Heteroscedasticity is indicated by a statistically significant value for $\hat{S}_4^2$.

For the simple regression example, $\hat{G}_4^2 = 13.92$, with $p=0.008$. Therefore, we reject H_0 and conclude that at least one of the assumptions does not hold. The specific component statistics had the following values:

$\hat{S}_1^2 = 0.0002, p=0.99$

$\hat{S}_2^2 = 0.0.147, p=0.70$

$\hat{S}_3^2 = 0.0.276, \, p=0.60$

$\hat{S}_4^2 = 13.490, \, p=0.0002$

Based on these results, we would conclude that the data are not homoscedastic (i.e., the variance is not homogeneous), but that normality and linearity do hold.

The traditional approach for assessing the assumption of normality of residuals involves using the QQ plot and Shapiro–Wilk test that we applied in the context of the *t*-test and ANOVA. The QQ plot appears in Figure 9.3.

Remember that we are looking to see whether the individual data points lie along the line and within the confidence interval, which they do in this case. This pattern leads us to conclude that the residuals are indeed normally distributed. The test for normality yielded a *p*-value of 0.05, which would lead

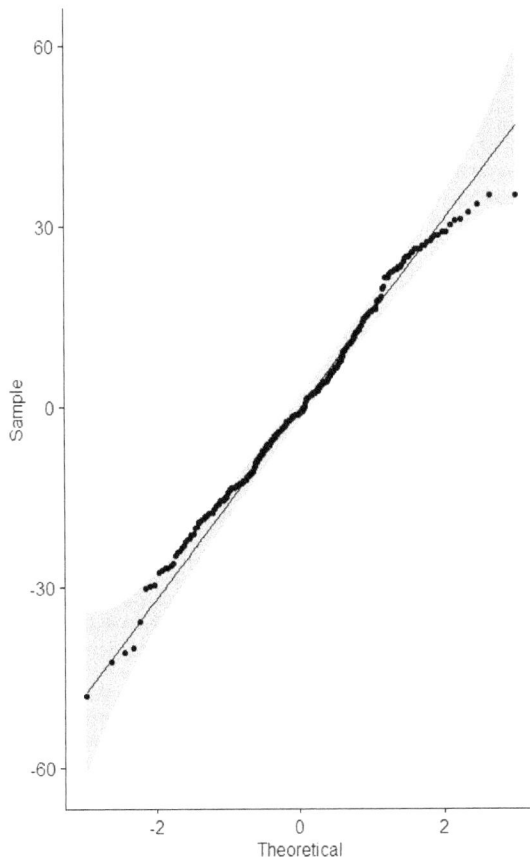

FIGURE 9.3
QQ plot of residuals for simple linear regression model.

us to reject the null hypothesis of normality if $\alpha=0.05$. Taken together, these results suggest that the assumption of normality may not hold, but if that is the case, the deviation is likely to be relatively small and therefore does not significantly impact the regression results.

We will assess the homogeneity of the variance assumption using a scatterplot of the model-predicted values of SOP and the residuals, looking to see whether the spread in residuals is relatively constant across values of the predicted scores, which would indicate that the homogeneity assumption holds. Conversely, if the spread in residuals varies across values of the predicted SOP scores, we would conclude that the assumption does not hold. The residual-predicted plot appears in Figure 9.4.

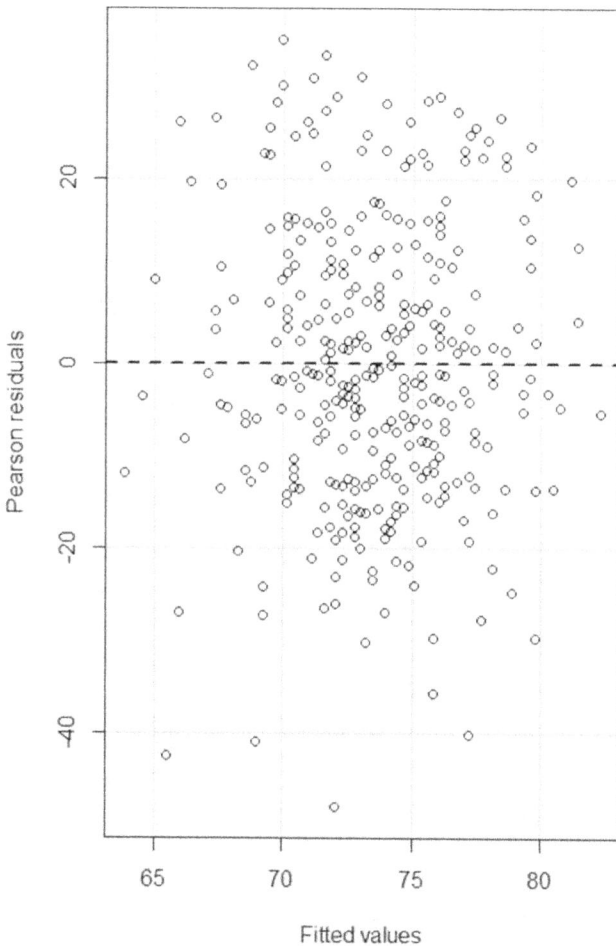

FIGURE 9.4
Residual-predicted plot for simple linear regression.

There does not appear to be a discernible pattern in the scatterplot, suggesting that the homogeneity of variance assumption does hold.

In addition to the residual by predicted scatterplot, we can also use two statistical tests of the null hypothesis that variance in the residuals is related to the mean of the model-predicted outcome (Breusch & Pagan, 1979) or to a linear combination of the independent variables (Cook & Weisberg, 1983). A statistically significant result for either of these tests would lead us to reject the null of homogeneous variances in the residuals across values of the predicted dependent variable. In the case of simple linear regression, these are identical null hypotheses, but this will not be the case for multiple regression, as we will see below. For this example, $p = 0.137$, leading us not to reject the null that the residual variance is unrelated to the mean of the model-predicted dependent variable. Thus, both the scatterplot and hypothesis tests support the assumption of homogeneous variance of the residuals across levels of the outcome variable.

We should stop for a moment and consider the full set of results regarding the assumptions of our model. In some respects, it appears that we have contradictory findings for both the assumption of normality of residuals and homoscedasticity of residuals. The Pena and Slate tests, as well as the QQ plot, suggest that the assumption of normality has likely held. However, the Shapiro–Wilk test was statistically significant, which would lead us to reject the null hypothesis that the residuals are normally distributed. How can we reconcile these differences? It will help to remember that all of these techniques are based on a sample and therefore are subject to the usual issues around sampling error. In addition, when we consider these results en toto, it appears that the departure from normality, if present, is very minor. We have two pieces of evidence suggesting that the residuals are normally distributed and one that might be thought of as borderline evidence against the assumption (Shapiro–Wilk test with $p = 0.05$). Thus, on the balance of evidence, it is probably reasonable to conclude that the residuals are normally distributed in the population. What about the homoscedasticity assumption? Again, we have conflicting evidence with a scatterplot, one hypothesis test suggesting that the residual variances are homoscedastic and another test suggesting that this is not the case. Again, when we consider the full set of evidence, any departures from homoscedasticity are likely to be minor. In addition, given the sample size of over 300 individuals, we can comfortably rely on the central limit theorem when applying inferential statistics to our results.

As was the case for the other statistical methods that we have discussed in this book, we will assess the assumption of independence with a thorough understanding of the sampling plan underlying our data. If the data are obtained using a simple random sampling scheme, then it is highly unlikely that there will be dependencies among the individuals. However, if the data come from a convenience sample, then we need to carefully consider from where individuals come and how they agree to participate with an eye toward identifying dependencies. In cases where data are collected in a longitudinal fashion, then serial dependency is likely to be present. We can assess this using the Durbin–Watson statistic. We will not discuss it here, given that it is used in very specific circumstances when data are collected longitudinally. However, we will learn more about this statistic and its application when we discuss models for serially correlated data in Chapter 12.

Finally, we need to assess whether there are unusual or outlying cases present in the data. Figure 9.5 displays individual values of Cook's D, studentized residuals, Bonferroni-corrected p-values associated with the studentized residuals, and the hat values that measure leverage.

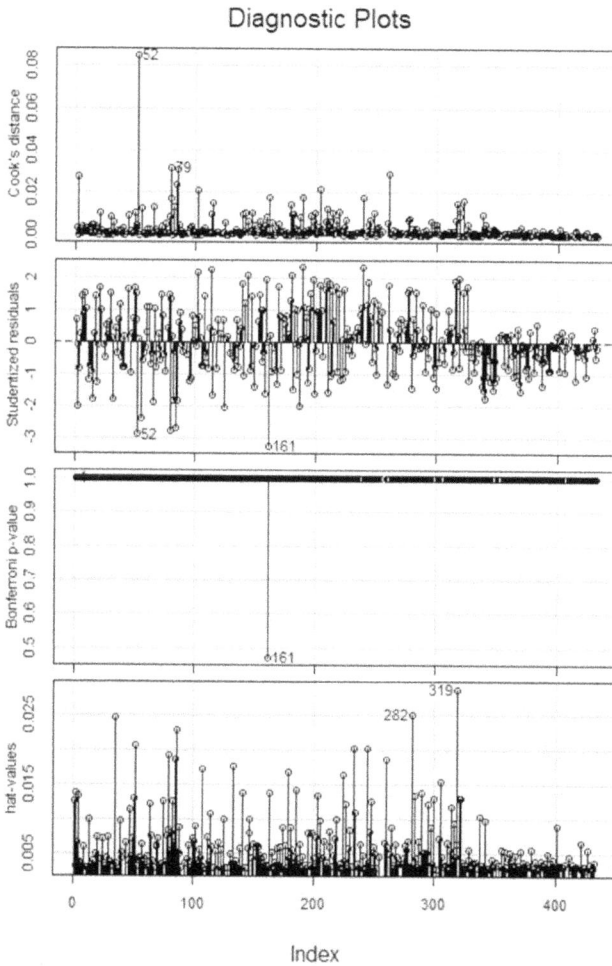

FIGURE 9.5
Index plots for Cook's distance, studentized residuals, Bonferroni-corrected p-values for studentized residuals, and hat values.

The Cook's D value for subject 52 is markedly larger than that of the other members of the sample, suggesting that this individual potentially influences the regression equation more than others in the sample. The studentized residuals for subjects 52 and 161 were unusually large when compared to the other members of the sample. However, the Bonferroni p-value for person 161, which was the smallest at 0.472, was not less than our alpha of 0.05. Thus, although the studentized residuals for individuals 52 and 161 were unusual when compared to the rest of the sample, we would not identify them as outliers with respect to the accuracy of the model prediction for the outcome once we correct for Type I error inflation. Finally, individuals 282 and 319 had the highest hat values in the sample, suggesting that they were the most unusual cases with respect to the independent variables. Using the 3*mean hat value cut-off as suggested by Fox (2015), individuals 35, 52, 80, 85, 86, 107, 133, 234, 245, 260, 282, and 319 had high-leverage/unusual values for the independent variable.

We'll conclude our investigation into influential individuals by examining the DFBETAS and DFFITS. Figure 9.6 displays a boxplot of the DFBETA values. We can see that several individuals have unusually large or small values, with case 52 being particularly notable in this regard.

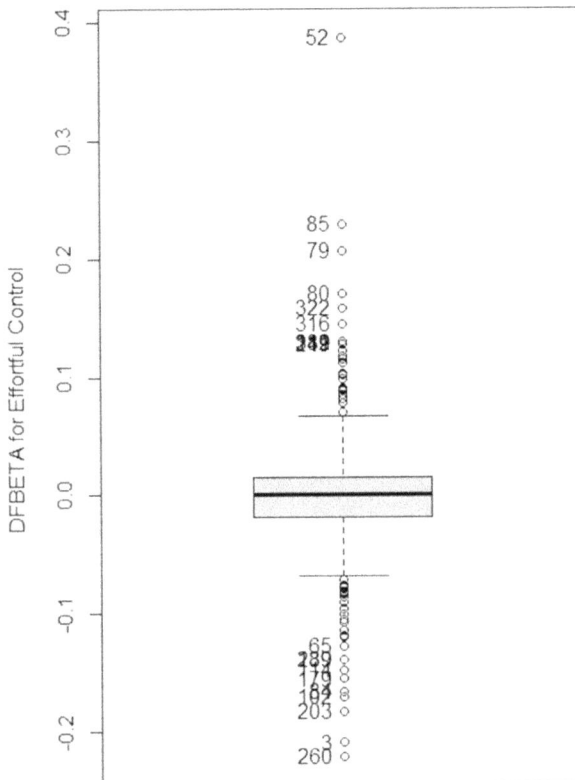

FIGURE 9.6
Boxplot of DFBETA values for the simple regression model.

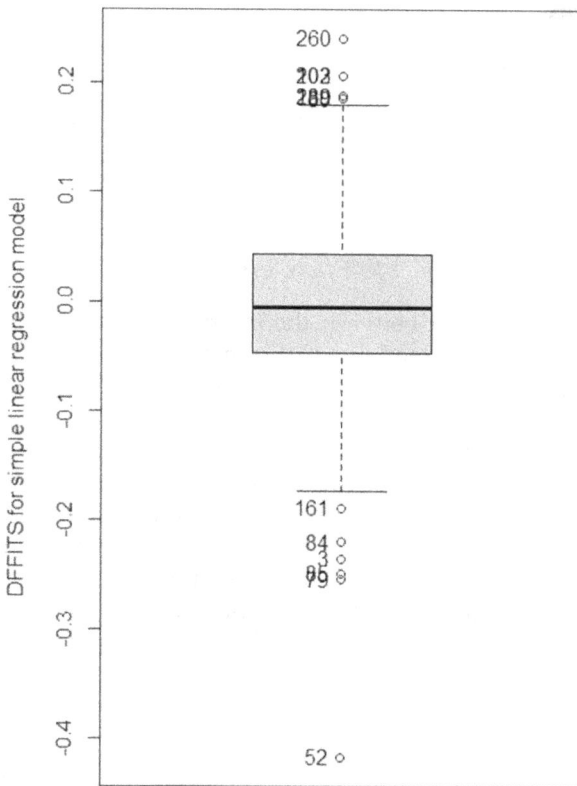

FIGURE 9.7
DFFITS for a simple linear regression model.

A boxplot for the DFFITS based on the simple linear regression model appears in Figure 9.7.

A number of the cases that had a marked impact on the regression coefficient in the simple linear regression model (based on large DFBETAS) also had large DFFITS, suggesting that they were distant from the rest of the sample with respect to the studentized residuals and leverage values. In particular, we can see that individuals 52, 260, and 203 stood out for both DFBETAS and DFFITS.

Taken together, we see that there are a few individuals who were unusual in some way when compared to others in the sample. However, the number of such individuals is not particularly large when considered in light of the full sample size. Thus, we may feel comfortable that our results should remain fairly stable whether these individuals are in the sample or not. As we will see in the next chapter, there are approaches to fitting the regression model that are designed to address the impact of outlying observations. We will see the extent to which results using these methods might differ from the standard OLS considered here.

Multiple Regression

The linear regression model can very easily be extended to allow for multiple independent variables at once, much as a one-way ANOVA can be easily extended to include several independent variables. In the case of two regressors, the model takes the form

$$y_i = \beta_0 + \beta_1 x_{1i} + \beta_2 x_{2i} + \varepsilon_i \tag{9.26}$$

The only major difference between the multiple regression model in (9.26) and the simple regression model in (9.1) is that each coefficient is interpreted while *holding constant* the value of the other regression coefficient. In particular, the parameters are estimated by b_0, b_1, and b_2, and inferences about these parameters are made in the same fashion with regard to both confidence intervals and hypothesis tests. The assumptions underlying this model are also the same as those described for the simple regression model. Despite these similarities, there are three additional topics regarding multiple regression that we need to consider here, including inference for the set of model slopes as a whole, adjusting the coefficient of determination for the more complex model, and collinearity among the independent variables.

With respect to model inference, for simple linear regression, the most important parameter is generally the slope. When there are multiple x variables in the model, the researcher may want to know whether the independent variables taken as a whole are related to y, as well as whether each independent variable separately is associated with the dependent variable. Therefore, some overall test of model significance is desirable. The null hypothesis for this test is that all of the slopes are equal to 0 in the population; i.e., none of the regressors are linearly related to the dependent variable. The test statistic for this hypothesis is calculated as

$$F = \left(\frac{n-p-1}{p}\right)\left(\left(\frac{R^2}{1-R^2}\right)\right) \tag{9.27}$$

Here terms are as defined in equation (9.7). This test statistic is distributed as an F with p and $n-p-1$ degrees of freedom. A statistically significant result would indicate that one or more of the regression coefficients are not equal to 0 in the population. Typically, the researcher would then refer to the tests of individual regression parameters, which were described above, to identify which were not equal to 0.

A second issue to be considered by researchers in the context of multiple regression is the notion of adjusted R^2. Adjustment of the standard coefficient of determination is necessary because the inclusion of additional independent variables in the regression model will always yield higher values of R^2, even when these variables are not statistically significantly related to the dependent variable. In other words, there is a capitalization on chance that occurs in the calculation of R^2 as more variables are entered into the regression model. As a consequence, models including many independent variables, each having a negligible relationship with y, may produce an R^2 suggesting that the model explains a great deal of variance in y, even when this isn't the case.

An alternative for measuring the variance explained in the dependent variable that accounts for this additional model complexity would be quite helpful to the researcher seeking to understand the true nature of the relationship between the set of independent and the dependent variables. Such a measure exists in the form of the adjusted R^2 value, which is commonly calculated as

$$R_A^2 = 1 - \left(1 - R^2\right)\left(\frac{n-1}{n-p-1}\right) \tag{9.28}$$

R_A^2 only increases with the addition of an x if it explains more variance than would be expected by chance. In addition, R_A^2 will always be less than or equal to the standard R^2. It is generally recommended to use this statistic in practice when models containing many independent variables are used.

A final important issue specific to multiple regression is that of collinearity, which occurs when one independent variable is strongly associated with one or more of the other independent variables. When this is the case, the regression coefficients and their corresponding standard errors can be quite unstable, resulting in strange or counterintuitive estimates and/or large standard errors. In turn, the result will be poor performance when it comes to statistical inference, such as hypothesis testing and confidence intervals. One very useful approach to investigate the issue of collinearity for a sample involves a statistic known as the variance inflation factor (VIF). To obtain the VIF for a given independent variable (x_j), we would regress all of the other independent variables onto x_j and obtain the $R_{x_j}^2$ value. We can then calculate

$$\text{VIF} = \frac{1}{1 - R_{x_j}^2} \tag{9.29}$$

The denominator of the VIF is known as the tolerance value. VIF is large when $R_{x_j}^2$ is near 1 (i.e., the tolerance is very small). Such a result occurs when x_j has very little unique variation when the other independent variables in the model are considered. That is, if the other $p-1$ regressors can explain a high proportion of the variability in x_j, then x_j does not add much to the model.

As we noted above, collinearity can lead to high sampling variation in b_j, resulting in large standard errors and unstable parameter estimates. Conventional rules of thumb have been proposed for determining when an independent variable is highly collinear with the set of other $p-1$ regressors. For example, the researcher might consider collinearity to be a problem if VIF>5 or 10 (Fox, 2016). The typical response to collinearity is to either remove the offending variable(s) or use an alternative approach to conducting the regression analysis, such as ridge regression or regression following a principal components analysis. We will discuss the first of these options in Chapter 12.

Multiple Regression Example

Now that we have considered how multiple linear regression differs from simple regression, let's return to our example involving SOP and EFFC. In addition, we will include negative affect (NEG), extraversion (EXT), and Male (1=male, 0=female). The regression coefficients, standard errors, test statistics, and p-values appear in Table 9.3.

Note that the Betas are the slopes estimated on standardized data. The results in Table 9.3 indicate that there are statistically significant positive relationships between SOP with EFFC, NEG, and EXT. In other words, higher levels of EFFC, negative affect, and extraversion were each associated with higher levels of SOP. In contrast, there was not a statistically significant difference in the mean SOP between male and female respondents.

Notice that the coefficient for EFFC and the intercept in the multiple regression model differ from the simple linear regression model values in Table 9.2. These differences are due to the impact of including additional independent variables in the model. Specifically, the 5.82 coefficient value for EFFC in the multiple regression model reflects its relationship with SOP *after accounting for NEG, EXT, and Male*. In contrast, the 4.47 value from the simple regression model is not adjusted for any other variables. This difference is important for us to keep in mind when we interpret the results in a multiple regression model. *The EFFC coefficient for the multiple regression model estimates the relationship between EFFC and SOP after accounting for an individual's scores on NEG, EXT, and Male.* It is not an estimate of the relationship between EFFC

TABLE 9.3

Coefficients, Standard Errors, Betas, Test Statistics, and p-Values for Multiple Regression Model

Variable	Coefficient	Standard Error	Beta	t	p	95% Confidence Interval
Intercept	8.62	9.10		0.95	0.34	−9.28, 26.53
EFFC	5.82	1.08	0.27	5.42	<0.001	3.71, 7.94
NEG	7.32	1.11	0.35	6.61	<0.001	5.14, 9.49
EXT	2.05	0.99	0.10	2.06	0.04	0.10, 4.01
Male	1.33	1.81	0.03	0.74	0.46	−2.23, 4.89

EFFC, effortful control; EXT, extraversion; N_AFF, negative affect.

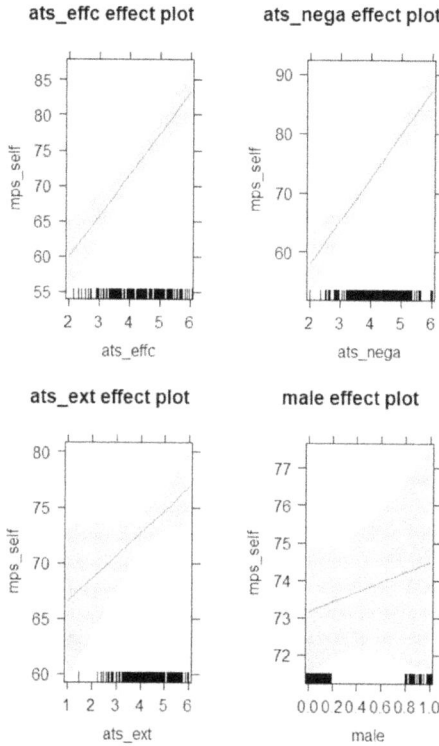

FIGURE 9.8
Effect plots for independent variables with SOP as the response. SOP, self-oriented perfectionism.

and SOP in isolation, which is the case for the simple linear regression model. Similar interpretations would be made for the other independent variables.

The effect plots for each independent variable appear in Figure 9.8.

We can see the positive relationships between EFFC, NEG, and EXT with SOP and the relatively narrow confidence band (represented as the shaded region). The plot for Male also shows a positive relationship with SOP but with a much wider confidence band.

The R^2 for the multiple regression model is 0.16, meaning that together EFFC, NEG, EXT, and Male account for approximately 16% of the variance in SOP. The adjusted R^2 is 0.15, suggesting that there was not a great deal of overlap in the variance in SOP associated with each of the independent variables. In other words, each of the four independent variables largely explained unique variance in SOP, given that when we adjusted for the model complexity, the coefficient of determination only declined by 0.01. Both the raw and adjusted R^2 values fall into the moderate range as defined above (Cohen, 1988).

As with the simple linear regression model, we need to assess the assumptions underlying the model. We can use the omnibus test developed for this purpose (Pena & Slate, 2006) just as we did for the simple linear regression problem. The omnibus test was statistically significant, indicating that at least one of the assumptions was not satisfied. The test of the null hypothesis for homoscedasticity was statistically significant ($\chi_1^2 = 16.366$, $p = 0.00005$). The tests for skewness, kurtosis, and linearity were not statistically significant, leading us to conclude that the normality and linearity assumptions have been met.

The QQ plot for the residuals appears in Figure 9.9.

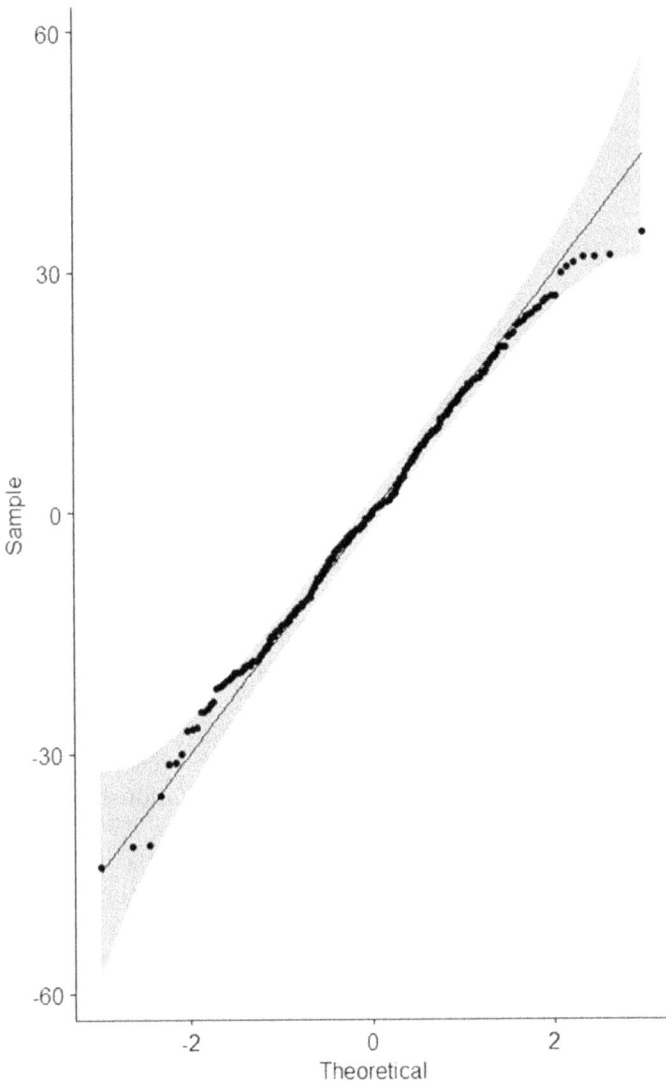

FIGURE 9.9
QQ plot of the multiple regression model residuals.

The individual residual values generally fall closely along the line denoting normality and within the 95% confidence band (shaded area). A density plot of the residuals reinforces that they are symmetrically distributed (Figure 9.10).

The skewness and kurtosis of the residuals are −0.09 and 2.90, respectively. These values indicate that the data are fairly symmetric but with a high peak,

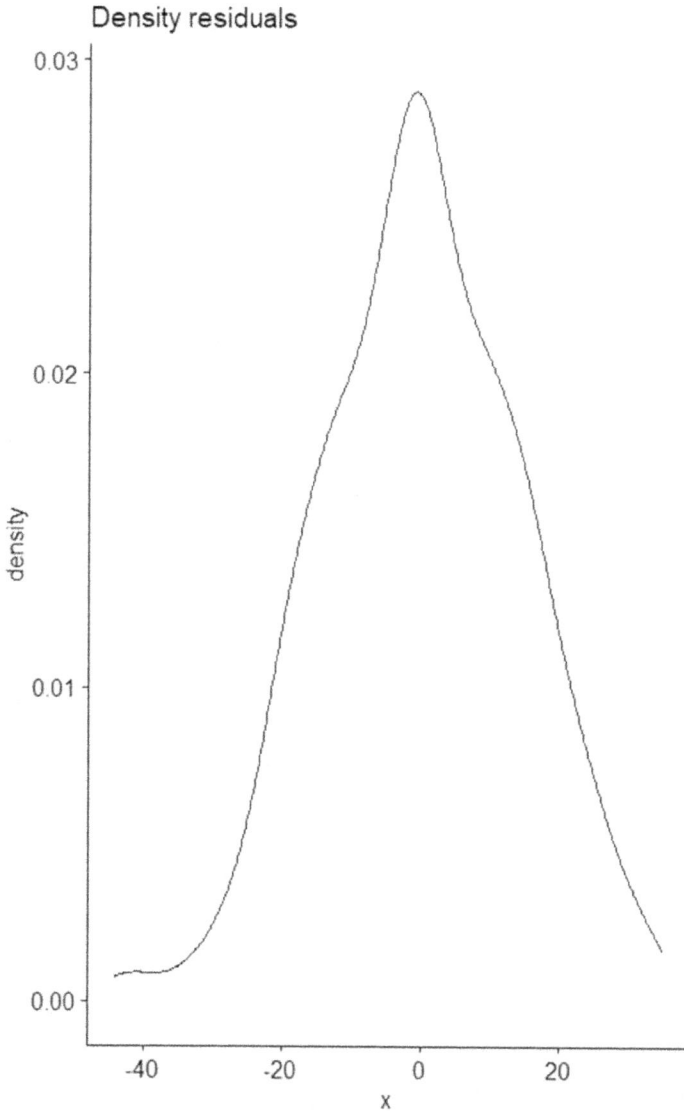

FIGURE 9.10
Density plot for the multiple regression model residuals.

which is exactly what we see in the density plot. The Shapiro–Wilk test was not statistically significant ($p=0.378$), meaning that we would not reject the null hypothesis that the residuals are normally distributed. Taken together, these results suggest that the residuals do appear to meet the normality assumption underlying OLS regression.

The scatterplot of the residuals and model-predicted values appears in Figure 9.11.

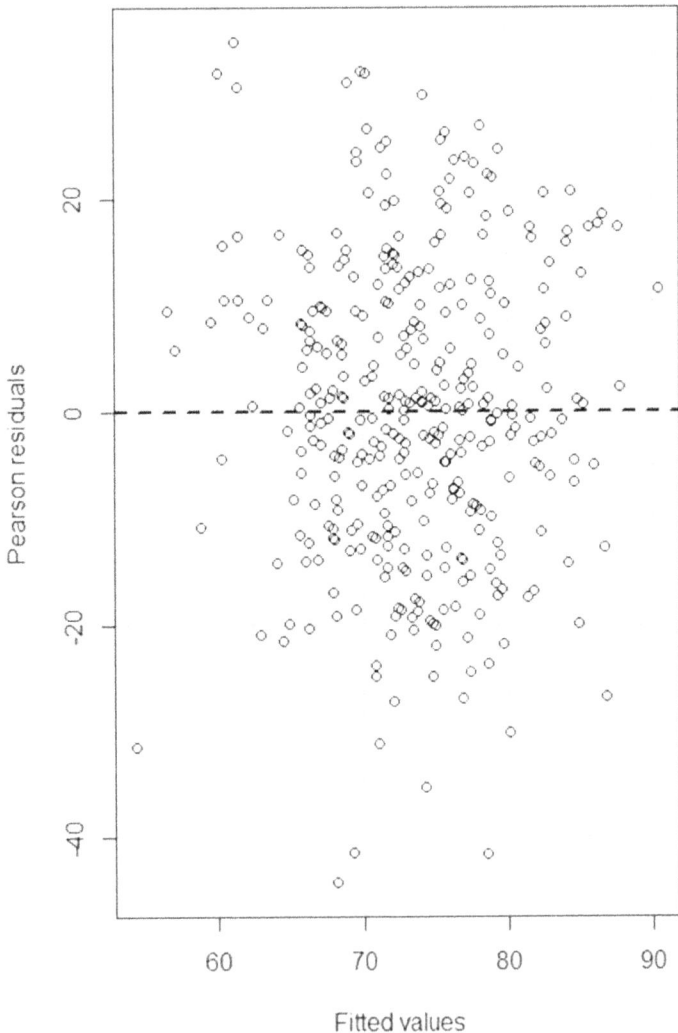

FIGURE 9.11
Scatterplot of residuals by model-predicted values for multiple regression model.

TABLE 9.4

VIF and Tolerance Values for Independent
Variables from Multiple Regression Model

Variable	VIF	Tolerance
EFFC	1.05	0.96
NEG	1.14	0.88
EXT	1.04	0.96
Male	1.11	0.90

EFFC, effortful control; EXT, extraversion; VIF,
variance inflation factor; NEG, negative affect.

There appears to be no discernible pattern in the plot with constant variance levels of the fitted values. The p-values associated with the Breusch–Pagan and Cook–Weisberg tests are 0.54 and 0.32, respectively. In conjunction with the residual by predicted plot, these results indicate that the residuals are homogeneously distributed. We refer the reader back to our discussion regarding how to consider the contradictory results for assumption checking that we discussed in the context of the simple linear regression model.

The VIF and tolerance values for the independent variables appear in Table 9.4.

None of the VIF values exceed the threshold of 5 suggested by Fox (2015). In addition, the tolerance values are all 0.88 or larger, indicating that none of the independent variable has more than 12% of its variance accounted for by the other independent variables.

Summary

Our focus in Chapter 9 has been linear regression analysis, which is one of the most important and widely used data analysis techniques in statistics. It provides researchers with a powerful tool to investigate relationships among variables and to obtain predictions for values of a dependent variable given one or more independent variables. In addition, we learned that regression and ANOVA are two aspects of the broader GLM family. Indeed, in Chapter 11, we will see that this framework can be further extended to situations in which the dependent variable is categorical, rather than continuous. Given that they come from the same modeling wellspring, it is not surprising that the foundational assumptions underlying ANOVA (normality, independence, and equal variances) also apply to linear regression models. In Chapter 9, we have learned how to assess these assumptions in the regression context, using a variety of tools. We have also discussed additional concerns that are more pertinent in regression, such as collinearity and identification of influential observations.

We are now ready to move on to an examination of alternative estimation techniques for regression models, which will be the focus of Chapter 10. We will see that many of the techniques that were developed in the context of ANOVA and correlation are also applicable with regression models. In Chapter 11, we will extend the regression paradigm to a variety of interesting applications, including model selection, model averaging, regularized estimation, modeling nonlinear relationships, identifying the most important variables in a multiple linear regression model, and the application of regression to more complex hypotheses. Next, however, we will consider how to estimate linear regression models when one or more of the assumptions underlying OLS have not been met.

10

Robust Linear Regression Models

In Chapter 9, we discussed ordinary least squares (OLS) linear regression, which was part of the broader family of the general linear model. This technique allowed us to estimate a model linking a dependent variable (y) to one or more independent variables (x) with a linear model of the form

$$y_i = \beta_0 + \beta_1 x_{1i} + \varepsilon_i \tag{10.1}$$

The model slope (β_1) characterizes the direction and magnitude of the relationship, whereas the intercept (β_0) reflects the mean of the dependent variable when the independent variable is set to 0. The impacts of all factors other than x are included in the error term, ε. Estimation of the parameters in equation (10.1) is done to minimize the sum of squared residuals as below:

$$\sum_{i=1}^{n} e_i^2 = \sum_{i=1}^{n} \left(y_i - \hat{y}_i\right)^2 \tag{10.2}$$

The model in (10.1) can be extended to include multiple independent variables in the form of multiple linear regression. A model with two independent variables would be expressed as

$$y_i = \beta_0 + \beta_1 x_{1i} + \beta_2 x_{2i} + \varepsilon_i \tag{10.3}$$

And, of course, a model with more than two independent variables would have a unique coefficient for each x.

In addition to the slope and intercept, we also learned about the R^2 statistic, which reflects the proportion of variability in the dependent variable that is associated with the set of independent variables. Finally, we saw that OLS rests on three primary assumptions: (1) the model residuals are normally distributed, (2) homogeneity of variance in the residuals across levels of the dependent variable, and (3) independence of the residuals after the independent variables are taken into account. When these assumptions are violated, the estimates and standard errors produced using OLS may not be accurate vis-à-vis the population values. In addition, we also saw that when there are strong relationships among the independent variables (i.e., collinearity), the standard OLS regression model parameters and standard errors may not be accurate.

Given the sensitivity of OLS to violations of these distributional assumptions, the focus of this chapter is on alternative approaches to estimating

DOI: 10.1201/9781003379324-10

regression model parameters when one or more of the assumptions are not met. We will discuss a wide array of such methods, many of which we have already been introduced to in other contexts, including rank-based methods, trimming, M estimators, inference based on the permutation or the bootstrap, and Bayesian estimation. In this chapter, we will investigate how these methods can be applied to the problem of estimating regression model parameters, with special attention on the slope, in situations when the assumptions outlined above have been violated. Throughout these discussions, we will use the example from Chapter 9 involving measures of personality obtained from a sample of college students. As a reminder, the dependent variable is a measure of self-oriented perfectionism (SOP) where higher scores reflect higher levels of perfectionism. In addition, we also have scores reflecting effortful control (EFFC) (ability to make plans and follow through with them), negative affect (NEG) (feelings of fear, frustration, etc.), extraversion (EXT), and whether one identifies as male (yes or no). Given that we devoted the first section of Chapter 10 to exploration of the data, we will not repeat those analyses here. The regression estimates for the various alternatives to OLS appear at the end of the chapter.

Rank-Based Regression Estimator

Throughout this book, we have seen the utility of rank-based approaches for dealing with data that do not conform to the distributional assumptions underlying many standard data analysis techniques. Ranks can also be used, in conjunction with least squares estimation, in the regression context, as well. The rank-based slope estimates are found by minimizing the function

$$\sum_{i=1}^{n} a\left(R\left(y_i - \hat{y}_i\right)\right)\left(y_i - \hat{y}_i\right) \tag{10.4}$$

where R is the rank and a is the score function, such as the Wilcoxon $\varphi_w\left[\dfrac{i}{n+1}\right]$

The first quantity in equation (10.4) involves the rank of the difference between the observed and model predicted values of the dependent variable; i.e., the rank of the residuals. The second component is the standard residual that we have become used to seeing.

The intercept for the rank-based model is then estimated separately, typically as the median of the residuals from the fitted model. Finally, we should note that score functions other than the Wilcoxon, such as normal scores, can be used in lieu of the Wilcoxon scores. Normal scores are taken as the inverse of the standard normal cumulative distribution function. The Wilcoxon scores are the most commonly employed in statistical software, however.

Least Absolute Deviation Regression

As we have mentioned multiple times now, OLS minimizes $\sum_{i=1}^{n} e_i^2$ across the full sample of n individuals to identify the optimal slopes and intercept. However, it is also true that outlying observations can have an outsized impact on this fit function and thus the slopes and intercept in the regression model. One approach that has been suggested (e.g., Birkes & Dodge, 1993) for dealing with this issue is the least absolute deviation (LAD) regression. Rather than minimizing the sum of the squared residuals, LAD minimizes the sum of the absolute value of the residuals (e):

$$\sum_{i=1}^{N} |e_i| \qquad (10.5)$$

By not squaring the residuals, LAD tends to reduce the impact of outliers on the estimation of model parameters (Birkes & Dodge, 1993).

Least Trimmed Squares Regression

The least trimmed squares regression (LTS) involves a very straightforward application of trimming as described in Chapter 2 and throughout the book. The goal underlying LTS is to minimize the sum of squares for the h smallest residuals (e_{ih}), where h represents the portion of the sample that remains after trimming the γ most extreme values. As an example, we have commonly set $\gamma = 0.2$, meaning that we trim the 0.2 most extreme high and 0.2 most extreme low values. When applied to the problem of linear regression, LTS minimizes $\sum_{i=1}^{h} e_i^2$ for the h individuals with the $1-\gamma$ least extreme residuals. In other words, those with the largest absolute value of residuals are first removed from the sample, and then OLS is applied to the remaining h individuals.

Least Trimmed Absolute Value Regression

The least trimmed absolute value (LTAV) regression estimator represents a combination of LTS and LAD. As with LTS, members of the sample with the h smallest residuals are used to determine the values of the slopes and intercept. However, whereas LTS minimizes the sum of the squared residuals for this subset, LTAV minimizes $\sum_{i=1}^{h} |e_i|$. Thus, cases with the γ largest and smallest residuals are first trimmed from the sample, and then the slope estimates are identified by minimizing the absolute value of residuals for the individuals who have not been trimmed. The core idea behind LTAV is that by first trimming the sample of outliers and then focusing on minimizing the absolute value rather than the square of the residuals, the impact of outlying observations is further reduced beyond what either trimming or only minimizing the absolute value alone can accomplish.

Least Median of Squares Regression

The least median of squares (LMS) technique finds the regression model parameter estimates that minimize the median of e_i^2 values across the n individuals in the sample. This approach stands in contrast to OLS, which minimizes the sum of the squared residuals, as mentioned above. As with several of the methods described here, LMS is designed to eliminate the impact of outlying residuals on regression parameter estimation.

Huber's *M* Estimator Regression

We discussed the theory behind M estimators of location in Chapter 2 and saw it applied to the problem of group mean comparisons in Chapters 3–5. As with trimmed estimators, this approach removes data points that are identified as outliers. However, whereas trimming removes a predetermined proportion of cases (γ) from each end of the data, M estimators use the data itself to identify potential outliers for removal. In the context of regression, the M estimators of the slopes and intercept are obtained by minimizing

$$\sum_{i=1}^{n} w_i e_i^2 \tag{10.6}$$

The weight value, w_i, is obtained using the following steps:

1. Estimate regression coefficients using OLS and obtain the residuals, e_i.
2. Calculate weights for each individual based on the residuals with the following:

$$w_i = \begin{cases} \dfrac{\psi\left(\dfrac{e_i}{\hat{\tau}}\right)}{\dfrac{r_i}{\hat{\tau}}} & \text{if } y_i \neq \hat{y}_i \\ 1 & \text{if } y_i = \hat{y}_i \end{cases} \tag{10.7}$$

$$\hat{\tau} = \frac{\text{Median of largest } n - p - 1 \text{ of } |e_i|}{0.6745}$$

where n is the sample size and p is the number of independent variables

$$\psi = \text{Max}\left[-K, \text{Min}(K, x)\right]$$

$$K = 2\sqrt{\frac{(p+1)}{n}}$$

3. Use OLS to estimate parameters minimizing $w_i e_i$ rather than e_i, where larger w_i indicates a greater contribution observation i in the estimation process

4. Repeat steps 1–3 until the change in regression estimates falls below a predetermined tolerance value.

A larger value of w_i means that individual i contributes more to the estimate of the slope(s) and intercept. As Wilcox (2012) notes, the M estimator was an important advancement in dealing with outliers, though there may be more effective approaches available now. Nonetheless, it continues to be widely used in practice and remains an important tool in the data analyst's toolbox.

Coakley–Hettmansperger Estimator Regression

A major limitation of the M estimator is its finite breakdown point of $2/n$. This means that it can successfully adjust the regression estimates for the presence of a small number of outliers. When many outliers are present, however, it loses its effectiveness. Coakley and Hettmansperger (1993) introduced an alternative based on the M estimation approach that has a breakdown point of nearly 0.5, meaning that it can work effectively even when nearly half of the sample consists of outliers (Coakley and Hettmansperger, 1993). As with the original M estimator, the Coakley–Hettmansperger (CH) technique involves a multi-step iterative process, which appears below.

1. Obtain initial estimates of b_0 and b_1 using LTS regression.

2. $$b_{1CH} = b_{1LTS} + \left(\underline{x}'\underline{B}\underline{x}\right)^{-1} \underline{x}'\underline{w}\psi\left(\frac{e_i}{w_i\hat{\tau}}\right)\hat{\tau} \tag{10.8}$$

$$w_i = \text{Min}\left[1, \left(b/(x_i - M_x)' C^{-1}(x_i - M_x)\right)^{9/2}\right]$$

where $\underline{w}$ is the diagonal of the weight matrix, $\underline{x}$ is the matrix of independent variables, $\underline{B}$ is the diagonal of $\psi'\left(\dfrac{e_i}{w_i\hat{\tau}}\right)$, ψ' is the derivative of ψ from Huber, C is the minimum volume ellipsoid (MVE) estimate of the covariance matrix of the independent variables, and M_x is the median from MVE of the independent variable x.

$$\hat{\tau} = 1.4826\left(1+5/\left(N-p\right)\right)M\left(\left|e_i\right|\right)$$

3. Iterate steps 1 and 2 until convergence is reached.

In their simulation study, Coakley and Hettmansperger (1993) showed that their estimator is particularly effective when the distribution of residuals is heavy-tailed. However, the CH estimator generally yields larger standard errors for the slopes and intercept than OLS, particularly as the distribution of residuals approaches normality (Wilcox, 2012).

S Estimator Regression

S estimators (Rousseeuw & Yohai, 1984) developed an approach to estimating the intercept and slopes that uses a measure of variability in residuals that are less impacted by outliers than the OLS approach. Specifically, S estimators minimize the percentage bend midvariance of the residuals, which is defined as

$$s_{PBMV}^2 = \frac{n\hat{\omega}_\beta^2\left(\sum_{i=1}^{N}\psi\left(y_i\right)\right)^2}{\left(\sum_{i=1}^{N}a_i\right)^2} \tag{10.9}$$

where $\hat{\omega}_\beta$ is the median of $|r_i - M_r|$, r_i is the residual for individual i, and M_r is the median of the residuals

$$y_i = \frac{r_i - M_r}{\hat{\omega}_\beta}$$

$$\psi = \text{Max}\left[-1, \text{Min}\left(1, r_i\right)\right]$$

$$a_i = \begin{cases} 1 & \text{if } |y_i| < 1 \\ 0 & \text{if } |y_i| \geq 1 \end{cases}$$

The key component of the S estimator that reduces the impact of outliers is the use of the residual median (M_r) and the weight $\hat{\omega}_\beta$. Individuals with residuals that are further from the median for the sample will have larger values of $\hat{\omega}_\beta$. In turn, larger values of $\hat{\omega}_\beta$ will yield smaller values of y_i and thus a reduced impact on s_{PBMV}^2. In other words, members of the sample whose residuals are further from the sample median will have less of an impact on the slope and intercept estimates.

Bayesian Regression Estimation

In Chapter 2, we discussed the framework underlying Bayesian estimation, including concepts such as prior distributions, Markov Chain Monte Carlo estimation, and assessment of model convergence. In subsequent chapters, we then applied these ideas to solving a variety of problems, including comparing central tendency across two or more groups and estimating correlations between variables. We can also apply Bayesian estimation to regression. And, as with other applications of Bayes, we can use either noninformative priors or informative priors for the slopes and intercepts. Indeed, for multiple regression with more than one independent variable, we can use informative priors for some slopes and noninformative priors for others. The Bayesian posterior parameter distribution for the regression coefficients can be expressed as

$$p(\beta\,|\,y,X) \propto L(\beta)p(\beta) \qquad (10.10)$$

where y is the dependent variable vector, X is the matrix of independent variables, β is the set of regression parameter estimates; e.g., slopes and intercept, $p(\beta\,|\,y,X)$ is the posterior distribution of regression model parameters given observed data, $L(\beta)$ is the likelihood-based estimates of regression parameters; e.g., OLS estimates from the observed data, and $p(\beta)$ is the prior distribution of regression parameters.

A common noninformative prior for the regression parameters is a uniform distribution over the interval $[-\infty,\infty]$ (Kaplan, 2014). A weakly informative prior would be the uniform distribution with a narrower range of potential values; e.g., uniform [–100,100]. Another common weakly informative prior is based on the standard normal distribution with a mean of 0 and a large standard deviation; e.g., normal (0, 100). Rather than a noninformative prior, we could instead elect to use an informative prior distribution that incorporates our knowledge of the likely parameter values from prior research and theory. For example, if several prior studies in the area have found that the slope relating to two variables ranges between 0.5 and 4.2, we might use the uniform [0,4] or the normal [2,1]. The choice of the prior will be based on the amount of available information about the relationships between the independent and dependent variables that the researcher brings to the study. When little or no previous work has been done in the area under study, the researcher will likely want to use noninformative priors, whereas when there is a great deal of previous research, that information can be incorporated into informative Bayesian priors and thus potentially improve the accuracy of the regression model. Even when noninformative priors are used and the likelihood (i.e., OLS) dominates the posterior distribution, the use of priors can help to stabilize the regression parameter estimates, particularly in situations involving small samples (Gelman et al., 2021). Put another way, Bayesian estimation is a potentially useful tool for researchers applying regression to small samples because the use of priors (informative or noninformative) can potentially yield more stable parameter estimates than OLS alone.

In addition to parameter estimation, the Bayesian approach can also be used to make inferences about hypotheses about the likely values of the model parameters. These inferences can be made using Bayes factors (BFs) much in the way that we used them in the context of group mean comparisons in Chapters 4, 5, and 6. As a reminder, the BF compares the posterior likelihood of one model against the posterior likelihood of another in the form of a ratio. In the case of regression, we might be most interested in comparing the likelihood of the full model including all of the independent variables (M_1) versus the likelihood of the null model in which only the intercept is included (M_0). The BF comparing these models is calculated as

$$BF_{01} = \frac{p(M_1 \mid y)}{p(M_0 \mid y)} \tag{10.11}$$

where $p(M_1 \mid y)$ is the posterior probability of model 1 including all independent variables and intercept and $p(M_0 \mid y)$ is the posterior probability of model 0 including only the intercept.

As we have discussed in previous chapters, larger values of the BF indicate a greater difference in the likelihood of one model vis-à-vis the other. We have adopted common guidelines for interpreting these values from Kaplan (2014), where $BF > 3$ or $BF < 1/3$ suggests that one model is more (or less) likely given the data and priors, whereas values between 1/3 and 3 fall into the region of indecision such that we cannot reach definitive conclusions about the relative likelihood of the models. In addition to comparing the posterior likelihoods for the full and null models, we can also compare the likelihoods of models that differ based on the sets of variables that are included.

In addition to comparing the likelihoods of the two models, we can also use Bayesian estimation to obtain the probabilities of regression model parameters taking specific values. This is analogous to the work that we did in Chapters 4 and 5 when we compared various hypotheses about the patterns of mean differences. For example, in the context of regression, we may wish to know the likelihood of the slope for a variable of interest exceeding 0, being less than 0, or being equal to 0. Likewise, we can use the posterior distribution to obtain probabilities for various orderings of the slopes; e.g., $\beta_1 > \beta_2 > \beta_3$ versus $\beta_2 > \beta_1 > \beta_3$ versus $\beta_1 = \beta_2 = \beta_3$. Such comparisons allow us to assess the relative importance of independent variables with respect to their relationships with the dependent variable. When making such comparisons, it is important that the independent variables be on the same scale, meaning that we will generally want to standardize them prior to conducting the analysis.

Permutation-Based Regression

The permutation-based approach to inference, which we discussed in the context of the t-test, one-way, and factorial ANOVA models, can also be applied in the context of simple and multiple regression analysis. The basic approach outlined in Chapter 2 is applicable in this case. Specifically, the regression model of interest is fit to all possible (or a large number) of data permutations, for each of which the slope and intercept estimates are estimated, creating the distribution for these parameters when the null hypothesis is true. Each coefficient for the observed data is then compared to the corresponding permutation distribution to obtain a p-value, which is defined as

$$p_{\text{permutation}} = 1 - p_b \qquad (10.12)$$

where p_b is the percentile for the coefficient in the permutation distribution.

The coefficient estimates are simply those that we obtain from OLS, and inference is done using this p-value from the permutations.

Regression Models for Dispersion, Skew, and Kurtosis

As we discussed in Chapter 9, the core assumptions underlying OLS estimation of the regression model include homogeneity of residual variance and normality of the residuals. When these assumptions are violated, OLS may yield biased parameter estimates (Fox, 2016). The focus in this chapter has been on methods that can provide unbiased estimates when the assumptions underlying OLS are not met. And, as we have seen, a plethora of options exist for this purpose. One commonality among these alternative estimators is that they are designed to reduce the impact of outlying data points that may be leading to the nonnormal and/or heteroscedastic data structure. These alternatives work by downweighting, removing, or transforming the potentially problematic values, or by creating an alternative reference distribution under the null hypothesis. Another approach to dealing with nonnormal and/or heteroscedastic data involves directly modeling the distributional characteristics that violate the OLS assumptions. This approach, referred to as a model for location, scale, and shape (MLSS), has been described by multiple authors, perhaps most notably Rigby and Stasinopoulos (2005).

The linear MLSS model for a single independent variable (x) can be expressed as

$$g_k(\theta_k) = \beta_{k0} + \beta_{k1}x_i \tag{10.13}$$

where θ_k is the distributional parameter k; i.e., mean, standard deviation, skewness, or kurtosis of the dependent variable, x_i is the value of independent variable for individual i, β_{k0} is the intercept for distributional parameter k, and β_{k1} is the coefficient linking independent variable.

The model represented in equation (10.13) allows us to model any or all of the four distributional parameters of the dependent variable. Standard linear regression only models the mean. In addition, the MLSS approach allows us to model the relationship between the distributional characteristics of the response and the independent variable(s). This means that we can account for and explore departures from normality (i.e., large skewness or kurtosis) and homoscedasticity (differences in the standard deviation) as a function of the independent variables. In so doing, we can both obtain less biased parameter estimates for the portion of the model focused on the mean, but we can also gain insights into why the assumption(s) have been violated (Rigby & Stasinopoulos, 2005). It should also be noted that this modeling approach can accommodate complex nonlinear relationships between the independent and dependent variables, though we will not delve into them here.

Given its ability to model not only the mean but also the distributional shape of the dependent variable, we can see that the MLSS framework can be useful in contexts where the assumptions of homogeneity of variance and/or normality are not tenable because we can model these characteristics directly and thereby account for them in parameter estimation. For example, if the data do not appear to be homoscedastic, we can include a term for the variance in our MLSS specification so that estimates associated with the mean (standard regression estimates) account for the heteroscedasticity. In addition, if we are interested in gaining insights into why homoscedasticity doesn't hold, we can include predictors for the variability in the model. Similarly, we can account for skewness and kurtosis with the MLSS and also include covariates in the model to predict them, if we wish. Thus, the MLSS approach allows us to both appropriately model our data by accounting for various distributional characteristics and to learn something about why the distribution appears as it does. As we will see below, in practice we will fit multiple MLSS models to the data and then use statistics such as the Akaike information criterion (AIC) to select the one that yields the best fit. The MLSS parameter estimates are obtained using a penalized likelihood function, details of which appear in Rigby and Stasinopoulos (2005).

Application of Robust Methods

Now that we have reviewed a number of robust approaches to estimating regression model parameters, let's apply them to the dataset that we introduced in Chapter 9. Recall that we are interested in assessing the relationships between SOP and several independent variables including EFFC, NEG, EXT, and whether the respondent is defined as male (Male). The OLS regression parameter estimates appear in Table 10.1.

The adjusted R^2 value for the model is 0.15, indicating that together the four independent variables account for approximately 15% of the variance in SOP.

The residuals for this model appear to have met the assumptions of normality and homoscedasticity, as the graphs below indicate (Figures 10.1 and 10.2).

There was no evidence of collinearity, given that the variance inflation factors were all below the commonly used cut-off value of 5 (Fox, 2015) (Table 10.2).

Despite the fact that the assumptions underlying OLS regression seem to have been met, we will still examine results for the estimation methods described in this chapter for the purposes of illustration. The multiple regression coefficients (standard errors) for OLS and the alternative estimators appear in Table 10.3.

The informative priors for the coefficients of the Bayesian estimator (INF) were drawn from a normal distribution with a variance of 1 and the following means: EFFC=4, NEG=6, EXT=1, and Male=1.5, respectively. These values were drawn from prior research. We can see that for most of the alternative estimators, the results are similar to those from OLS. Certainly, it seems to be the case that we would reach a qualitatively similar decision regarding the magnitudes of the relationships between the independent and dependent variables, as well as whether these relationships are statistically significant. Such concordance among the methods was not universally the case, as we see with LMS and the S estimators. For these latter techniques, the EXT

TABLE 10.1

Coefficients, Standard Errors, Betas, Test Statistics, and p-Values for Multiple Regression Model

Variable	Coefficient	Standard Error	Beta	t	p	95% Confidence Interval
Intercept	8.62	9.10		0.95	0.34	−9.28, 26.53
EFFC	5.82	1.08	0.27	5.42	<0.001	3.71, 7.94
NEG	7.32	1.11	0.35	6.61	<0.001	5.14, 9.49
EXT	2.05	0.99	0.10	2.06	0.04	0.10, 4.01
Male	1.33	1.81	0.03	0.74	0.46	−2.23, 4.89

EFFC, effortful control; NEG, negative affect; EXT, extraversion.

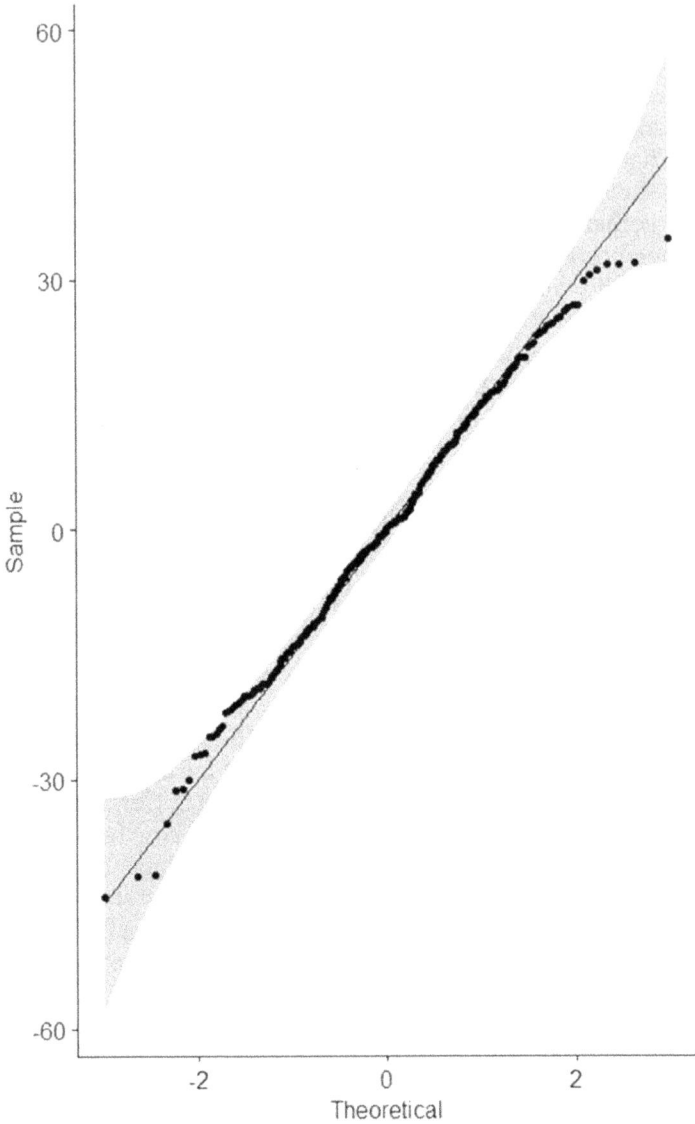

FIGURE 10.1
QQ plot of the multiple regression model residuals.

coefficient was larger than that of the other approaches. In addition, we see that the LMS estimate of the EFFC coefficient was lower than that of the others, whereas the estimate from S for this variable was somewhat larger.

As we noted above, in addition to the parameter estimation and inference about them, Bayesian estimation also allows the data analyst to assess the

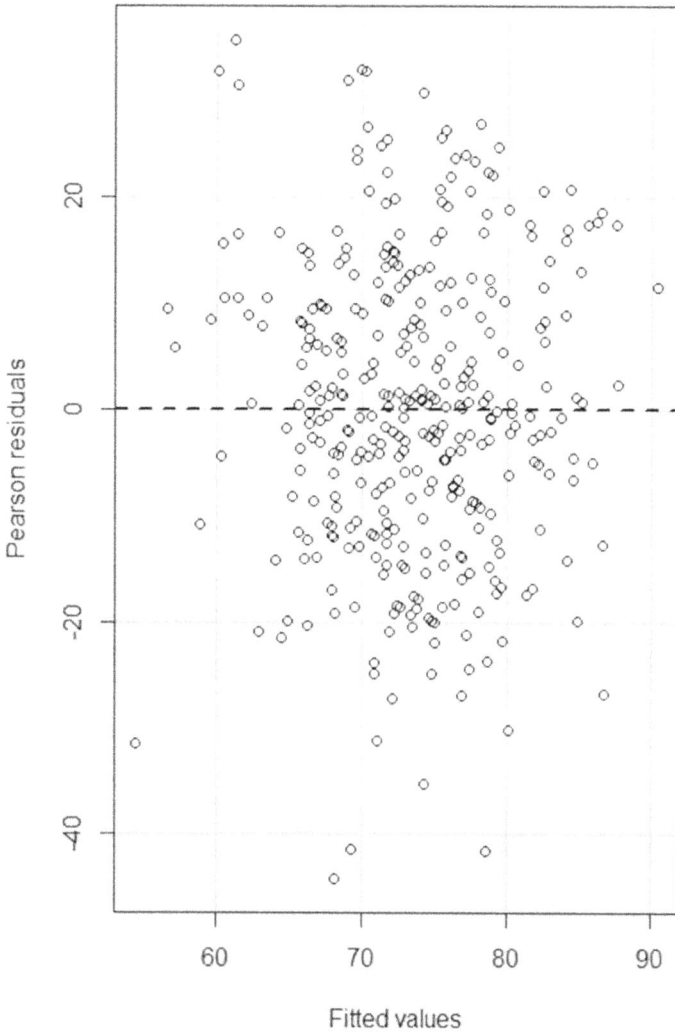

FIGURE 10.2
Scatterplot of residuals by model predicted values for multiple regression model.

likelihood of various potential regression outcomes. For example, we can use the posterior distribution of the model parameters to reach conclusions about the probability that each parameter is greater than, equal to, or less than 0. The values for each parameter in the multiple regression example we've been working are shown in Table 10.4.

These probabilities are drawn directly from the posterior distributions for each parameter. Thus, we can conclude that across the 10,000 points in the

TABLE 10.2

VIF and Tolerance Values for Independent
Variables from Multiple Regression Model

Variable	VIF	Tolerance
EFFC	1.05	0.96
NEG	1.14	0.88
EXT	1.04	0.96
Male	1.11	0.90

VIF, variance inflation factor; EFFC, effortful
control; NEG, negative affect; EXT, extraversion.

TABLE 10.3

Coefficients (Standard Error) for OLS and Alternative Estimators

Estimator	Intercept	EFFC	NEG	EXT	Male
OLS	8.62 (9.10)	5.82[a] (1.08)	7.32[a] (1.11)	2.05[a] (0.99)	1.33 (1.81)
Rank	7.27 (9.32)	5.84[a] (1.10)	7.37[a] (1.13)	2.31[a] (1.02)	1.49 (1.85)
M	5.61 (9.32)	5.90[a] (1.10)	7.58[a] (1.13)	2.42[a] (1.02)	1.56 (1.86)
CH	7.53	5.55	7.33	2.57	1.28
LTS	11.99 (8.65)	5.11[a] (1.02)	7.01[a] (1.05)	2.43[a] (0.94)	1.47 (1.70)
LTAV	10.00	5.34	7.00	2.54	1.37
LMS	1.59 ()	2.25 ()	6.24 ()	8.07 ()	−4.63 ()
LAD	13.93 ()	5.41[a] ()	5.89[a] ()	2.67[a] (0.64)	−0.0003 ()
S	10.81 ()	7.50[a] ()	2.22[a] ()	4.62[a] ()	−3.66 ()
Bayes (naïve)	8.62 (9.13)	5.82[a] (1.08)	7.32[a] (1.11)	2.05[a] (0.99)	1.33 (1.81)
Bayes (INF)	9.11 (7.59)	5.78[a] (0.99)	7.28[a] (1.02)	2.02[a] (0.93)	1.31 (1.76)
Permutation	8.62	5.82[a]	7.32[a]	2.05[a]	1.33

[a] Statistically significant at $\alpha=0.05$.
OLS, ordinary least square; EFFC, effortful control; EXT, extraversion; NEG, negative affect; LTS,
least trimmed square; LTAV, least trimmed absolute value; LMS, least median of square; LAD,
least absolute deviation.

TABLE 10.4

Posterior Probability of Model Parameter
Being Less Than 0, 0, and Greater Than 0

Variable	P (<0)	P (0)	P (>0)
Intercept	0.02	0.88	0.10
EFFC	0	0	1.00
NEG	0	0	1.00
EXT	0.01	0.58	0.41
Male	0.02	0.90	0.08

EFFC, effortful control; EXT, extraversion;
NEG, negative affect.

posterior distribution for the EFFC coefficient, 100% were above 0. In contrast, for the posterior distribution of the coefficient for Males, only 8% fell above 0 and 90% fell below 0. These results provide us with insights into the likely values of the parameter in the population above and beyond simple dichotomous decisions about statistical significance. For example, it is extremely likely that the coefficients for EFFC and NEG are greater than 0, whereas it is less clear whether the EXT coefficient is 0 or greater than 0. We can be quite sure that the coefficient for Males is likely to be 0 in the population. These results provide us with greater information than the hypothesis test because they show likelihoods for a full set of possible outcomes rather than a simple dichotomous choice (statistically significant or not).

In addition to the likelihood of various values for individual model parameters, we can also use the Bayesian paradigm to assess hypotheses about potential patterns of the regression parameters. For example, we saw how this could be used for hypotheses about the ordering of group means in the context of analysis of variance (Chapter 5). In the context of regression, we might be interested in hypotheses about the relative magnitudes of the coefficients. As we noted above, when assessing these hypotheses we need to ensure that the variables are on the same scale so that their values are comparable. Thus, it is recommended that we standardize the variables prior to using BFs to assess these hypotheses. Once we've done so, we can compare the likelihoods of two hypotheses versus one another, as well as the probability of a given hypothesis within a set of hypotheses. Again, this is very much akin to the assessment of various hypotheses about the means that we examined in Chapters 4 and 5.

As an example of using BFs to compare various hypotheses and to determine the probability of each, let's consider the three possible orderings of the regression coefficients:

$$H_0 : \beta_{\text{EFFC}} = \beta_{\text{NEG}} = \beta_{\text{EXT}} = \beta_{\text{Male}} = 0$$

$$H_1 : \beta_{\text{EFFC}} > 0 \ \& \ \beta_{\text{NEG}} > 0 \ \& \ \beta_{\text{EXT}} > 0 \ \& \ \beta_{\text{Male}} = 0$$

$$H_2 : \beta_{\text{NEG}} > \beta_{\text{EFFC}} > \beta_{\text{EXT}} > \beta_{\text{Male}}$$

These hypotheses are drawn from prior research and theory and reflect questions of interest to the researcher. The corresponding BFs and probabilities for each of the hypotheses appear in Table 10.5.

Of the three hypotheses, the most likely hypothesis is H_1, which states that the coefficients for EFFC, NEG, and EXT are all greater than 0 in the population, whereas the coefficient for Males is 0. In other words, the most likely hypothesis states that higher EFFC, NEG, and EXT are associated with higher degrees of SOP, and there are no differences in perfectionism between male and female students. Of the three hypotheses, this one has a probability of 0.7.

TABLE 10.5

Bayes Factors and Likelihoods Associated with H_0, H_1, H_2, and H_3

Hypothesis	BF	Probability
$H_0 : \beta_{\text{EFFC}} = \beta_{\text{NEG}} = \beta_{\text{EXT}} = \beta_{\text{Male}} = 0$	0.00	0.00
$H_1 : \beta_{\text{EFFC}} > 0$ & $\beta_{\text{NEG}} > 0$ & $\beta_{\text{EXT}} > 0$ & $\beta_{\text{Male}} = 0$	49.45	0.70
$H_2 : \beta_{\text{NEG}} > \beta_{\text{EFFC}} > \beta_{\text{EXT}} > \beta_{\text{Male}}$	20.79	0.30

BF, Bayes factor.

In addition to assessing the probability of each hypothesis, we can also compare their relative likelihoods by calculating the ratios of the various BFs. Given that the BF for H_0 is 0, we will not include it in these calculations. The relative likelihood of H_1 to H_2 is calculated as

$$\frac{49.45}{20.79} = 2.38$$

This value falls within the region of indeterminacy, meaning that we cannot state with confidence that H_1 is more likely than H_2.

We will conclude our demonstration of alternatives to OLS regression with the MLSS. Recall that this model allows us to fit relationships between individual covariates and the mean, standard deviation, skewness, and kurtosis of the dependent variable. As we discussed above, traditional regression only assesses relationships between the independent variables and the mean of the dependent variable. Thus, the MLSS approach will be ideal for situations in which the homoscedasticity and/or normality assumptions underpinning the standard linear model do not hold. In addition, we can gain insights into the variable(s) that might be associated with the location and spread in the data as well as its shape.

To simplify our discussion for pedagogical purposes, we will demonstrate the use of the MLSS approach using SOP as the dependent variable and NEG as the only independent variable. Three models were fit to the data, one with no independent variables (M_0), the second including NEG as a predictor of the mean and standard deviation of SOP (M_1), and the third including NEG as a predictor of the mean, standard deviation, skewness, and kurtosis of SOP (M_2). To ascertain the optimal model, we will consider the AIC and Bayesian information criterion (BIC) values of each, along with the R^2. Recall that AIC and BIC are measures of model fit with a penalty added to account for model complexity. Models with smaller values of these statistics are taken to provide a better fit to the data after accounting for their complexity. These values for each MLSS model appear in Table 10.6.

The AIC was minimized for M_2, and for BIC, M_1 had the lowest value. The largest R^2 value was associated with M_2, though it was quite close in value to that of M_1. Given that overall, these model fit values favor M_2, we will consider the estimates for it, which appear in Table 10.7.

TABLE 10.6

Model Fit Statistics for MLSS Models

Model	AIC	BIC	R^2
M_0	2703.95	2711.13	0.00
M_1	2676.68	2691.82	0.09
M_2	2661.95	2737.62	0.11

AIC, Akaike information criterion; BIC, Bayesian information criterion.

TABLE 10.7

Parameter Estimates for NEG from MLSS Model M_2

Term	Coefficient	Standard Error	t	p
Mean intercept	59.70	1.44	41.51	<0.001
Mean NEG	3.17	0.97	3.26	0.001
SD intercept	1.62	0.39	4.17	<0.001
SD NEG	0.27	0.10	2.75	0.006
Skew intercept	−0.78	0.47	−1.64	0.101
Skew NEG	0.20	0.15	1.35	0.18
Kurtosis intercept	−1.98	0.78	−2.53	0.012
Kurtosis NEG	0.69	0.21	3.31	0.001

NEG, negative affect; MLSS, model for location, scale, and shape.

There is a statistically significant relationship between NEG and the mean, standard deviation, and kurtosis of SOP. For each of these distributional features, the coefficient for NEG was positive, meaning that higher levels of NEG were associated with higher levels of SOP (mean), higher variability in perfectionism, and greater kurtosis in perfectionism. However, skewness in perfectionism scores was not associated with the level of NEG. From a practical perspective, we can conclude that individuals with greater levels of NEG tended to be more perfectionistic, but that the variability in perfectionism was also greater for these individuals. The higher kurtosis for individuals with larger NEG scores result means that the tails of the SOP distribution were heavier for these individuals.

Summary

Our goal in Chapter 10 was to introduce a variety of alternative methods for fitting a linear regression model to the data. As has been the case throughout this book, these alternatives include techniques based on ranks, trimming, downweighting outliers, and using permutations to obtain a null distribution for the parameters of interest. In addition, we also learned about an approach to modeling the variability and shape of the distribution for the dependent variable. This latter approach can be used to both account for departures from the assumptions underlying OLS regression when estimating coefficients and also to gain insights into the cause of these departures. In this chapter, we also explored how the use of Bayesian estimation can help when we have small samples, when we have prior information that could prove useful in parameter estimation, and when we would like to know the relative likelihoods of various hypotheses.

Given this plethora of options for fitting regression models, a natural question that we might have is, which method should we use when? First of all, if the assumptions of OLS are met, it is always the optimal method to use. OLS will yield the most accurate and efficient estimates of the slopes and the intercept. Second, when the assumptions of OLS are not met, there is no single optimal approach for estimating regression parameters. However, based on prior research, there are some guidelines that researchers and data analysts can use to make the best choice for their particular scenario.

Prior research has demonstrated that LAD is very robust to outliers in the dependent variable, but not to those in the independent variable or to bivariate outliers. In addition, it has been shown to be particularly useful when the dependent variable has heavy tails. Therefore, if an investigation of the data using the methods for identifying influential cases that we discussed in Chapter 9 reveals that there are outliers with respect to the dependent variable but not to the independent variables or influential cases in terms of model fit, then LAD is a good alternative to OLS. In contrast to LAD, regression methods based on trimming the data (this includes M estimates) are optimal if there are many outliers for *either* the dependent or independent variables (or both). However, it must also be noted that trimmed data techniques tend to have lower power than most of the other methods that we have discussed here. This lower power is undoubtedly due to the reduction in sample size associated with trimming.

In the presence of outliers, M estimates are also generally preferable to those from OLS. A better option in the presence of outliers might be CH. It has been shown to be a better alternative to OLS than the M estimator in such cases, given its superior ability to control the deleterious impact of outliers. Indeed, CH may well be a strong competitor in general when it comes to fitting regression models in all contexts, though more study of it needs to be done. It is particularly attractive because it works well even with normally distributed data, which is not generally true of the robust methods, especially when N is large. However, when N is small, CH has lower power than some of the other methods and thus may not be very useful in those cases. In summary, a good rule of thumb is that when the data do not conform to the assumptions underlying OLS, the researcher should try several of the methods described in this chapter and compare the results with one another. If the parameter estimates are relatively similar, then we can have some confidence that their consensus provides a clear picture of what is likely the case for the population. However, if these estimates differ, then the researcher should use the guidelines outlined above, as well as current research, to decide which approach is optimal given the structure of their data.

11

Regression for Dichotomous Dependent Variables

In the previous two chapters, we considered regression models for situations in which the dependent variables were continuous. However, in many applications, the outcome variable of interest is categorical, rather than continuous. For example, a researcher might be interested in predicting whether or not an incoming university freshman is likely to graduate from college in 4 years, using high school grade point average and admissions test scores as the independent variables. Here, the outcome is the dichotomous variable graduation in 4 years (yes or no) and will almost surely not produce normally distributed model errors. As we have seen in Chapter 9, the standard ordinary least squares (OLS) regression model operates under the assumption of normally distributed and homoscedastic residuals. As such, it will likely not be appropriate for situations in which the dependent variable is dichotomous. Thus, we might consider applying one of the robust alternatives that were described in Chapter 10. However, while not resting on the assumption of normality or homoscedasticity, these techniques do tacitly assume that the dependent variable is continuous in nature. As an example, consider the rank-based estimation procedure that we studied in Chapter 10. Attempting to assign ranks to the raw values of a dependent variable that is coded as 0 and 1 would yield an identical distribution of values, where all of the individuals with a 0 receive a rank of 1, and those with a 1 receive a rank of 2. Similarly, attempting to rank the residuals from such models would be similarly unfruitful. The trimming and other similar approaches to robust estimation would also not be particularly helpful in this case. For example, how would we trim individuals from a sample when only two dependent variable values (or a very small range of residuals) are present?

Despite the aforementioned limitations associated with using the regression models that we have discussed to cases where the dependent variable is dichotomous, alternative models for such variables are available. In this chapter, we will only consider situations in which the dependent variable is dichotomous. However, there exist models for use when the dependent variable consists of three or more ordered categories (ordinal logistic regression), three or more unordered categories (nominal logistic regression), and counts (Poisson regression). Taken together, these alternatives for categorical outcome variables are referred to as generalized linear models (GLiMs). GLiM is a broad category that includes not only the categorical response models

DOI: 10.1201/9781003379324-11

listed above but also GLMs such as linear regression and ANOVA. In the remainder of this chapter, we will describe logistic regression for a dichotomous dependent variable, followed by some alternatives that can be used with this type of data and that may be particularly appropriate in particular contexts.

Logistic Regression Model for a Dichotomous Outcome Variable

To begin our discussion of logistic regression for a dichotomous dependent variable, let's consider an example involving a sample of 20 men, 10 of whom have been diagnosed with coronary artery disease and 10 who have not. Each of the 20 individuals was asked to walk on a treadmill until they became too fatigued to continue. The outcome variable in this study was the diagnosis, and the independent variable was the time walked until fatigue; i.e., the point at which the subject requested to stop. The goal of the study was to find a model predicting coronary artery status as a function of time walked until fatigue. If an accurate predictive equation could be developed, it might be a helpful tool for physicians to use in helping to diagnose heart problems in patients. In the context of Chapter 9, we might consider applying a linear regression model to these data, as we found that this approach was useful for estimating predictive equations. However, recall that there were a number of assumptions upon which appropriate inference in the context of linear regression depends, including that the residuals would be normally distributed. Given that the outcome variable in the current problem is a dichotomy (coronary disease or not), the residuals will almost certainly not follow a normal distribution. Therefore, we need to identify an alternative approach for dealing with dichotomous outcome data such as these.

Perhaps the most commonly used model for linking a dichotomous outcome variable with one or more independent variables (either continuous or categorical) is logistic regression. The logistic regression model takes the form

$$\ln\left(\frac{p(Y=1)}{1-p(Y=1)}\right) = \beta_0 + \beta_1 x \tag{11.1}$$

Here, y is the outcome variable of interest, taking the values 1 or 0, where 1 is typically the outcome of interest. (Note that these dichotomous outcomes could also be assigned other values, though 1 and 0 are probably the most commonly used in practice). This outcome is linked to an independent variable, x, by the slope (β_1) and intercept (β_0). Indeed, the right side of this equation should look very familiar, as it is identical to the linear regression model from Chapter 9. However, the left side is quite different from what we see in linear regression and is called the logistic link function, or the logit. Within the parentheses are the odds that the outcome variable will take the value of 1. For our coronary artery example, 1 is associated with having coronary artery disease and 0 is with not having it. To make the relationship between the odds and the independent variable (time walking on the treadmill until fatigue) linear, we need to take the natural log. Thus, the logit link for this problem is the log of the odds of an individual having coronary artery disease. Interpretation of the slope and intercept in the logistic regression model is the same as interpretation in the linear regression context. A positive value of β_1 indicates that the larger the value of x, the greater the log-odds of the target outcome occurring. The parameter β_0 is the log-odds of the target event occurring when the value of x is 0. While mathematically this result is just fine, it is a bit difficult to make sense of it in application, because we are dealing with log units and often the independent variable cannot be 0. We faced very similar issues around the intercept for linear regression in Chapter 9.

As with standard OLS regression, we can make inferences about population relationships using our sample. For example, based on our sample estimates, we can test the null hypothesis that an independent variable is not related to the dependent variable in the population. More formally, the null and alternative hypotheses are:

$$H_0 : \beta_1 = 0$$

$$H_A : \beta_1 \neq 0$$

We can test this hypothesis using Wald's χ^2 statistic.

$$\chi^2 = \left(\frac{\hat{\beta}_1}{\text{SE}_{\hat{\beta}_1}} \right)^2 \tag{11.2}$$

where $\text{SE}_{\hat{\beta}_1}$ is the standard error of the coefficient.

The χ^2 statistic is then referenced to the chi-square distribution with 1 degree of freedom to obtain a p-value. And, as with the linear regression models in Chapter 9, a p-value less than our α (e.g., 0.05) leads us to reject H_0 and conclude that there is likely to be a relationship between the independent variable and the log-odds of the event occurring. The sign of β_1 tells us about the nature of the relationship between the independent and dependent variables. A positive slope means that as x increases in value, the likelihood of the event occurring also increases, whereas a negative slope indicates that larger x values are associated with a lower likelihood of the event.

In addition to the slope itself, we can make the interpretation of our results somewhat more intuitive by exponentiating β_1, thereby converting it to an odds ratio. To provide context for this calculation, let's return to our heart disease example. In addition to measuring the time that the subjects can walk on the treadmill before becoming fatigued, let's also assess whether they have completed a formal exercise program in the last year (yes or no). In that case, the odds ratio reflects the relative odds that an individual who completed an exercise program in the prior year was diagnosed with heart disease. Formally, we can express these odds as:

$$\frac{P_{\text{exercise program}}\left(\text{heart disease}\right)\big/ 1-P_{\text{exercise program}}\left(\text{heart disease}\right)}{P_{\text{no exercise program}}\left(\text{heart disease}\right)\big/ 1-P_{\text{no exercise program}}\left(\text{heart disease}\right)} \tag{11.3}$$

In the numerator, we see the ratio of the probability that those in the exercise program are diagnosed with heart disease versus the probability of those in the exercise program not being diagnosed with heart disease. Similarly, the denominator is the ratio of those not in the exercise program who have been diagnosed with heart disease versus those in the exercise program not diagnosed with heart disease. A value greater than 1 in the numerator would indicate that for those in the exercise program, the probability of being diagnosed with heart disease is greater than not being so diagnosed. A value less than 1 reflects the opposite outcome; i.e., those in the exercise program have a lower probability of being diagnosed with heart disease than being diagnosed with it. A similar interpretation can be applied to the denominator for those who have not participated in an exercise program.

Now that we understand the meaning of each term in (11.3), we can put them together and consider the meaning of the odds ratio itself. An odds ratio value greater than 1 would mean that individuals in the exercise program group have greater odds of being diagnosed with heart disease than those in the no-exercise group. And, conversely, a value less than 1 would indicate that those who have participated in an exercise program have lower odds of being diagnosed with heart disease than those who have participated in such a program. For this particular example, the slope estimate relating exercise program participation and heart disease diagnosis is –0.353. We can then exponentiate this value to obtain the odds ratio:

$$e^{\hat{\beta}_1} = e^{-0.353} = 0.703$$

We can interpret this value as follows: The odds of an individual who has participated in an exercise program during the previous year being diagnosed with heart disease is 0.703 times that of an individual who has not participated in an exercise program over the last year. For continuous independent variables, we interpret the odds ratio as the change in the odds of the event occurring (e.g., being diagnosed with heart disease) for every 1-unit increase in the independent variable value. We can also construct a confidence interval around the odds ratio. If 1 falls within the interval, we would conclude that there is no relationship between the independent and dependent variables because it is possible that the odds of the event occurring are equal in the population.

In Chapter 9, we learned that a very useful way to describe the strength of the relationship between the independent variables in our model and the outcome variable is with the proportion of variation in the dependent variable that is accounted for by the independent variable. This statistic, R^2, ranges between 0 and 1, with larger values indicating a stronger relationship between the set of independent variables and the dependent variable. In the context of logistic regression, we do not have a direct analog for R^2, but we can calculate a pseudo-R^2 value that provides us with information regarding how much better our model fits the data than a model with no independent variables at all. There are three commonly used pseudo-R^2 statistics, each of which appears below:

Cox and Snell R^2:

$$R_{CS}^2 = 1 - \left(\frac{LL_0}{LL_M} \right)^{2/n} \tag{11.4}$$

Nagelkerke R^2:

$$R_N^2 = \frac{R_{CS}^2}{1 - (LL_0)^{2/n}} \tag{11.5}$$

McFadden R^2:

$$R_L^2 = \frac{-2LL_0 - (-2LL_M)}{LL_0} \tag{11.6}$$

where LL_0 is the log-likelihood value for the null model (no independent variables), LL_M is the log-likelihood value for the target model, and n is the sample size.

These Pseudo-R^2 values are not, strictly speaking, true R^2. However, they do provide us with information about how much the unexplained variation in the outcome variable, which is expressed through the LL, declines when the independent variables are added to the model. The larger the pseudo-R^2, the more "explanatory power" the independent variables have with respect to the dependent variable.

Logistic Regression Example

To demonstrate the application of logistic regression in practice, we will consider an example involving a student course satisfaction survey from a college science class. The outcome of interest is an item asking, "Overall, were you satisfied with this course?" with responses of yes (1) or no (0). The independent variables are measures of how many hours per week students spend studying for all courses in which they're enrolled, the number of hours per week spent studying for this science class, and their self-reported study habits scored on a 4-point Likert scale from very poor (1) to very good (4). We are interested in understanding the degree to which each of these independent variables is related to the overall satisfaction with the class. To investigate these relationships, we will fit a logistic regression model to the data.

The results of the logistic regression model appear in Table 11.1.

There was a statistically significant negative relationship (0 is not in the confidence interval) between each study for all courses and study habits. The negative result associated with Study for all reveals that the more time a student spends studying for all courses, the lower the likelihood that they will report being satisfied with this course. Likewise, students who reported having better study habits were also less likely to indicate that they were satisfied with this class.

As we discussed earlier in the chapter, the odds ratio reflects how the odds of the outcome of interest (e.g., being satisfied with the course) change in conjunction with changes in the independent variable. As an example, for every additional hour a student studied for all of their classes, the odds of being satisfied were 0.79 times as large. Thus, a student who reports studying 10 hours per week is only 0.79 times as likely to be satisfied with the science class as those who spend 9 hours studying per week. We can also express this relationship in terms of how much *more* likely an individual is to be satisfied if they study 1 hour less per week by taking the inverse of the odds ratio: $1/0.79 = 1.27$. Thus, for every 1 hour *less* a person spends studying for all of their courses, they are 1.27 times more likely to report being satisfied with this class.

TABLE 11.1

Logistic Regression Results for Course Satisfaction

Variable	Coefficient (95% CI)	Standard Error	Odds Ratio (95% CI)
Intercept	1.87 (1.60, 4.48)	0.45	20.36 (4.93, 88.04)
Study for all	−0.14 (−0.43, −0.03)	0.06	0.79 (0.65, 0.97)
Study for this	−0.14 (−0.53, 0.08)	0.10	0.80 (0.59, 1.09)
Study habits	−0.31 (−0.85, −0.18)	0.11	0.60 (0.43, 0.84)

Finally, as with OLS regression (Chapter 9), we can assess the overall relationship of the independent variables with the outcome using a measure of shared variance; i.e., R^2. As we described above, there are a number of options, with McFadden, Cox–Snell, and Nagelkerke being among the most widely used. For the course satisfaction data, these three values were 0.02, 0.03, and 0.04, respectively. These results indicate that the set of independent variables accounted for a small proportion of the variance in the level of student satisfaction with their courses. Put another way, a student's study habits and practices did not have a strong collective relationship with the degree to which students were associated with the science class.

Bayesian Logistic Regression

In Chapter 10 we talked about the use of Bayesian estimation in the context of linear regression. The same concepts that we reviewed there, including details of Markov Chain Monte Carlo (MCMC) such as the length of the chain, the burnin and thinning periods, and priors, are applicable in the context of logistic regression as well. We will not devote space here to these issues, given that they've been discussed elsewhere in the book, particularly in Chapter 2. We will, however, describe the application of Bayesian estimation to fitting logistic regression models and demonstrate how making adjustments to some of the Bayesian settings does (and does not) impact the results.

For the course satisfaction example, we will begin by fitting the logistic regression model using a Bayesian estimator with a set of weakly informative priors. Specifically, the prior mean for the coefficients had a mean of 0 with standard deviations of 2.5, 3.3, and 4.0 for study for all courses, study for this course, and study habits. The total number of chains for the MCMC estimator was 10,000, with the first 5,000 serving as the burnin period and with four separate chains. The mean of the posterior distribution for each coefficient, the Bayesian standard error (standard deviation of the posterior distribution), the odds ratio, and the 95% credibility intervals for the coefficients and odds ratios appear in Table 11.2.

TABLE 11.2

Logistic Regression Results for Course Satisfaction Using Bayesian Estimator with Noninformative Priors

Variable	Coefficient (95% CI)	Standard Error	Odds Ratio (95% CI)
Intercept	3.03 (1.58, 4.54)	0.75	20.89 (4.85, 93.69)
Study for all	−0.23 (−0.43, −0.03)	0.10	0.79 (0.65, 0.97)
Study for this	−0.22 (−0.54, 0.08)	0.15	0.80 (0.58, 1.08)
Study habits	−0.51 (−0.85, −0.18)	0.03	0.60 (0.43, 0.83)

The results are qualitatively quite similar to those from the frequentist logistic regression model (Table 11.1). The amount of time that students spent studying overall and their self-reported study habits were negatively associated with course satisfaction. For every 1 hour increase in time spent studying per week for all of their courses, the odds of being satisfied with the class being assessed were only 0.79 times as large as for those who spent 1 hour per week less time studying. Similarly, for every 1 point higher (on a 4-point scale) that a student rated their study habits, their odds of being satisfied with the class were only 0.60 times as large as for a student with a 1-point lower self-rating. There was not a statistically significant relationship between the amount of time that the students spent studying for this class and their course satisfaction.

We can assess the convergence of the model parameter estimation using several tools, much as we did for the Bayesian linear regression model (Chapter 10). First, we can examine density plots of the parameter estimates across the retained MCMC samples and the trace plots of these values for the four chains. Recall that ideally, the density plots should be symmetric around a single point, and the chains should be centered on a single value (and the same single value for all four chains). These plots appear in Figure 11.1.

The density and trace plots both show that the MCMC algorithm reached convergence for all of the model parameters. In addition, the Gelman-Rubin value for each parameter estimate is 1.0, indicating that the variability in estimates across the four chains is equal to that within the chains. In other words, the four chains independently converged to very similar values across the MCMC replications.

Finally, we can assess model fit by comparing the predicted value for the outcome based on the estimated model parameters with the actual outcomes. In the sample itself, the probability of an individual being satisfied with the class is 0.59, whereas the model-predicted probability of being satisfied is 0.60. In other words, the model appears to yield comparable outcomes to the observed data, suggesting that it may be working well. We can visualize the observed and model-predicted (y_{rep}) values in Figure 11.2. The greater the overlap between y and y_{rep}, the better the model fits the data. Based on the results in Figure 11.2, it seems clear that there is a high degree of overlap between the observed and model-predicted values, providing further evidence for the reasonable fit of the data.

Next, let's refit the logistic regression model, but now with strongly informative priors. Imagine that prior research in mathematics courses found coefficient linking the amount of time that a student spent studying and their log-odds of course satisfaction was −0.3. We use this value as the basis of an informative prior from the normal distribution with a mean of −0.3 and a standard deviation of 0.1 for Study for this course. The results appear in Table 11.3.

There are clear differences in the coefficients, standard errors, and credibility intervals for each model parameter when we use an informative prior for

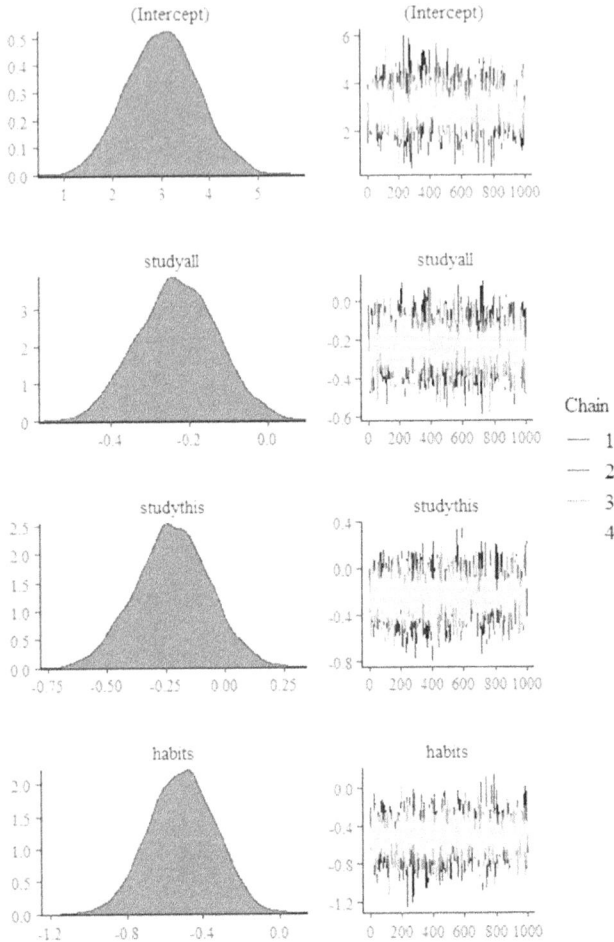

FIGURE 11.1
Density and trace plots for MCMC estimates of the course satisfaction logistic regression model. MCMC, Markov Chain Monte Carlo.

Study for this course versus when we used the weakly informative priors in the previous analysis. Of particular interest is that the 95% credibility interval for Study for this course no longer includes 0 (compared to Table 11.2), which leads us to conclude that there is a negative relationship between satisfaction with the science class and the number of hours that they spent studying for it. The other results were qualitatively the same for the naïve and informative prior models.

With respect to model fit and convergence, Figure 11.3 supports that the logistic regression model with informative priors did reach convergence for each model parameter, given the largely symmetric density plots and stable overlapping values across the four chains.

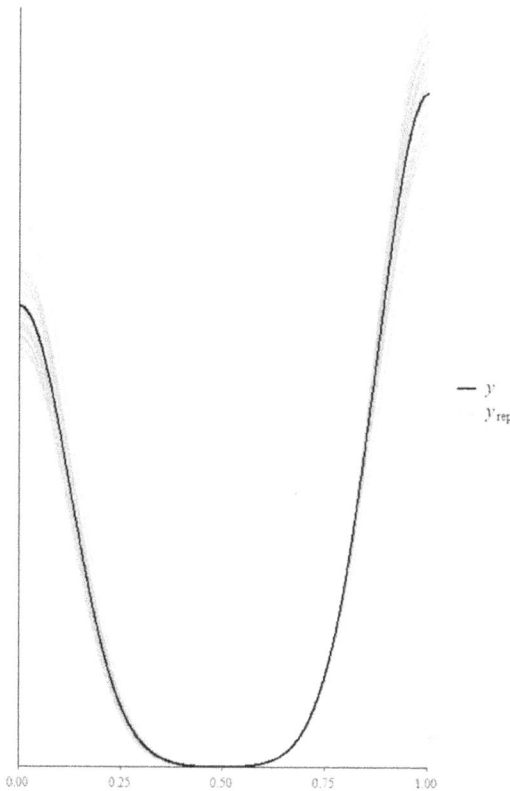

FIGURE 11.2
Observed and model-predicted distributions of course satisfaction.

TABLE 11.3

Logistic Regression Results for Course Satisfaction Using Bayesian
Estimator with Strongly Informative Priors

Variable	Coefficient (95% CI)	Standard Error	Odds Ratio (95% CI)
Intercept	2.73 (1.98, 3.49)	0.39	15.32 (7.25, 32.79)
Study for all	−0.24 (−0.39, −0.11)	0.07	0.78 (0.68, 0.90)
Study for this	−0.24 (−0.39, −0.07)	0.08	0.79 (0.68, 0.93)
Study habits	−0.37 (−0.54, −0.21)	0.08	0.69 (0.53, 0.81)

The Gelman-Rubin value was 1.0 for each parameter estimate, also suggesting estimation convergence. Finally, the probability of course satisfaction for the data simulated using the model parameter estimates is 0.6, which is quite similar to the actual value of 0.59. In addition, Figure 11.4 displays a high overlap between the observed and model-predicted outcome values. Thus, we would conclude that, like the model with weakly informative priors, the model with a highly informative prior for Study for this course yields a good fit to the data.

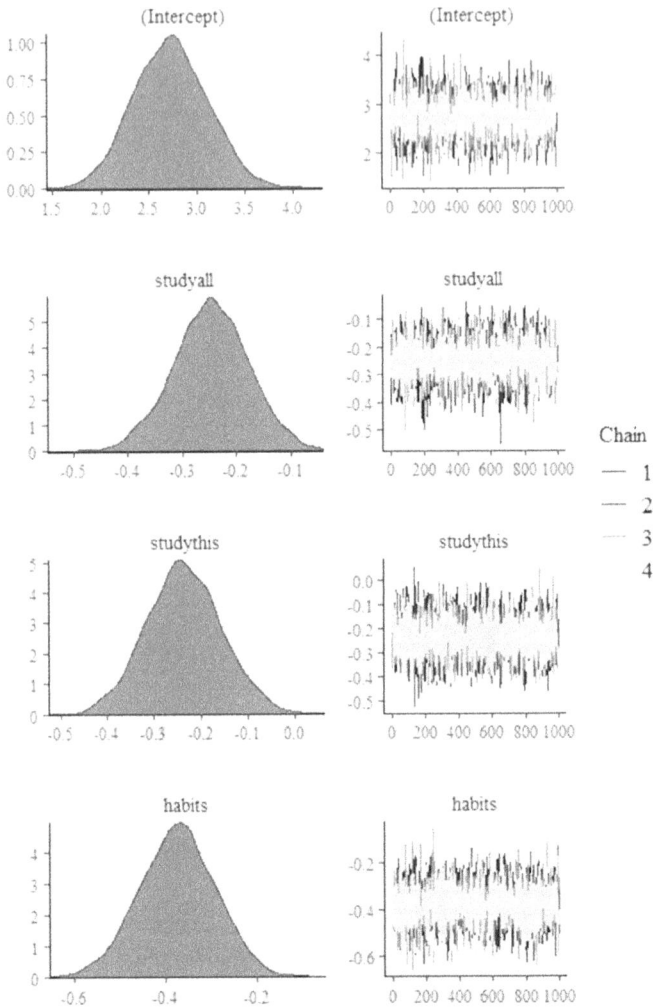

FIGURE 11.3
Density and trace plots for MCMC estimates of course satisfaction logistic regression model with informative priors. MCMC, Markov Chain Monte Carlo.

Other Models for Dichotomous Outcomes

There are several alternative regression models for dichotomous data that might be more appropriate than logistic regression in some situations. These models rely on different assumptions about the mechanisms and distributions underlying the observed dichotomous outcome. We will review each of these models briefly and then discuss when they may be more appropriate than standard logistic regression. Finally, we will present examples as a way of summarizing our discussion and demonstrating how you might make use of these techniques.

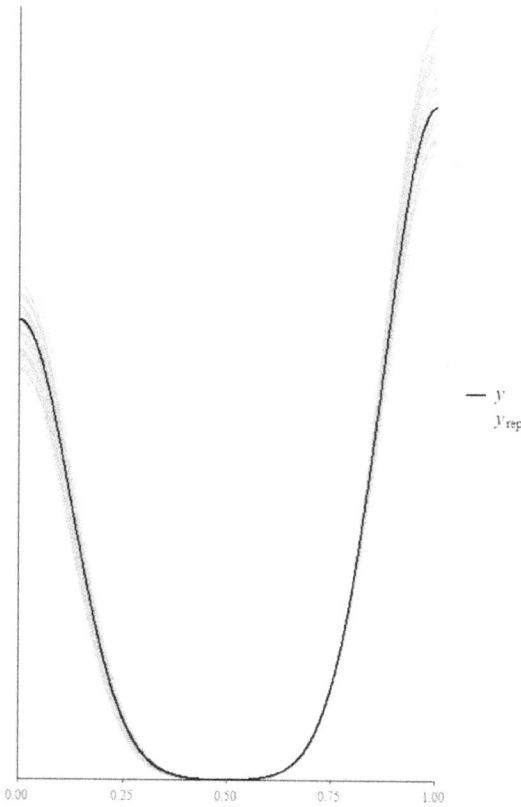

FIGURE 11.4
Observed and model-predicted distributions of course satisfaction with informative priors.

The first alternative model that we will examine is Probit regression. In equation (11.1), we learned that the function linking the dichotomous outcome variable with the set of predictors takes the form $\ln\left(\dfrac{p(Y=1)}{1-p(Y=1)}\right)$. This logistic link rests on the tacit assumption that the variable is truly dichotomous. Examples of true dichotomies would include pregnant/not pregnant, treatment/control, dead/alive, and the like. In many situations, however, it may be possible that there is a continuous variable underlying the observed dichotomy. For example, we might classify students as reading at grade level based on a test score. The observed outcome is reading at grade level or not, but underlying this dichotomy is a score on the reading test. Thus, even though we might treat grade-level reading as a dichotomous variable, in reality, it is a simplification of a more complex continuous score. Similarly, a psychologist might classify a client as being clinically depressed based on clinical judgment and scores on standardized instruments.

However, in reality, depression is a more complex process lying on a continuum that underlies our dichotomous diagnosis. In such a situation, we can use a regression model based on the Probit link function, as in equation (11.7).

$$\phi^{-1}\left(\pi\left(y_{ij}\right)\right) = \beta_0 + \beta_1 x_i \qquad (11.7)$$

where ϕ^{-1} is the inverse of the normal distribution and $\pi\left(y_{ij}\right)$ is the probability of the event of interest; e.g., depression diagnosis, meeting reading standard.

The Probit link function converts the probability of the event of interest to a standard normal (z) score. For example, if the probability of a student meeting the reading standard is 0.85, the transformed value is $\phi^{-1}(0.85) = -1.036$. Thus, the response value in Probit regression is the continuous z-score, rather than the log of the likelihood of the event, as with logistic regression. Probit regression operates under the assumption that the process underlying the observed dichotomous variable, y, is represented by the normally distributed variable y^*. The observed dichotomy comes from comparing the value of y^* with a threshold. When it falls above the threshold, the observed value of y is 1, and when y^* falls below the threshold, y is 0. It is important to keep in mind that we only observe y. y^* is an unobserved variable that we assume underlies our observed response. Under this assumption, we can express the Probit model as

$$y^* = \beta_0 + \beta_1 x_i + \varepsilon \qquad (11.8)$$

where ε is the random error with mean of 0 and standard deviation of 1.

Notice that equation (11.8) is identical to the linear regression model in Chapter 9. As we will see, we can interpret β_1 as the change in y^* given a 1-unit change in the independent variable, x. This means that the model parameter estimates of logistic and Probit regression models are not on the same scale and thus cannot be directly compared.

The normality assumption for y^* that underlies Probit regression may be limiting in situations where the probability of an event might be either very small or very large. In such cases, the probability distribution (and therefore the distribution of y^*) will be skewed. The complementary log–log (CLL) link function and associated regression model may be a worthwhile alternative to consider in such cases. The CLL model takes the form

$$\ln\left(-\ln\left(1 - \pi\left(y_{ij}\right)\right)\right) = \beta_0 + \beta_1 x_i \qquad (11.9)$$

As we noted above, the CLL link is particularly useful when $\pi\left(y_{ij}\right)$ is very low (or very high) for most values of x, but then increases (or decreases) rapidly after a particular value of x. For this reason, the CLL model is sometimes known as the extreme value model.

A final alternative to logistic regression that we will consider in this chapter is Beta regression. This modeling paradigm was described by Ferrari and Cribari-Neto (2004) for situations in which the data analyst is working with continuous instantiations of categorical outcomes, such as proportions and rates for an outcome of interest. For example, rather than modeling individual reading test outcomes (meets standard/did not meet the standard) for students in multiple schools, we might instead consider the proportion of individuals who met the standard across each school. In this case, we are interested in the collective test performance for schools, rather than performance at the individual student level.

Ferrari and Cribari-Neto (2004) point out that standard logistic regression has several limitations that might limit its utility in many situations. These limitations include the fact that the model coefficients are expressed in the log-likelihood scale, rather than the original scale of the data. In addition, they noted that in many applications the outcomes are skewed (i.e., most members of the sample have relatively high or relatively low likelihoods of the outcome of interest), and the variability in the outcomes tends to be higher near the mean than at the extremes.

These limitations, in addition to the desire in some cases to model proportions directly, rather than the dichotomous variables that underlay them, led Ferrari and Cribari-Neto (2004) to develop an alternative model framework for what we typically think of as dichotomous data. Specifically, they proposed the Beta regression model for situations in which the outcome is a proportion or a rate. This model looks quite similar to the other model frameworks that we have discussed in this chapter, with a link function being used to transform the dependent variable value and a set of model coefficients associated with the independent variables. The logit link function is the most common link, leading to the following form for the Beta regression model:

$$\ln\left(\frac{\mu}{1-\mu}\right) = \beta_0 + \beta_1 x_i \tag{11.10}$$

where μ is the mean of the dependent variable y.

As an example, y might be the proportion of examinees in a school who have met the reading test score standard. In this case, μ would be the mean proportion of students in a school meeting the reading test standard for individuals at a given level on the independent variable, x. In general, y must fall between 0 and 1, meaning that μ will as well. We will work through an example of Beta regression later in the chapter.

Examples Using Models with Alternative Models for Dichotomous Data

In the case of course satisfaction, it would be reasonable to assume that underlying the dichotomous rating (satisfied or not satisfied) is a continuous variable reflecting the extent to which an individual is satisfied with the

TABLE 11.4

Probit Regression Results for Course Satisfaction

Variable	Coefficient (95% CI)	Standard Error
Intercept	1.87 (0.99, 2.75)	0.45
Study for all	−0.14 (−0.27, −0.02)	0.06
Study for this	−0.14 (−0.33, 0.05)	0.10
Study habits	−0.31 (−0.52, −0.11)	0.11

course. If the value of this unobserved trait for an individual student lies above a threshold, the course will receive a positive rating, whereas if the value of the trait is below the threshold, the rating will be negative. In such a case, we can use Probit regression as an alternative to logistic regression. The coefficients for this model appear in Table 11.4.

As we noted above, it is important to remember that the dependent variable in the Probit regression model is the latent satisfaction and not the observed dichotomy. Therefore, the coefficients in Table 11.4 are not comparable to those in Tables 11.1–11.3; i.e., they reflect relationships on a different scale. The confidence intervals for Study for all and Study habits did not include 0, meaning that these two variables had a statistically significant negative relationship with the continuous satisfaction variable underlying the observed dichotomy. This would lead to a similar conclusion as was derived from logistic regression, namely, that students who spent more time studying for all of their courses and who rated their study habits relatively higher had lower levels of latent course satisfaction. The McFadden, Cox–Snell, and Nagelkerke R^2 values for the Probit model were 0.02, 0.03, and 0.04, which were nearly identical to those from the logistic regression model. And, as with that model, these relatively low values indicate that the Probit regression model does not account for much of the variability in the latent course satisfaction variable.

Finally, we will finish our analysis of the course satisfaction data with the CLL model. Recall that it is most commonly recommended for use with a highly skewed dichotomous outcome variable. Such cases arise when one of the possible values is much more likely than the others. For the current example, this is actually not the case, given that approximately 59% of students indicated that they were satisfied with the course as opposed to 41% who were not satisfied. Nonetheless, for pedagogical purposes, we will apply the CLL model to the course satisfaction problem described above. The parameter estimates for this model appear in Table 11.5.

The results of the CLL model can be interpreted in much the same way that we do for logistic regression. We can see in Table 11.5 that the amount of time students spend studying for all of their courses and their self-reported study habits are both negatively associated with course satisfaction, just as was the

TABLE 11.5

Complementary Log–Log Regression Results for Course Satisfaction

Variable	Coefficient (95% CI)	Standard Error	Odds Ratio (95% CI)
Intercept	1.61 (0.70, 2.52)	0.46	5.00 (2.02, 12.37)
Study for all	−0.15 (−0.28, −0.02)	0.07	0.86 (0.76, 0.98)
Study for this	−0.24 (−0.37, 0.05)	0.10	0.85 (0.69, 1.05)
Study habits	−0.32 (−0.53, −0.11)	0.11	0.72 (0.59, 0.90)

case for logistic (and Probit) regression; i.e., their 95% confidence intervals do not include 0. Also, as we saw with logistic regression, the amount of time that students spent studying for the target course was not associated with course satisfaction. The odds ratios can also be obtained for the CLL model and are interpreted just as with logistic regression. Thus, for example, a student who studies for all courses 1 hour longer than her classmate is 0.86 times as likely to say that they were satisfied with this science class. The pseudo-R^2 values for the CLL model are

McFadden=0.02

Cox–Snell=0.03

Nagelkerke=0.04

These values are essentially identical to those for both the logistic and Probit regression models, suggesting that the study variables do not account for a large portion of the variance in the extent to which students are satisfied with the science class.

We will conclude our examination of statistical models for dichotomous outcome variables with Beta regression. In our discussion above, we noted that Beta regression is particularly useful for cases in which the dependent variable lies between 0 and 1, such as with proportions and rates. What differentiates this context from the other modeling frameworks that we have reviewed in this chapter is that the dependent variable can take *any* value between 0 and 1. In contrast, for logistic, Probit, and CLL regression, the observed outcome will be either 0 or 1. Thus, these models are not appropriate for situations in which y can be something other than 0 or 1, leading us to the need for Beta regression.

The example that we will use to demonstrate Beta regression involves passing rates on a statewide high school graduation exam for 98 schools in an American state. In addition to the proportion of high school juniors who achieved a passing score, the dataset also includes the proportion of students whose family's income fell above the median for the state as a whole. The researcher who collected the data is interested in whether there is a relationship between the two variables. The specific hypothesis is that there is

TABLE 11.6

OLS Regression Results for Graduation Exam Passing Rate

Variable	Coefficient (95% CI)	Standard Error
Intercept	0.13 (0.02, 0.29)	0.08
SES	1.04 (0.60, 1.47)	0.22

OLS, ordinary least squares.

a positive relationship between the proportion of students with household incomes above the state median and the proportion of students who received a passing score on the graduation exam.

Without knowing about Beta regression, we might view this as a standard OLS regression problem. The outcome variable is continuous on the scale between 0 and 1, and we are interested in the relationship between it and another continuous variable. The results of the OLS model appear in Table 11.6.

Because the confidence interval does not include 0, we would conclude that there is a statistically significant positive relationship between the proportion of students with household incomes above the state median and the proportion who passed the graduation exam. This result supports our hypothesis. The R^2 for this model is 0.19, meaning that approximately 19% of the variation in the proportion of students passing the exam was associated with the proportion of student households having incomes above the state median.

An examination of the residual plots (Figures 11.5–11.7) appears to suggest that the assumption of normal residuals has not been met.

This conclusion regarding distributional assumption violations for the residuals is further supported by the fact that the omnibus test for the assumptions was statistically significant ($p < 0.0001$), leading us to reject the null hypothesis that the assumptions have been met. Further exploration shows that the tests for both skewness ($p < 0.0001$) and kurtosis ($p < 0.0001$) were statistically significant, leading us to reject the null hypotheses of symmetric and non-kurtotic data. The skewness estimate for the residuals is 1.62. In summary, the residuals do not appear to be normally distributed.

Given that the dependent variable is a proportion and thus falls between 0 and 1, we can use Beta regression as an alternative to OLS. As we discussed earlier in the chapter, Beta regression accounts for distributional issues associated with data such as proportions and rates, including skewness such as that in evidence here (Ferrari & Cribari-Neto, 2004). The coefficients, confidence interval, and standard errors for the Beta regression model appear in Table 11.7.

To address our research hypothesis about the relationship between the proportion of students with household incomes in excess of the state median and the graduation exam passing rate, we will focus on the coefficient for SES. We see that the coefficient is positive and that the confidence interval does not include 0, leading us to the conclusion that there is a statistically significant positive relationship between the two variables. In other words,

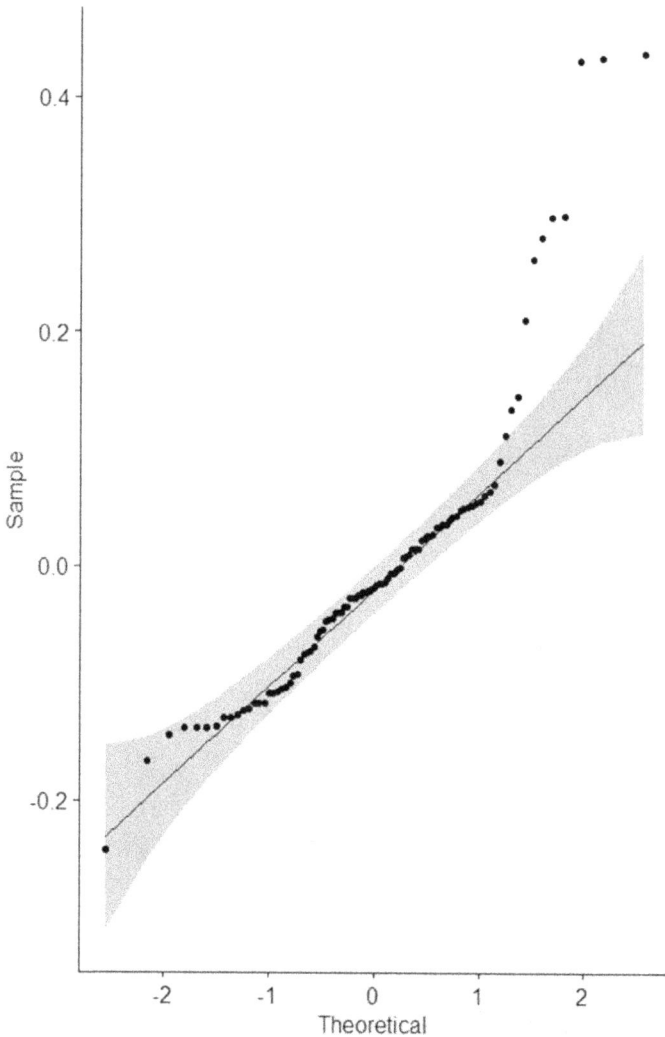

FIGURE 11.5
Normal QQ plot for school graduation exam passing rate data.

higher values of the socioeconomic status (SES) variable were associated with a higher passing rate, which supports the researcher's hypothesis. We can convert the coefficient for SES to the original proportion scale by back-transforming the logit in equation (11.10), yielding a value of 0.75. This result indicates that if a school were to go from 0% to 100% of students with household income above the state median, the proportion of students passing the exam will increase by 75%. In a perhaps more realistic example, if the proportion of students with household incomes above the state median were

Density residuals

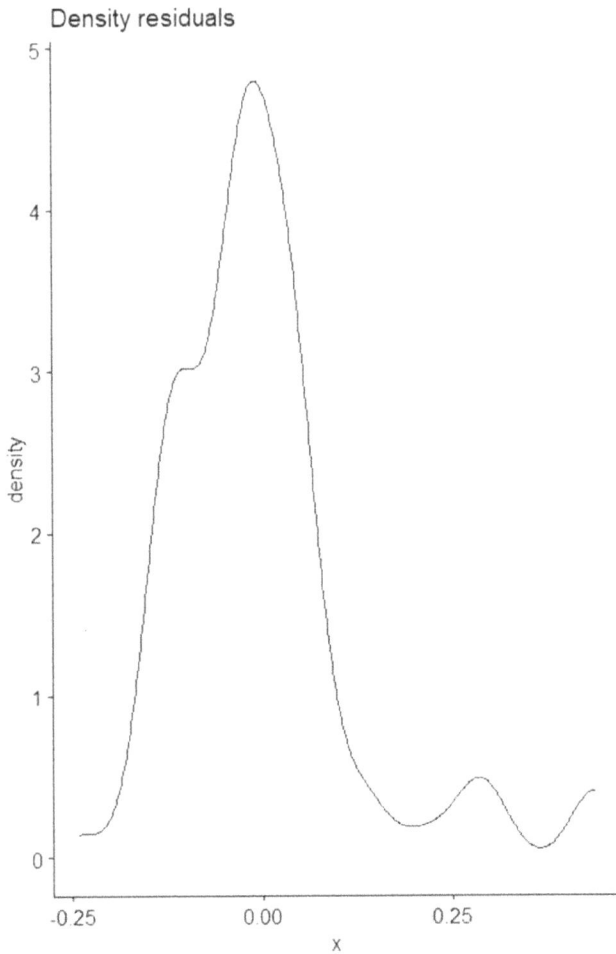

FIGURE 11.6
Density plot for residuals school graduation exam passing rate data.

to increase by 0.1 (i.e., 10%), the proportion of students passing the graduation exam would increase by $0.1 \times 0.75 = 0.075$, or 7.5%. Finally, the pseudo-R^2
value is 0.14, meaning that the proportion of students with household income
greater than the state median was associated with approximately 14% of the
variability in the proportion of students at the school who pass the graduation exam. Finally, Phi is a measure of precision in the Beta distribution, with
larger values indicating a greater spread in the data around the mean. It is
not particularly informative in this case, as we are primarily interested in
the relationship between the independent variable and the mean (proportion
passing the test), rather than the variance around the mean.

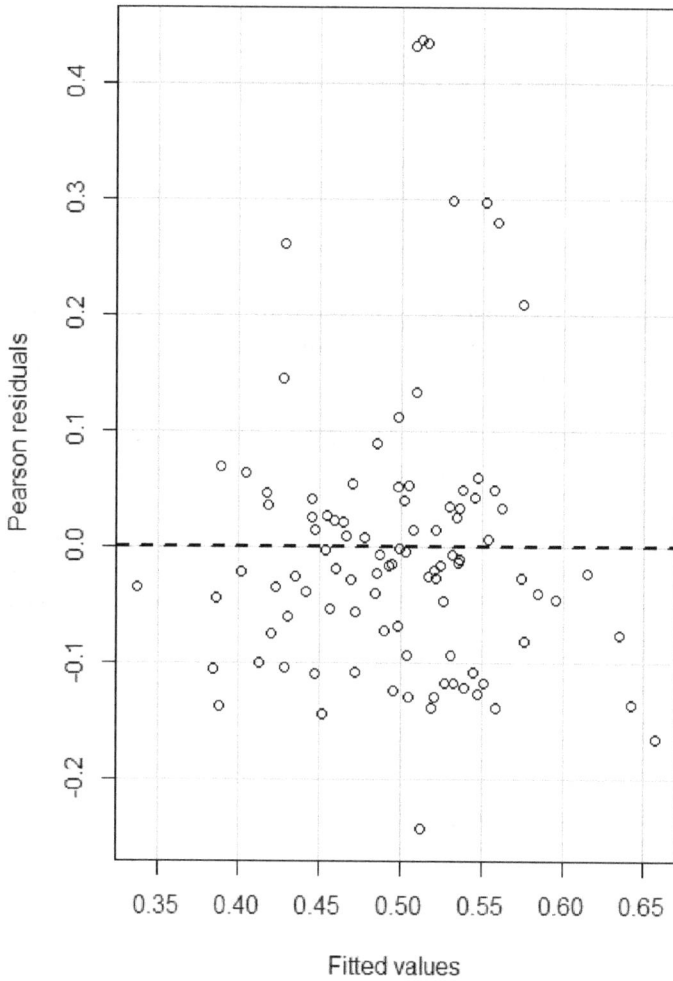

FIGURE 11.7
Residual by the predicted plot for school graduation exam passing rate data.

TABLE 11.7

Beta Regression Results for Graduation Exam Passing Rate

Variable	Coefficient (95% CI)	Standard Error
Intercept	−1.47 (−2.18, −0.76)	0.36
SES	4.24 (2.27, 6.22)	1.01
Phi	11.68 (8.54, 14.82)	1.60

Summary

In this chapter, we have considered several models for use when the dependent variable is dichotomous. Certainly, the most commonly used of these models is logistic regression. And indeed, in the majority of cases that a researcher will encounter in practice, it will be the optimal approach to use. Such is particularly the case when the outcome is a true dichotomy (e.g., treatment/control, alive/dead). However, if the observed dichotomous variable is likely based upon an underlying continuity (e.g., achieves reading mastery; depressed), then the researcher should consider using the Probit or CLL models. If the response is highly skewed, the CLL is preferable, whereas for other cases the Probit would be suitable. In cases where the dependent variable is a proportion or rate, Beta regression is most likely the optimal approach to use, particularly if the assumptions underlying OLS regression are not met. Finally, researchers should not forget about the opportunities offered by Bayesian estimation. It works well with small samples and allows the researcher to incorporate prior knowledge into the estimation process. Therefore, when researchers have relatively strong hypotheses about one or more of the model parameters based on earlier research and/or theory, Bayesian estimation with informative priors is a worthy alternative for the researcher to consider.

12

Advanced Issues in Regression Modeling

Introduction

In Chapter 9, we were introduced to the topic of linear regression. We learned that the linear regression model related one or more independent variables with a dependent variable, expressing these relationships with slopes. The slope for a given independent variable provides us with information about both the direction and magnitude of its relationship with the outcome variable. The multiple linear regression models take the form

$$y_i = \beta_0 + \beta_1 x_{1i} + \beta_2 x_{2i} + \cdots + \beta_j x_{ji} + \varepsilon_i \tag{12.1}$$

where y_i is the dependent variable value for individual i, x_{ji} is the value of independent variable j for individual i, β_j is the slope relating independent variable j to the dependent variable, β_0 is the intercept, and ε_i is the random error for individual i.

In parametric regression, the model parameters are estimated using ordinary least squares (OLS), which minimizes the sum of the squared residuals (the difference between the observed and model-predicted values of the dependent variable). OLS rests on the assumption that the residuals are normally distributed, homoscedastic (have constant variance), and independent. When these assumptions are not met, OLS will likely not produce trustworthy parameter estimates. In such cases, we learned in Chapter 10 that there are a plethora of alternative techniques that can be used to obtain regression parameter estimates. And in Chapter 11, we considered a regression model that is useful for cases when the dependent variable is dichotomous, rather than continuous as was the case for the models in Chapters 9 and 10.

Our goal in Chapter 12 is to learn about a number of extensions to the regression model that will be useful for a variety of specific research scenarios that we may encounter in practice. These extensions will allow us to identify subsets of independent variables that are associated with the dependent variable from among a larger set. In a related concept, we will also see

DOI: 10.1201/9781003379324-12

that Bayesian estimation can be used to average results across a number of models, including different subsets of the variable, thereby providing a potentially more holistic view of the relationships between the independent and dependent variables. We will then describe models that go beyond the linear regression framework and allow us to estimate nonlinear relationships among variables. Our attention will then turn to more complex regression models in which systems of relationships among variables can be explored through mediation and moderation. We will conclude this chapter by examining Dominance analysis (DA), a statistical technique for ordering the independent variables based on the strength of their relationships with a dependent variable, and by modeling time-dependent data in the regression framework. The overarching goal of this chapter is to provide you with an introduction to a set of data analysis options that can be used to address research questions that lie beyond the scope of linear regression. In addition, we hope that learning about this set of tools will broaden and deepen your statistical toolbox. It is important to note here that our coverage of these topics in this chapter is not exhaustive. Each topic covered herein warrants a full chapter (or even a book) of its own. However, hopefully, this chapter will pique your desire to seek out more in-depth treatment of one or more of the topics covered here.

Variable Selection

In some cases, researchers are faced with a situation in which they have a large number of independent variables to consider and need to reduce this set to a more manageable size. Ideally, decisions about the variables to include in a regression model are based primarily on theory. Statistical considerations should not be the driving force determining the variables to be included in a regression model. One good rule of thumb for researchers to use in this regard is that for each variable in the model, there should be one or ideally multiple citations in the literature supporting its inclusion. However, it is also the case that the theory underlying a particular analysis may not be particularly well developed such that the researcher cannot easily make choices about which variables to exclude from an analysis. In such cases, they may wish to make use of a variable selection technique to identify the independent variables that are most strongly related to the dependent for inclusion in the model.

Such variable selection techniques are not without controversy because they are inherently atheoretical in nature. The variables that are selected for inclusion in the model are not necessarily the most theoretically meaningful, but rather they exhibit the strongest relationships with the dependent variable for the sample at hand. It is easy to see why this is potentially problematic. Rather than building a theoretically sound model, we may end up with one that includes the most statistically important predictors of the outcome but does not tell a coherent story about what is happening in the real world. In addition, the variable selection is based on a single sample. Therefore, if this sample is not representative of the population, we may end up with a model that does not generalize in a meaningful way. For these reasons, the use of variable selection techniques is generally not popular among statisticians. Nonetheless, given its popularity in the applied research literature, it is important for you to be familiar with some of the more popular approaches for variable selection.

Stepwise Regression

One of the most common methods for variable selection is stepwise regression. There are two approaches for carrying out stepwise variable selection: forward and backward regression. The forward stepwise algorithm starts with all of the independent variables (x) outside of the model. Each is then entered into a simple regression model one by one and the hypothesis test results, as well as the R^2 indicating the proportion of variance in y accounted for by this model including x, are saved. The independent variable that accounts for the largest R^2 value and that also has a statistically significant coefficient estimate is then retained in the model. If none of the independent variables have a statistically significant relationship with y, the algorithm stops and an empty model is returned. Assuming that at least one of the independent variables is retained, in the next step of the algorithm, each of the $p-1$ remaining independent variables is then entered in turn into the model, including the first retained x, and the R^2 is saved. The x from this step, yielding the greatest increase in R^2 and having a statistically significant coefficient value, is then retained in the model. The forward stepwise algorithm continues until none of the variables outside of the model exhibit a statistically significant relationship with the dependent variable.

Backward stepwise regression begins with all of the independent variables included in the model, and a total R^2 is generated for this full model. Next, each variable is removed in turn, and the resulting model R^2 is calculated. The variable that reduces R^2 the least, with no statistically significant change in R^2 is then removed from the model. The algorithm repeats the step with the remaining variables and removes the variable that reduces R^2 the least with a non-significant change in R^2. This process is repeated until there are no variables that can be removed without resulting in a statistically significant reduction in the value of R^2.

Best Subsets Regression

An alternative variable selection approach is the best subset regression. With best subsets regression, each possible combination of the independent variables (from single predictor models to the full model containing all predictors) is fit to the data, and one or more measures of statistical fit [e.g., model complexity adjusted R^2, Mallow's C_p, Bayesian information criterion (BIC)] are used to select the optimal model. Adjusted R^2, Mallow's C_p, and the BIC are calculated as follows:

$$R_A^2 = 1 - \frac{\left(1 - R^2\right)(N-1)}{N-p-1} \tag{12.2}$$

$$C_p = \frac{\sum_{i=1}^{N}\left(y_i - \hat{y}_i\right)^2}{S^2} - N + 2(p+1) \tag{12.3}$$

$$\text{BIC} = -2\,\text{LL} + \ln(N)(p+1) \tag{12.4}$$

where R^2 is the proportion of variance in the dependent variable explained by the independent variable, S^2 is the residual variance, and LL is the model log-likelihood.

Model selection is based on optimizing one of these fit statistics. For example, we could select the model that maximizes R_A^2 or the one that minimizes C_p and/or BIC. In practice, researchers often consider all of these fit statistics and select the model that optimizes the majority of them.

Regularized Estimation

In some research contexts, the number of variables that can be measured (p) approaches, or even exceeds, the number of individuals on whom such measurements can be made (N). For example, researchers working with gene assays may have thousands of measurements that were made on a sample of only ten or 20 individuals. The consequence of having such small samples coupled with a large number of measurements is known as high-dimensional data. In such cases, standard statistical models often do not work well, yielding biased standard errors for the model parameter estimates (Bühlmann & van de Geer, 2011). These biased standard errors in turn can lead to inaccurate Type I error and power rates for inferences made about these parameters. High dimensionality can also result in parameter estimation bias due to the presence of collinearity (Fox, 2016). Finally, when p exceeds N, it may not be possible to obtain parameter estimates at all using standard estimators.

In the context of standard single-level data structures, statisticians have developed estimation methods that can be used with high-dimensional data, with one of the more widely used such approaches being regularization or sometimes known as shrinkage methods. Regularization involves the application of a penalty to the standard estimator (e.g., OLS) such that the coefficients linking the independent variables to the dependent variables are made smaller, or shrunken toward 0. The goal of this technique is that only those variables that are most strongly related to the dependent variable are retained in the model, whereas the others are eliminated by having their coefficients reduced (shrunken) to 0. This approach should eliminate from the model-independent variables that exhibit weak relationships to the dependent variable, thereby rendering a reduced model.

One of the most popular regularization approaches is the least absolute shrinkage and selection operator (lasso; Tibshirani, 1996), which can be expressed as

$$e^2 = \sum_{i=1}^{N} \left(y_i - \hat{y}_i\right)^2 + \lambda \sum_{j=1}^{p} \hat{\beta}_j \tag{12.5}$$

where y_i is the observed value of the dependent variable for individual i, $\hat{y}_i$ is the model-predicted value of the dependent variable for individual i, $\hat{\beta}_j$ is the sample estimate of the coefficient for independent variable j, and λ is the shrinkage penalty tuning parameter.

The tuning parameter, λ, is used to control the amount of shrinkage, with larger λ values corresponding to greater shrinkage of the model; i.e., a greater reduction in the number of independent variables that are likely to be included in the final model. If λ is set to 0, the resulting estimates are equivalent to those produced by the standard estimator; i.e., least squares, ML.

The key aspect of using the lasso is the determination of the optimal λ value. The most common approach to finding the appropriate tuning parameter value is through the use of cross-validation. With standard cross-validation, the researcher divides the full sample into k subsamples using random selection. One of these subsamples is then designated as the training set, and the others are known as the test sets. The lasso is applied to the training set for a variety of λ values, and the resulting $\hat{\beta}$ estimates are applied to each of the test samples to obtain predicted values of y_i for each individual. The mean square error for test set k with tuning parameter value λ ($MSE_{k\lambda}$) is then calculated for each of the test samples as

$$\mathrm{MSE}_{k\lambda} = \frac{\sum_{i=1}^{N} \left(y_{ik} - \hat{y}_{ik\lambda}\right)^2}{N_k} \tag{12.6}$$

where y_{ik} is the dependent variable value for subject i in test set k, $\hat{y}_{ik\lambda}$ is the model-predicted dependent variable value for subject i in test set k using λ, and N_k is the sample size for test set k.

The $\mathrm{MSE}_{k\lambda}$ values are then averaged across the K test samples for each value of λ, and the optimal value of λ is the one yielding the lowest mean $\mathrm{MSE}_{k\lambda}$.

If the sample is too small to be divided into training and cross-validation samples, a variation called leave-one-out or jackknife cross-validation can be used instead. With this method, the lasso model is fit to the data leaving out one individual and then applying the cross-validation method described above to compare that individual's actual and predicted values of y. This individual is then placed back into the sample, another individual is removed, the lasso model fits the data, and model parameters are applied to the data of the newly removed individual to obtain a cross-validation estimate of the value in equation (12.6). This approach is repeated for each individual in the sample so that the $\mathrm{MSE}_{k\lambda}$ is calculated involving all members of the sample. Regardless of which method of cross-validation is used, the optimal value of λ is the one that corresponds to the smallest MSE.

In summary, the goal of the lasso estimator is to eliminate from the model those independent variables that contribute very little to the explanation of the dependent variable by setting their $\hat{\beta}$ values to 0, while at the same time retaining independent variables that are important in explaining y. The optimal λ value is specific to each data analysis problem. A number of approaches for identifying it have been recommended, including the use of cross-validation to minimize the mean squared error (Tibshirani, 1996) or the selection of λ that minimizes the BIC. This latter approach was recommended by Schelldorfer et al. (2011), who showed that it works well in many cases. Zhao and Yu (2006) also found the use of the BIC for this purpose to be quite effective. With this approach, several values of λ are used, and the BIC values for the models are compared. The model with the smallest BIC is then selected as being optimal.

The Ridge Estimator

An alternative to the lasso that uses a very similar penalty function is the ridge estimator. As with the lasso, the ridge penalty is based on the LS estimator. However, rather than using the sum of the absolute values of the coefficients for the penalty term as with the lasso, the sum of the squared coefficients (rather than the absolute values) is used instead. The ridge algorithm seeks to minimize the following fitting function:

$$e_i^2 = \sum_{i=1}^{N}(y_i - \hat{y}_i)^2 + \lambda \sum_{j=1}^{p} \beta_j^2 \qquad (12.7)$$

Although the penalty used in ridge regression looks very similar to that used with the lasso, in fact, they have one primary difference, which is that the lasso serves as both a variable selection procedure and a parameter shrinkage method (Tibshirani, 1996). In practice, this means that the lasso will yield a model in which multiple independent variables may be effectively excluded due to having coefficients of 0. However, the ridge estimator will drive the coefficients of some variables down to near 0 but will typically not make these coefficients 0 exactly. The result is that ridge models will tend to be less parsimonious than those produced by the lasso (Hastie, Tibshirani, & Wainwright, 2015). Nonetheless, as with the lasso, the ridge estimator seeks to minimize the number of independent variables that effectively contribute to the prediction of the dependent variable and thereby reduce problems associated with high-dimensional data, such as unstable parameter estimates and collinearity. Determination of the optimal tuning parameter value is done using cross-validation as with the lasso.

The Elastic Net Estimator

The ridge and lasso estimators can be combined in the form of the elastic net (Zou & Hastie, 2005). The elastic net fitting function uses both the absolute squared values and squared coefficients, as well as a second tuning parameter, α:

$$e_i^2 = \sum_{i=1}^{N}(y_i - \hat{y}_i)^2 + \lambda \sum_{j=1}^{p}\left(\alpha\beta_j + (1-\alpha)\beta_j^2\right) \qquad (12.8)$$

When $\alpha = 0$, equation (2.4) simplifies to the ridge regression model, whereas when $\alpha = 1$, it is the lasso model. The values of α and λ can be selected using cross-validation in the same manner as described above.

Example of Variable Selection

As an example of the application of variable selection, let's consider a situation in which we are interested in treating self-oriented perfectionism (SOP) as our dependent variable, and we have 13 independent variables that reflect measures of various personality traits. The results of a multiple regression analysis appear in Table 12.1.

TABLE 12.1

Results of Multiple Regression Analysis for SOP and Personality Measures

Variable	Estimate	Standard Error	t	p
Intercept	17.18	10.18	1.69	0.09
Fear	0.86	0.96	0.89	0.37
Frustration	2.66	0.98	2.71	0.01
Sadness	1.30	0.87	1.50	0.14
Disinhibition	0.89	0.87	0.98	0.33
Activation control	5.38	0.92	5.84	<0.01
Attention control	−0.81	0.90	−0.91	0.36
Inhibitory control	0.43	1.02	0.42	0.67
Sociability	−0.49	0.77	−0.65	0.52
High intensity pleasure	1.38	0.95	1.46	0.15
Positive affect	0.12	0.89	0.14	0.89
Neutral perception	0.77	0.93	0.83	0.41
Affective perception	1.72	0.96	1.79	0.07
Associative perception	−1.24	0.95	−1.30	0.19

SOP, self-oriented perfectionism.

TABLE 12.2

Results of Backward Stepwise Regression Analysis for SOP and Personality Measures

Variable	Estimate	Standard Error	t	p
Intercept	19.81	5.84	3.39	<0.01
Frustration	3.10	0.84	3.67	<0.01
Sadness	1.98	0.77	2.59	0.01
Activation control	5.29	0.79	6.72	<0.01
Affective perception	1.67	0.75	2.24	0.03

SOP, self-oriented perfectionism.

Based on this regression model, we would conclude that frustration and activation control are positively related to SOP. The adjusted model had an R^2 of 0.21, indicating that this set of variables accounts for approximately 21% of the variation in SOP.

Applying backward stepwise regression yields the following model.

The adjusted $R^2 = 0.22$. In other words, the reduced model accounted for as much variance in SOP as did the full regression model. The forward selection and best subsets using Mallow's Cp both identified the same four-variable model as did backward regression in Table 12.2.

The results for the three regularization techniques appear in Table 12.3. Notice that variables not included in the table had 0 regression coefficient values.

TABLE 12.3

Results of the Lasso and Elastic Net for SOP
and Personality Measures

Variable	Lasso	Elastic Net
Fear	0.96	0.96
Frustration	2.83[a]	2.84[a]
Sadness	1.53	1.52
Activation control	4.42[a]	5.21[a]
Neutral perception	0.61	0.61
Affective perception	1.35	1.35

[a] Statistically significant for $\alpha = 0.05$.
SOP, self-oriented perfectionism.

The lasso and elastic net methods identified the same set of variables as having non-zero coefficients. In addition, the results were quite similar for the two methods. For both, only frustration and activation control were found to have statistically significant relationships with SOP.

Bayes Model Averaging

Bayes model averaging (BMA) represents an alternative to variable selection techniques. Rather than identifying an optimal set of independent variables based upon some criterion, as with stepwise, best subsets, or regularization, BMA takes information from a large set of models and then averages the results, weighted by the model posterior likelihoods. A primary advantage of this approach is that it does not require the elimination of independent variables from the analysis but rather allows for all of the potential predictors to be included in the analysis, shrinking the impact of those that are consistently associated with low-likelihood models (Raftery, 1995). Thus, rather than making dichotomous decisions about whether each variable should be included in the analysis or not, we fit all possible models for our set of predictors and then use a weighted average of the coefficient value for each variable as our estimate of its relationship with the dependent variable.

To illustrate how BMA works, let's consider a regression problem in which we have four independent variables. We should note here that BMA is most useful when we have many independent variables, but to clearly illustrate how it works, we will stay with a fairly simple model. The full regression model, including all of the variables, would take the form:

$$y_i = \beta_0 + \beta_1 x_{i1} + \beta_2 x_{i2} + \beta_3 x_{i3} + \beta_4 x_{i4} + \varepsilon_i \qquad (12.9)$$

where y_i is the dependent variable value for individual i, β_0 is the intercept, β_k is the slope for independent variable k, x_{i1} is the value of independent variable j for individual i, and ε_i is the random error for individual i.

With four independent variables, there are $2^4 = 16$ different models that could be fit to the data, using all combinations of the predictors. Each of these models has an associated posterior likelihood value, as we discussed in Chapter 9. These likelihoods can be summed across the 16 possible models to create a likelihood value for the full model space. With BMA, these values are then used to calculate the posterior probability for model j, $P(M_j \mid D)$, for each of the 16 models:

$$P(M_j \mid D) = P(D \mid M_j) \frac{P(M_j)}{\sum_{j=1}^{16} \left(P(D \mid M_j) P(M_j) \right)} \qquad (12.10)$$

where $P(D \mid M_j)$ is the probability of the data given model j and $P(M_j)$ is the prior probability of model j.

The $P(M_j \mid D)$ is then used to calculate the BMA posterior means $(E\left[\hat{\beta} \mid D\right])$ and variance $(V\left[\hat{\beta} \mid D\right])$ of the coefficients for independent variable k:

$$E\left[\hat{\beta} \mid D\right] = \sum_{j=1}^{16} \hat{\beta} P(M_j \mid D) \qquad (12.11)$$

$$V\left[\hat{\beta} \mid D\right] = \sum_{j=1}^{16} \left(\text{Var}\,\hat{\beta} \mid D, M_j + \hat{\beta}^2 \right) P(M_j \mid D) - E\left[\hat{\beta} \mid D\right]^2 \qquad (12.12)$$

where $\hat{\beta}$ is the estimate of regression coefficient for model j and $\text{Var}\,\hat{\beta} \mid D, M_j$ is the variance of regression coefficient for model j.

The result of the BMA for each independent variable is then reported as $E\left[\hat{\beta} \mid D\right]$ and inference based on $\text{Var}\,\hat{\beta} \mid D, M_j$ for each independent variable. In other words, the slope estimate from BMA for a given independent variable is the weighted average of the slopes for that variable from each of the 16 possible models. Thus, models that are more likely, given the prior distribution and the data, will contribute more to the value of the average slope than will models with lower relative likelihoods.

The results of BMA for our SOP and personality variables problem appear in Table 12.4.

The variables that were most strongly associated with SOP for the other approaches that we've considered here, namely, Activation Control, Frustration, and Sadness, were also the most important when we averaged across all of the possible models.

TABLE 12.4

BMA Results for SOP and Personality
Measures Model

Variable	Estimate
Fear	0.29
Frustration	2.95
Sadness	1.56
Disinhibition	0.05
Activation control	5.48
Attention control	−0.11
Inhibitory control	0.01
Sociability	−0.01
High intensity pleasure	0.10
Positive affect	0.01
Neutral perception	0.19
Affective perception	0.91
Associative perception	−0.02

SOP, self-oriented perfectionism; BMA,
Bayes model averaging.

Nonlinear Model Estimation

A primary assumption underlying the regression models that we discussed
in Chapter 9 is that the model is linear in nature. Mathematically, the linear
model for a sample can be expressed by the following equation:

$$y = b_0 + b_1 x \tag{12.13}$$

where y is the dependent variable, x is the independent variable, b_0 is the
intercept, and b_1 is the slope.

In some situations, we may be interested in assessing whether the underlying relationship between x and y is nonlinear in nature. One very simple
nonlinear model includes a squared term for the independent variable. This
is known as the quadratic model and takes the form:

$$y_i = b_0 + b_1 x + b_2 x^2 \tag{12.14}$$

where b_2 is the coefficient for the squared time term.

This model captures the nonlinear relationship through the inclusion of
the quadratic term, x^2. When the relationship between x and y is actually

linear, the coefficient b_2 in equation (12.14) will be 0. However, when the relationship is nonlinear, the estimate of the linear relationship in equation (12.13), b_1, will be adjusted for the presence of the nonlinearity through the inclusion of $b_2 x^2$ (Fox, 2016). In contrast, when the standard linear model in equation (12.13) is used with nonlinear data, the estimate of b_1 will be biased (Fox; i.e., not statistically accurate).

When making use of the nonlinear regression model in equation (12.14), we are making a fairly strict assumption about the nature of the model underlying the relationship between the independent and dependent variables. Namely, we are assuming that this relationship involves a linear and a quadratic term. While useful in many situations, this model is also fairly limited in the types of relationships that can be fit to the data. Thus, it would be helpful if we could accommodate a more complex type of relationship when appropriate. One very flexible approach that we can use for this purpose is the locally weighted scatterplot smoother (LOWESS). This technique fits a regression model between an independent and dependent variable and obtains predicted values for each data point. This is not particularly notable, as we can do this with any regression model. What differentiates LOWESS from standard regression is the fact that predictions of the dependent for a given individual ($\hat{y}_i$) involve only individuals with values of the independent variable that are close to (in the neighborhood of) our target individual. In addition, those with values of the independent variable that are closer to our individual will be weighted greater than those who are further away. These distances are calculated only for those individuals who are within the selected subset of cases around the target point. For example, the selected subset might include the 50% of the sample that is closest to the target observation based on the values of the independent variable.

There are a number of possible weighting schemes that have been associated with LOWESS estimation. One of the more common approaches, the tricube, takes the following form:

$$w_{ij} = \left(1 - \left| \frac{d_{ij}}{\max(d_{ij})} \right|^3 \right)^3 \tag{12.15}$$

where $d_{ij} = |x_i - x_j|$, $\max(d_{ij})$ is the maximum distance from the target data point within the selected subset, x_i is the independent variable value for target data point, and x_j is the independent variable value for individual j.

For individuals outside of the selected subset, $w_{ij} = 0$. Once the w_{ij} values are calculated, weighted least squares are conducted in which the sum of the squared weighted residuals is minimized to obtain slope and coefficient estimates. The weighted least squares criterion is

$$\sum_{i=1}^{n} w_{ij} (y_i - \hat{y}_i)^2 \tag{12.16}$$

where y_i is the observed dependent variable value for individual i and $\hat{y}_i$ is the LOWESS model-predicted dependent variable value for individual i.

Typically, the results of a LOWESS model will be expressed in the form of a scatterplot for the dependent and independent variable values. We will examine such a plot below.

Another option for fitting nonlinear relationships in our data comes in the form of generalized additive models (GAMs), which link an outcome variable, y, with one or more independent variables, x, using smoothing splines. Splines are piecewise polynomials for which individual functional sections are joined at locations in the data known as knots (Hastie & Tibshirani, 1990). A commonly used approach is the cubic spline, which is simply a set of cubic polynomial functions joined together at the knots, for which each section between the knots has unique model parameter values:

$$y = \beta_0 + \beta_1 x + \beta_2 x^2 + \beta_3 x^3 \tag{12.17}$$

where y is the dependent variable, x is the independent variable, and β_j is the coefficient for model term j.

The cubic spline essentially fits different versions of the model between each pair of adjacent knots. The more knots in the GAM, the more piecewise polynomials that will be estimated, and the more potential detail about the relationship between x and y will be revealed; i.e., the more complex the nonlinearity will be. GAMs take these splines and apply them to a set of one or more predictor variables as in equation (12.18):

$$y_i = \beta_0 + \sum f_j\left(x_{ji}\right) + \varepsilon_i \tag{12.18}$$

where y_i is the value of dependent variable for individual i, x_i is the value of independent variable j for individual i, f_j is the smoothing spline for independent variable j, and ε_i is the random error for individual i.

Each independent variable has a unique smoothing function, and the optimal set of smoothing functions is found by minimizing the penalized sum of squares criterion (PSS):

$$\text{PSS} = \sum_{i=1}^{N}\left\{y_i - \right\}^2 + \sum_{j=1}^{p}\lambda_j\int f_j''(t_j)^2\, dt_j \tag{12.19}$$

Here, y_i is the value of the response variable for subject i, and λ_j is a tuning parameter for variable j such that $\lambda_j \geq 0$. The researcher can use λ_j to control the degree of smoothing that is applied to the model. A value of 0 results in an unpenalized function and relatively less smoothing, whereas values approaching ∞ result in an extremely smoothed (i.e., linear) function relating the outcome and the predictors. The GAM algorithm works in an iterative fashion, beginning with the setting of β_0 to the mean of Y. Subsequently, a smoothing function is applied to each of the independent variables in turn, minimizing the PSS. The iterative process continues until the smoothing functions for the various predictor variables stabilize, at which point final model parameter estimates are obtained. Based upon empirical research, a recommended value for λ_j is 1.4 (Wood, 2006), and as such, it will be used in this study.

Example of Nonlinear Models

We will investigate nonlinear relationships between a measure of anxiety and SOP from our variable selection example. In this case, anxiety is the dependent variable and SOP is the independent. First, let's consider the quadratic model in which SOP and SOP squared are the predictors. The coefficient for the linear term is 0.03 ($p<0.001$), and the quadratic coefficient is 0.041 ($p<0.001$). Thus, we can conclude that there is a nonlinear relationship between anxiety and SOP, as well as a linear relationship after accounting for the quadratic term. The results of the quadratic model appear in Figure 12.1.

We can see that the anxiety is lowest for individuals with SOP values in the midrange and highest for those with low or high perfectionism.

The LOWESS curve with a bandwidth of 0.5 appears in Figure 12.2.

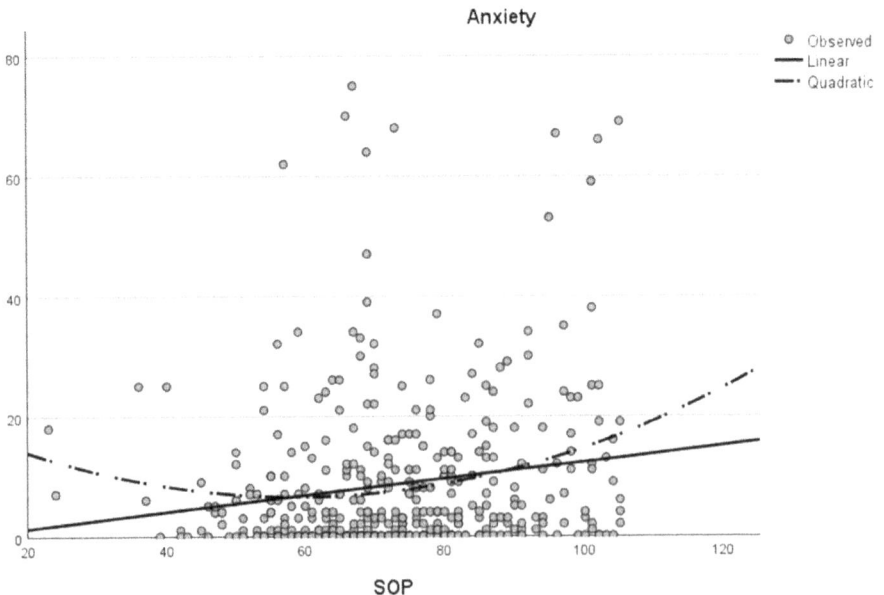

FIGURE 12.1
Linear and quadratic model results for the relationship between anxiety and SOP. SOP, self-oriented perfectionism.

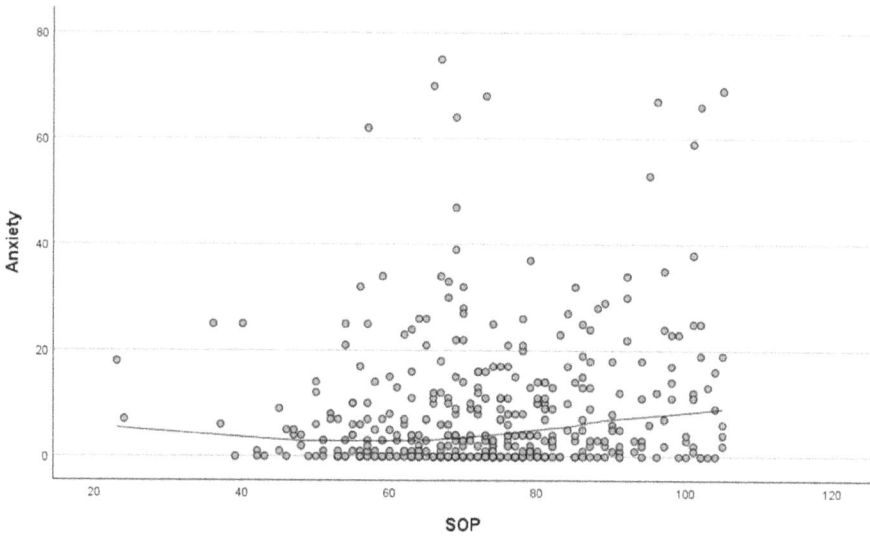

FIGURE 12.2

LOWESS curve for the relationship between anxiety and SOP. LOWESS, locally weighted scatterplot smoother; SOP, self-oriented perfectionism.

Similar to the results for the quadratic model, the LOWESS curve suggests that anxiety levels are higher for individuals with either low or high levels of perfectionism.

Finally, we will consider the results for the GAM, which appear in Figure 12.3.

The results of the GAM are generally similar to those for the quadratic and LOWESS models. However, we do see that the GAM-based curve is generally flat for most values of SOP and then takes an upward trajectory when the perfectionism score nears 100. This result suggests that higher levels of anxiety are generally associated with higher levels of perfectionism but not with lower levels, in contrast to the results for the quadratic and LOWESS approaches.

Moderation and Mediation

Mediation analysis is an important aspect of social science research. It allows a researcher to explore the relationship between a predictor variable (x) and an outcome (y) through, or as mediated by, a third variable (m). This partially mediated relationship can be viewed as Figure 12.4.

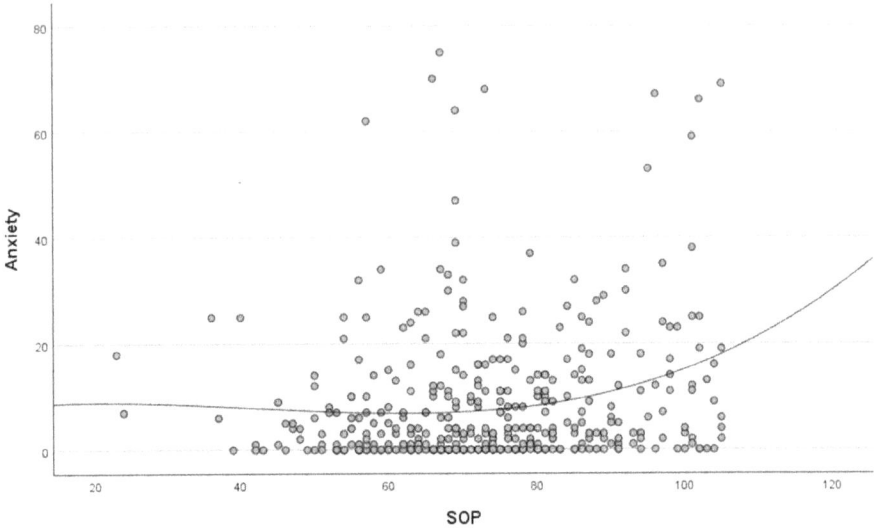

FIGURE 12.3
GAM curve for the relationship between anxiety and SOP. GAM, generalized additive model;
SOP, self-oriented perfectionism.

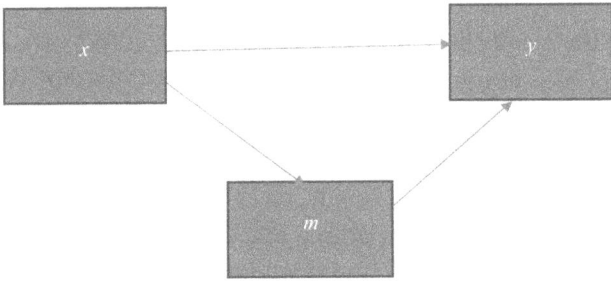

FIGURE 12.4
Partial mediation model.

This path model can be expressed by the following set of equations:

$$m_i = a_0 + a_1 x_{1i} + e_i \tag{12.20}$$

where m_i is the value of mediator for individual i, x_{1i} is the value of independent variable for individual i, a_0 is the mediator model intercept, a_1 is the coefficient linking independent variable to mediator, and e_i is the random error for individual i in the mediator model

$$y_i = b_0 + b_1 m_{1i} + c_1 x_{1i} + f_i \tag{12.21}$$

where y_i is the value of outcome variable for individual i, b_0 is the outcome variable model intercept, b_1 is the coefficient linking mediator to outcome variable, c_1 is the coefficient linking independent variable to the outcome variable, and f_i is the random error for individual i in the outcome variable model

Parameter estimates from models (12.20) and (12.21) can be used to calculate the indirect relationship between x and y, as well as the total relationship between the two variables. The indirect effect is expressed as

$$\text{IND} = a_1b_1 \tag{12.22}$$

The total relationship between x and y can then be written as

$$\text{TOT} = \text{IND} + c_1 = a_1b_1 + c_1 \tag{12.23}$$

Researchers have shown that the indirect effect does not follow a normal distribution (MacKinnon et al., 2002) and that the bootstrap offers a viable alternative for obtaining confidence intervals for this effect (Falk, 2018).

To demonstrate a mediation example, we will return to the dataset that we used in Chapter 9, which was collected from college students. For this example, we will use a measure of mastery approach (MAP) to learning as the dependent variable, effortful control (EFFC) as the primary independent variable of interest, and SOP as the mediator. Higher scores on MAP indicate that a learner is more likely to learn for learning's sake, whereas higher scores on EFFC indicate that an individual has a greater ability to regulate their own behavior and cognition. SOP refers to an individual's level of perfectionism that is drawn from within. Our hypothesis is that the relationship between EFFC and MAP is mediated by SOP.

To obtain the standard error and confidence interval for the mediation model, 5,000 bootstrap draws were used. The coefficients, standard errors, and 95% confidence intervals for the direct and indirect effects in the proposed mediation model appear in Table 12.5.

The mediation model had an R^2 of 0.15, indicating that SOP, EFFC, and the indirect effect accounted for approximately 15% of the variability in MAP. In addition, we see that the bootstrap confidence intervals for the direct and

TABLE 12.5

Coefficients, Standard Errors, and 95% Confidence Intervals for the Mediation Model

Relationship	Coefficient	Standard Error	95% Confidence Interval
Intercept	10.33	0.90	8.56, 12.10
EFFC>MAP	1.02	0.18	0.67, 1.37
SOP>MAP	0.04	0.01	0.02, 0.06
EFFC>SOP>MAP	0.17	0.06	0.07, 0.31

EFFC, effortful control; MAP, mastery approach; SOP, self-oriented perfectionism.

indirect effects did not include 0. Thus, we can conclude that both EFFC and SOP are related to MAP in a positive manner, such that individuals with higher EFFC and SOP were more likely to report wanting to learn for learning's sake. In addition, the relationship between EFFC and MAP was also mediated through SOP. Taken together, these results tell us that there is a partially mediated relationship between SOP and MAP with both a direct and indirect component.

Whereas mediation focuses on relationships between two variables through a third, moderation involves exploration of how a variable impacts the relationship between two variables. We discussed this idea in the context of interactions in the analysis of variance (ANOVA) in Chapter 6. With ANOVA, an interaction occurs when the difference in group means for one variable depends on the value of another variable. For example, if a reading instruction treatment impacts reading test scores differently for boys and girls, we would conclude that there is an interaction between gender and treatment conditions. Typically, we use the term interaction when discussing ANOVA. However, we could refer to this as moderation, as gender moderates the impact of the treatment on reading test scores. This notion can be directly applied to linear regression such that we can assess whether one independent variable moderates the relationship between another independent variable and the response in the form of an interaction between the two independent variables. This model takes the form

$$y_i = \beta_0 + \beta_1 x_{1i} + \beta_2 x_{2i} + \beta_3 x_{1i} x_{2i} + \varepsilon_i \tag{12.24}$$

where y_i is the dependent variable value for individual i, x_{ji} is the value of independent variable j for individual i, β_0 is the model intercept, β_j is the coefficient for model term j, and ε_i is the random error for individual i.

The model in equation (12.24) is quite similar to the multiple linear regression model that was the focus of Chapter 9. The model can be estimated using OLS or the robust estimation techniques described in Chapter 10. The estimated model is

$$y_i = b_0 + b_1 x_{1i} + b_2 x_{2i} + b_3 x_{1i} x_{2i} \tag{12.25}$$

Of particular interest for our discussion here is the interaction between the two independent variables, expressed as $x_{1i} x_{2i}$. Let's assume the theory suggests that x_2 moderates the relationship between x_1 and y. The strength and nature of this moderation effect are reflected in β_3. If this coefficient is statistically significant, we can conclude that x_2 moderates the relationship between x_1 and y. In other words, the strength of the relationship between x_1 and y is dependent upon the value of x_2. In addition, β_3 reflects the change in the relationship between x_1 and y for a 1-unit change in x_2, known as the conditional effect.

If the interaction term is found to be statistically significant, perhaps the most common approach to interpreting the nature of the moderation is through the calculation of conditional effects. The conditional effect reflects the relationship between x_1 and y at a given level of x_2. It is calculated as

$$\theta_{x_1y} = b_1 + b_3 x_{2i} \qquad (12.26)$$

The algebra used to obtain the conditional effect from equation (12.26) appears in Darlington and Hayes (2018). Typically, researchers select values of x_2 representing relatively high, medium, and low values. These might be the 75th, 50th, and 25th percentile values, or one standard deviation above the mean, the mean, and one standard deviation below the mean. However, a researcher with a special interest in specific values of x_2 could also use those when calculating the conditional effects. We will see an example of these calculations in the example below.

A potential problem with the conditional effects approach to probe interactions is that the results are dependent upon the values of the moderator that the researcher selects. And different researchers who select different moderator values can come to different conclusions regarding the nature of the interaction effects. Thus, an alternative approach for gaining insights into interaction effects is the Johnson–Neyman technique (JN; Johnson & Neyman, 1936). The JN method identifies the values of the moderator for which the conditional effect (θ_{x_1y}) transitions from being statistically significant to statistically non-significant. We are then left with regions of the moderator for which the target variable has a statistically significant relationship with the dependent variable. The JN technique is too complex for a researcher to readily do by hand but can be carried out using software, specifically the PROCESS macro in the SPSS, SAS, and R software environments (Hayes, 2023).

To demonstrate mediation, we will return to the example from Chapter 10 involving SOP. Recall that this dataset was drawn from the population of college students and included a measure of SOP where higher scores indicated a greater level of perfectionism derived from their own internal standards. In addition, this dataset also includes a measure of sadness from the adult temperament scale (SAD) and a measure of chronic stress (STRESS), where higher scores indicate greater stress. We are interested in ascertaining whether SOP moderates the relationship between STRESS and SAD. The population regression model reflecting this set of relationships is

$$\text{STRESS} = \beta_0 + \beta_1 \text{SAD} + \beta_2 \text{SOP} + \beta_3 \text{SAD*SOP} + \varepsilon_i$$

The sample estimates for this model appear in Table 12.6.

The interaction between SAD and SOP was statistically significant ($p = 0.04$), leading us to conclude that perfectionism does appear to moderate the

TABLE 12.6

Regression Parameter Estimates for Moderation Model

Variable	Coefficient	Standard Error	t	p
Intercept	−33.72	12.55	−2.69	0.01
SAD	8.87	2.82	3.15	0.002
SOP	0.46	0.20	2.38	0.02
SAD * SOP	−0.09	0.04	−1.99	0.04

SOP, self-oriented perfectionism; SAD, sadness.

TABLE 12.7

Conditional Effects for the Relationship of SAD with STRESS at Selected Percentiles of SOP

SOP Value	Coefficient	Standard Error	t	p
49 (25th percentile)	4.48	0.81	5.55	<0.001
59 (50th percentile)	3.58	0.60	6.00	<0.001
72 (75th percentile)	2.42	0.76	3.18	0.002

SOP, self-oriented perfectionism; SAD, sadness.

relationship between sadness and chronic stress. In addition, the $R^2 = 0.10$ for the model means that approximately 10% of the variability in STRESS is associated with SAD, SOP, and the interaction of SAD and SOP. The conditional effects for the relationship of SAD with STRESS at the 25th, 50th, and 75th percentiles of SAD appear in Table 12.7.

We could calculate these coefficients by hand using the equation for the conditional effect. For example, for SOP=49, the conditional relationship between SAD and STRESS is $\theta_{SAD,STRESS} = 8.87 - 0.09 * 49 = 4.48$. Similar calculations can be done to obtain the conditional effect coefficients in Table 12.6. These results show that the relationship between SAD and STRESS is stronger when SOP is smaller. In other words, the lower the SOP that a person has, the stronger the relationship between their level of sadness and their chronic stress. In all cases, this relationship was positive, meaning that higher levels of sadness were associated with higher levels of chronic stress.

The JN procedure identified the region of significance for SAD and STRESS as lying from an SOP value of 78.01 and below. This region of the data includes 92.9% of the sample, and the coefficients are positive, as seen in Table 12.8.

We can see that the relationship between SAD and STRESS is positive across levels of SOP, but that the coefficient declines as SOP increases in value.

There is one final point that we need to make with respect to both mediation and moderation. In this chapter, we have focused almost exclusively on the statistical aspects of these modeling strategies. While such issues are obviously of key importance when conducting these analyses, perhaps even

TABLE 12.8

Conditional Effects for the Relationship of SAD with STRESS in the JN Region of Significance

SOP	b	SE	t	p
25.0000	6.6296	1.7355	3.8200	0.0002
28.5000	6.3160	1.5882	3.9769	0.0001
32.0000	6.0024	1.4430	4.1597	0.0000
35.5000	5.6888	1.3007	4.3735	0.0000
39.0000	5.3752	1.1624	4.6242	0.0000
42.5000	5.0616	1.0296	4.9160	0.0000
46.0000	4.7480	0.9048	5.2476	0.0000
49.5000	4.4344	0.7918	5.6008	0.0000
53.0000	4.1208	0.6962	5.9190	0.0000
56.5000	3.8072	0.6262	6.0795	0.0000
60.0000	3.4936	0.5910	5.9113	0.0000
63.5000	3.1800	0.5967	5.3294	0.0000
67.0000	2.8664	0.6422	4.4634	0.0000
70.5000	2.5528	0.7200	3.5454	0.0004
74.0000	2.2392	0.8210	2.7273	0.0067
77.5000	1.9256	0.9378	2.0534	0.0407
78.0141	1.8796	0.9559	1.9663	0.0500
81.0000	1.6120	1.0651	1.5136	0.1310
84.5000	1.2984	1.1996	1.0824	0.2798
88.0000	0.9848	1.3391	0.7355	0.4625
91.5000	0.6712	1.4822	0.4529	0.6509
95.0000	0.3577	1.6280	0.2197	0.8262

SOP, self-oriented perfectionism; SAD, sadness; JN, Johnson–Neyman.

more important are the theoretical considerations underlying the research questions and hypotheses that we would like to investigate using mediation or moderation. Of particular importance is that there be a strong theory based on prior literature to support the models under consideration. Without such a theory to support them, the results of statistical analyses cannot be taken as substantively meaningful. For example, without prior theory to suggest that it is reasonable (and even likely) for the relationship between EFFC and learning motivation to be mediated by SOP, the results of statistical analysis cannot tell us anything meaningful about the real world. Only when there is strong conceptual grounding to support the research hypothesis *preceding* the collection and analysis of data we can use the results from our models to make statements about the real world. It is extremely important for researchers to keep in mind this need for a strong conceptual basis when working with any statistical models, including mediation and moderation.

Dominance Analysis

Researchers using regression analysis may be interested in ascertaining which variables included in the model are the most important. Much research has been devoted to the development of methods for assessing relative variable importance in the context of observed variable regression models. Perhaps the most prominent of these methods is DA, which was described by Azen and Budescu (2003). This technique, which is described in more detail below, compares the change in variance explained for the outcome variable by regression models including subsets of the independent variables of interest. Variables for which this change is largest are said to be relatively more important than those for which the change is smaller.

Researchers interested in gaining insights into the relative importance of predictor variables with respect to an outcome often turn to DA (Azen & Budescu, 2003). The overarching goal of DA is to rank the predictors in terms of the amount of variance in the dependent variable that each explains. More specifically, DA involves the calculation of ΔR_j^2, the change in the variance explained when variable j is included in the model versus when it is removed. For example, to ascertain whether variable j is relatively more important than variable l, with respect to the dependent variable, every possible subset of the remaining k variables is used in regression analysis. For each of these subsets, a model without either variable k or l is fit, and the R^2 value is retained. Then a model involving this variable subset, as well as variable j, is fit to the data, and the resulting R^2 is saved. Finally, a model involving the original variable subset, also including variable k, is fit to the data, and the R^2 value is retained. The change in R^2 for each variable is then calculated (i.e., ΔR_j^2 and ΔR_l^2). This approach is applied to all possible subsets and variable combinations possible for a given dataset.

There are three types of dominance possible in the context of DA:

1. Complete dominance occurs when ΔR^2 is always larger for variable j than variable l for all possible model subsets.

2. Conditional dominance occurs when, for any size of the subset model, the mean ΔR^2 for variable j is greater than that of variable l. But this doesn't mean that for each model, ΔR^2 is larger for variable 1.

3. General dominance occurs when across all possible models the average ΔR^2 for variable j is larger than that of variable l.

Of the aforementioned types of dominance, general dominance is perhaps the most widely used in practice (Azen & Traxel, 2009). For a given independent variable, x_j, the general dominance statistic (Azen & Budescu, 2003) is calculated as

$$d_j = \frac{1}{J} \sum_{m=0}^{J-1} \frac{1}{C(J-1,m)} \sum_{qm}^{C(J-1,m)} R^2_{x_j|S_{qm}(x_j)} \qquad (12.27)$$

where m is the size of model prior to entry of x_j, $C(J-1,m)$ is the number of combinations when selecting m elements from $J-1$ possibilities, $S_{qm}(x_j)$ is the subset of variables to which x_j can be added, given m, and $R^2_{x_j|S_{qm}(x_j)}$ is the squared semipartial correlation of the dependent variable and x_j given $S_{qm}(x_j)$.

The statistic d_j is the mean increase in the squared semipartial correlation $(R^2_{x_j|S_{qm}(x_j)})$ for the variable x_j across all independent variable subsets. Independent variables with larger values of d_j are said to exhibit general dominance over those variables with smaller values.

Luo and Azen (2013) extended DA to the context of hierarchical (sometimes referred to as multilevel) linear models. As those authors noted, the multi-level model framework brings with it a complexity that is not found with single-level regression models such as those described above. In particular, the assessment of the variance accounted for by a set of predictors is not a settled issue, with multiple statistics available for this purpose and the added complexity of variables contributing to explained variance at multiple levels of the model. Luo and Azen demonstrated that the measure of explained variance to be used when assessing the relative importance of predictors in a multilevel model is dependent on the level at which the predictor vari-ables appear and the level of the response variable for which DA should be conducted.

One approach to inference that has been suggested for DA involves the bootstrap, which accounts for the sampling variability inherent in these val-ues (e.g., Braun, Converse, & Oswald, 2019; Azen & Traxel, 2009; Tonidandel et al., 2009; Azen & Budescu, 2003). The bootstrap can be used to obtain stan-dard errors and/or confidence intervals for d_j. These in turn can be used to ascertain the ordering of the dominance statistics once sampling variability has been taken into account. A second approach for assessing the relative importance of the individual predictors in a DA involves the use of Bayes Factors (Gu, 2021). This technique, which is discussed in more detail in the *Bayesian Inference* section below, allows the researcher to compare hypotheses about the ordering of the independent variables based on d_j. For example, the researcher may have two possible hypotheses of interest, differing based on the relative importance of x_1 and x_2.

$$H_1 : d_1 > d_2 > d_3 > d_4$$
$$H_2 : d_2 > d_1 > d_3 > d_4 \qquad (12.28)$$

where d_j is the general dominance value for predictor j.

TABLE 12.9

Dominance Analysis Results for Relationship of
SOP with Personality Measures

Variable	Dominance Coefficient
Activation control	0.104
Sadness	0.029
Frustration	0.025
Affective perception	0.022
Fear	0.018
Neutral perception	0.016
Disinhibition	0.009
Inhibitory control	0.004
High intensity pleasure	0.004
Associative perception	0.004
Attentional control	0.03
Positive affect	0.002

SOP, self-oriented perfectionism.

The Bayes Factor can be calculated to determine which of these hypotheses is more likely and thereby whether x_1 or x_2 is a more important predictor of the dependent variable.

Example of Dominance Analysis

We will demonstrate DA using the SOP and personality measures from earlier in the chapter. Specifically, we are interested in ordering the set of independent variables based on the strength of their relationship with SOP. Table 12.9 includes the variables ordered by the dominance coefficients

The most important variable was activation control. When added to the model, it contributed an average increase of 0.104 to the model R^2. The next most important variables, sadness, frustration, and affective perception, contributed to the R^2 on average between 0.02 and 0.03. These results make it clear that activation control has the strongest relationship with SOP.

Regression with Time-Dependent Data

One of the foundational assumptions of many statistical analyses is that the errors associated with the dependent variable for individuals in the sample are independent of one another. In other words, once we have accounted for the relationship(s) of the independent and dependent variable(s), whatever is left over in the dependent variable is uncorrelated from one data point to another. We have seen this assumption applied to virtually every analysis that has been discussed in this text. This independence assumption can be very difficult to check, unlike other common assumptions that we may be familiar with (e.g., normality, homogeneity of errors).

The primary impact of correlated errors is manifested in the standard errors of the model parameters, such as regression coefficients. In general, correlated errors will lead to an underestimation of the model standard errors. In turn, underestimated standard errors lead to overestimated test statistics associated with the model parameters. In turn, when these values are overestimated, the p-value will be smaller than it should be, leading to an inflation of the Type I error rate.

There are a number of potential causes for correlated errors, some of which are idiosyncratic and not amenable to a statistical solution (e.g., inclusion of a pair of siblings in a sample of test examinees). However, there are causes that are perhaps more obvious and which can be dealt with using statistical models. These alternative causes of correlated error include the nesting of individual students within teachers or schools, which we will address in Chapter 13. In addition, the collection of observations on the same individual over time (time series) can also lead to correlated residuals. When the time series is fairly short, we can use standard statistical models such as repeated measures ANOVA, as in Chapter 7. For longer time series, it may be more beneficial to turn to models specifically designed for such serially correlated data. Time series is a very broad area of statistics that is widely used in many fields, particularly economics, business, and climate research. We will be touching on only a very, very small subset of these models in this chapter. The model that will be our focus might prove most useful in the context of social science or health science research. Specifically, we are going to learn how to fit standard regression models with serially correlated errors.

A common description of such data is that it is autoregressive in nature, meaning that a value at the current time is directly related to a value at a prior time. As an example, let's take as an example the rate of a particular cancer diagnosis in the United States between 1950 and 2004. We would like to know how strongly related gender is with this diagnosis. To investigate this relationship, we could use a standard simple linear regression model with diagnosis rate as the dependent variable and gender as the independent variable. The simple linear regression model for addressing our research question would be

$$y_t = \beta_0 + \beta_1 x_t + \varepsilon_t \tag{12.29}$$

where y_t is the dependent variable value for time t, x_t is the independent variable value for time t, β_0 is the intercept, β_1 is the coefficient relating x and y, and ε_t is the error at time t.

In standard OLS regression models, we assume that the ε are normally distributed and independent of one another. However, with time series data the error term actually has the following form:

$$\varepsilon_t = \phi_1 \varepsilon_{t-1} + \phi_2 \varepsilon_{t-2} + \cdots + w_t \tag{12.30}$$

where ϕ_1 is the autocorrelation for lag 1, ε_{t-1} is the error at lag 1, and $w_t \sim \text{iidN}(0, \sigma^2)$; sometimes referred to as white noise.

By contrast, the error term when no time dependence exists takes the following form:

$$\varepsilon_t \sim \text{iidN}(0, \sigma^2) \tag{12.31}$$

To properly fit the error term with serially correlated data, we can account for the lagged errors using an autoregressive operator term $(\Phi^{-1}(B))$ so that our model becomes:

$$y_t = \beta_0 + \beta_1 x_t + \Phi^{-1}(B) w_t \tag{12.32}$$

$\Phi^{-1}(B)$ essentially models the impact of the lagged error terms on the value of the current error term.

When considering the application of time series regression, we first need to determine whether there is in fact autocorrelation within the residuals. We can make this determination by fitting standard regression models and then plotting the autocorrelation for the residuals from that model. If the autocorrelation is not near 0, then we know it is an issue for our data. Typically, autocorrelation is assessed using autocorrelation function (ACF) and partial autocorrelation (PACF) plots. The ACF reflects the relationship between lags not controlling for other lags, whereas PACF reflects this relationship controlling for earlier lags. For autoregressive models, the ACF dies slowly, and the PACF stops at the level of autoregression.

For moving average models, the PACF dies slowly, and the ACF stops at the moving average level. The results for the cancer diagnosis data suggest that we have an AR(1) process (Figure 12.5).

In addition to the ACF and PACF, we can also assess whether there is an autoregressive process present in our data using the Durbin–Watson statistic (Durbin & Watson, 1951). It tests the null hypothesis that the lag-1 autoregression coefficient is 0; i.e., that there is no relationship between measurements that are adjacent in time. This statistic takes the form

$$D = \frac{\sum_{t=2}^{n} (E_t - E_{t-1})^2}{\sum_{t=1}^{n} E_t^2} \tag{12.33}$$

where E_t is the residual for time t and E_{t-1} is the residual for time $t - 1$.

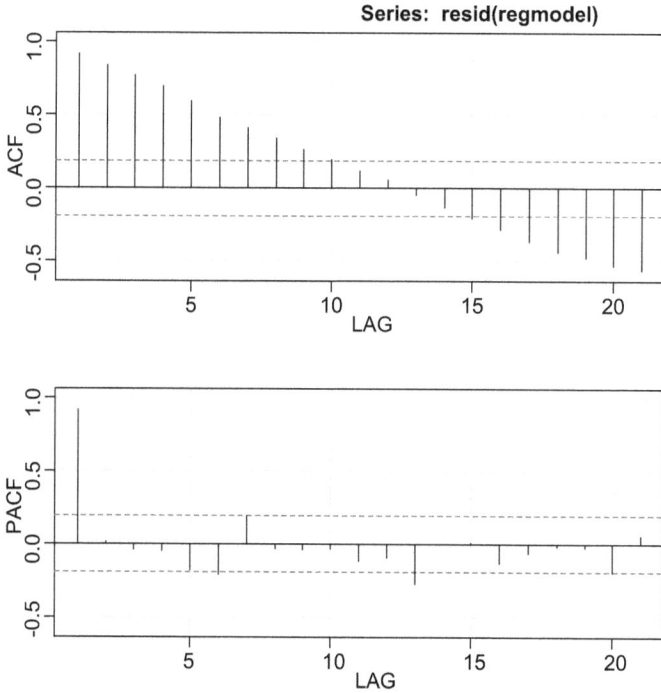

FIGURE 12.5
ACF and PACF graphs for cancer diagnosis by year data. ACF, autocorrelation function; PACF, partial autocorrelation.

When assessing the independent errors assumption of a regression model, we can consider both the autocorrelations and D.

- Prior to looking at the results of an AR(1) model, let's be sure that this is the appropriate model.
- The Ljung–Box Q statistic assesses H_0: the residuals are independent.

$$Q = n(n+2) \sum_{k=1}^{h} \frac{\rho_k^2}{n-k}$$

where n is the sample size, k is the lag, and ρ is the autocorrelation.

- The results here show that the residuals have little to no autocorrelation, suggesting that we have successfully modeled the data.

Let's consider an example of time series regression. The state of California collected data on the number of highway fatalities per million miles driven, and we will examine data for the years between 1983 and 1997. Figure 12.6 displays the number of traffic fatalities per million miles driven in the state of California between 1983 and 1997.

FIGURE 12.6
Number of traffic fatalities per million miles driven for California: 1983–1997.

We can see that the fatality rate per million miles driven declined over time. A researcher is interested in investigating whether there is a relationship between the safety belt use rate in the state and the fatality rate. This question is of interest because there has been research showing that at the individual level, wearing a seatbelt is associated with a higher survival rate in automotive accidents. Thus, it is of interest to determine whether this micro-level relationship translates into lower fatality rates if the safety belt use rate increases, which is precisely what occurred in California in the 1980s and 1990s.

Figure 12.7 is a scatterplot showing the relationship between seatbelt use and the fatality rate.

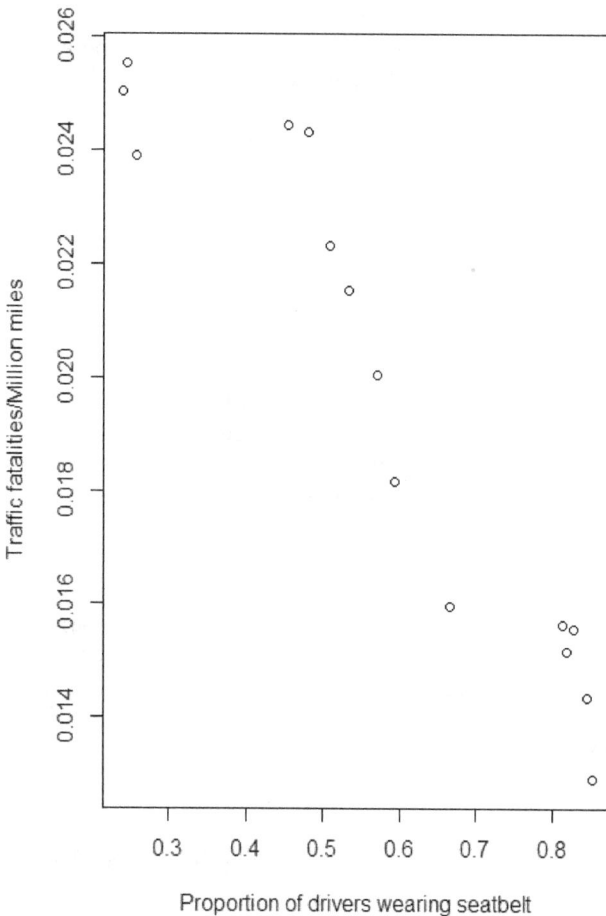

FIGURE 12.7
Scatterplot for relationship between seatbelt use rate and fatality rate in California: 1983–1997.

It does appear that the fatality rate is lower when the seatbelt use rate is larger. The correlation between the two variables is −0.94, indicating that this relationship is quite strong.

OLS linear regression would seem to be a useful tool for further examining the relationship between seatbelt use and fatality rates. The regression results appear in Table 12.10.

The regression analysis reinforces our finding that there is a negative relationship between the proportion of drivers wearing seatbelts and the fatality rate.

An important issue for us to consider in this example is the potential serial correlation of the fatality rate due to the fact that the data are collected over time. As we discussed above, it is possible that the OLS assumption of independent errors can be violated due to relationships of measurements that are adjacent in time. We can assess this assumption using a variety of tools, including the ACF, PACF, and Durbin–Watson statistic. For the seatbelt proportion model, $D = 1.16$, $p = 0.028$, leading us to reject the null hypothesis that the first-order autocorrelation is 0. The first-order ACF and PACF were 0.36, suggesting a moderate-sized (Cohen, 1988) relationship between fatality rates in adjacent years. The second-order ACF and PACF values were −0.06 and −0.22, respectively. Thus, once we account for the first-order autocorrelation (PACF), the relationship between fatality rates that are 2 years apart is smaller than for those in adjacent years.

One approach for attempting to deal with autocorrelation is to include the year as an independent variable in the model to account for the time effect. As we will see, this approach does not actually deal with this problem, but it is often used in practice. The results of this model appear in Table 12.11.

The relationship between seatbelt use proportion and the fatality rate is no longer statistically significant when we include the year, which has a

TABLE 12.10

Linear Regression Results for the Model Relating the Seatbelt Use Proportion and the Fatality Rate Per Million Miles Driven in California: 1983–1997

Variable	Coefficient	Standard Error	t	p
Intercept	0.03	0.001	27.13	<0.001
Seatbelt proportion	−0.02	0.002	−10.41	<0.001

TABLE 12.11

Linear Regression Results for the Model Relating the Seatbelt Use Proportion and Year with the Fatality Rate Per Million Miles Driven in California: 1983–1997

Variable	Coefficient	Standard Error	t	p
Intercept	2.30	0.55	4.19	0.001
Seatbelt proportion	0.003	0.006	0.631	0.54
Year	−0.001	0.0003	−4.14	0.001

significant negative relationship with the fatality rate. Therefore, we would conclude that the fatality rate per million miles driven declined over time and that the safety belt use rate is not associated with the fatality rate. The Durbin–Watson results were $D = 0.84$, $p = 0.01$, with ACF and PACF values of 0.49 for lag-1 and 0.13/−0.15 for lag-2. Thus, the assumption of independent errors does not seem to have been met simply through the inclusion of the year variable.

As we described above, an alternative approach to dealing with serial correlation, such as that in the current dataset, is to fit an autoregressive model, which includes the lagged error term as a predictor of the outcome variable. Given the results from the OLS model, there definitely appears to be a lag-1 autoregression and perhaps one for lag-2, as well, though this is less certain. The results for the lag-2 autoregression model appear in Table 12.12.

The estimate of the autocorrelation is 0.49, though it was not statistically significant. In addition, the seatbelt use proportion was found to be statistically significantly related to the fatality rate with a coefficient of −0.02. Thus, we would conclude that after accounting for the first-order autocorrelation in the residuals due to time, a higher rate of seatbelt use in the state was associated with a decrease in the number of fatalities per million miles driven.

We will conclude this analysis by fitting the autoregression model with both the lag-1 and lag-2 term2 (AR2 model). The results for this model appear in Table 12.13.

The second-order autoregressive term (AR2) was not statistically significant, though the first-order term (AR1) was. Thus, we would conclude that

TABLE 12.12

Results for the AR1 Model Relating the Seatbelt Use Proportion and Year with the Fatality Rate Per Million Miles Driven in California: 1983–1997

Variable	Coefficient	Standard Error	t	p
Intercept	0.03	0.002	12.02	<0.001
Seatbelt proportion	−0.02	0.004	−4.47	0.001
AR1	0.49	0.35	1.38	0.19

AR1, first-order autoregressive term.

TABLE 12.13

Results for the AR2 Model Relating the Seatbelt Use Proportion and Year with the Fatality Rate Per Million Miles Driven in California: 1983–1997

Variable	Coefficient	Standard Error	t	p
Intercept	0.29	0.17	1.70	0.12
Seatbelt proportion	−0.01	0.005	1.70	0.12
AR1	1.32	0.25	5.19	0.0003
AR2	−0.43	0.26	−1.67	0.12

AR1, first-order autoregressive term; AR2, second-order autoregressive term.

there is no need to include AR2, and we will rely on the AR1 model results to understand the relationship between seatbelt use proportion and the fatality rate after accounting for the autocorrelation due to time.

As we said above, the area of time series analysis is quite broad and encompasses a wide variety of models and applications, particularly in the area of forecasting. In this chapter, we have only touched on the very basics of one small piece of this major area of statistics. For many basic applications, what we have discussed here will be sufficient. However, it is important to remember that this rich area of statistical practice and theory exists so that if you are in need of it, you will know how to access it.

Summary

The purpose of this chapter was to provide you with an introduction to several extensions of the regression model that might prove useful for specific research scenarios that are frequently encountered in practice. These situations include the selection of independent variables to retain in our model when we are faced with a large possible number of them to consider. We saw that we can make such decisions based on optimizing fit statistics or by using regularization techniques in which model coefficients are driven toward 0 for all variables so that only those having the strongest relationships with the dependent variable are retained. In addition, we learned that BMA offers an alternative that would not require us to choose the variables to retain in the model. Rather, all of the variables are included, and their relationships with the dependent variable are averaged across all possible models. We then considered techniques for estimating nonlinear relationships between the independent and dependent variables using polynomial regression, LOWESS, and smoothing splines. Systems of regression models were then featured in the form of mediation and moderation. These models allow us to consider more complex relationships among the variables of interest, including scenarios in which the relationship of an independent and dependent variable goes through a mediator, or when the relationship differs depending on the value of a moderator. We concluded the chapter by considering how we might order the independent variables in terms of the strength of their relationships with the outcome and with an examination of modeling time-dependent data.

The purpose of this chapter was simply to provide an introduction to some very important topics in the area of regression. This coverage was not intended to be exhaustive, and indeed each of these topics could easily warrant a chapter of its own. However, it is hoped that the material presented here will spark the interest of many readers and lead them to do further investigations on their own from among the many excellent resources available that describe these techniques. And, for relatively straightforward problems, the discussion in this chapter should be sufficient for conducting some basic analyses. There are many exciting and useful applications of regression models, and we hope that this chapter has made that point clear.

13

Multilevel Modeling

Throughout this book, we have considered a variety of models and data analysis strategies that researchers can use to address hypotheses about populations of interest. For example, in Chapters 4–7 we learned how the *t*-test and analysis of variance (ANOVA) can be used to compare the mean of a dependent variable across two or more groups. We then turned our attention to questions about the relationships among variables using correlation coefficients and linear regression. We extended this work to include situations in which the outcome variable of interest is dichotomous (Chapter 11). In the previous chapter, we examined a variety of applications for and extensions to the regression modeling framework. In this chapter, we will consider yet another modeling framework built upon the general linear model that can be used when the individual observations are clustered together. Such clustering can come in the form of students being sampled from schools, employees from businesses, or patients from health clinics, among others. As we will see, such clustering leads to a violation of the independence assumption, which is fundamental to the correct use of most of the techniques that we have discussed in the previous chapters.

The assumption of independence essentially means that there are no relationships among individuals in the sample for the dependent variable, *once the independent variables in the analysis are accounted for*. In the examples we have presented in the previous chapters, this assumption has always been met. However, in many cases, the method used for selecting the sample does lead to correlated responses among individuals. For example, a researcher interested in the impact of a new teaching method on student achievement might randomly select schools for placement in either a treatment or control group. If school A is placed into the treatment condition, all students within the school will also be in the treatment condition. It would be reasonable to assume that the school itself, above and beyond the treatment condition, would have an impact on the performance of the students. This impact would manifest itself as correlations in achievement test scores among individuals attending that school. Thus, if we were to use a simple one-way ANOVA to compare test score means for the treatment and control groups with such cluster sampled data, we would likely be violating the assumption of independent errors because a factor beyond the treatment condition (in this case the school) would have an additional impact on the outcome variable.

We typically refer to the data structure described above as nested, meaning that individual data points at one level (e.g., student) appear in only one level of a higher-level variable such as school. Thus, students are nested within the

DOI: 10.1201/9781003379324-13

school. Such designs can be contrasted with a crossed data structure whereby individuals at the first level appear in multiple levels of the second variable. In our example, students might be crossed with after-school organizations if they are allowed to participate in more than one. For example, a given student might be on the basketball team as well as in the band. The focus of this book has been exclusively on crossed data structures up to this point. In this chapter, we will turn our attention to nested data, such as in our example.

When researchers apply standard statistical methods to multilevel data, such as the regression model described in Chapter 1, the assumption of independent errors is violated. For example, if we have achievement test scores from a sample of students who attend several different schools, it would be reasonable to believe that those attending the same school will have scores that are more highly correlated with one another than they are with scores from students attending other schools. This within-school correlation would be due, for example, to having a common set of teachers, a common teaching curriculum, coming from a common community, and a single set of administrative policies, among numerous other reasons. The within-school correlation will in turn result in an inappropriate estimate of the standard errors for the model parameters, which in turn will lead to errors of statistical inference, such as p-values smaller than they really should be and the resulting rejection of null effects above the stated Type I error rate, regarding the parameters. Recalling our discussion in Chapter 1, the test statistic for the null hypothesis of no relationship between the independent and dependent variable is simply the regression coefficient divided by the standard error. If the standard error is underestimated, this will lead to an overestimation of the test statistic, and therefore statistical significance for the parameter in cases where it should not be; i.e., a Type I error at a higher rate than specified. Indeed, the underestimation of the standard error will occur unless τ^2 is equal to 0.

In addition to the underestimation of the standard error, another problem with ignoring the multilevel structure of data is that we may miss important relationships involving each level in the data. Recall that in our example, there are two levels of sampling: students (level 1) are nested in schools (level 2). Specifically, by *not* including information about the school, for example, we may well miss important variables at the school level that help to explain performance at the examinee level. Therefore, beyond the known problem with misestimating standard errors, we also proffer an incorrect model for understanding the outcome variable of interest.

Intraclass Correlation

In cases where individuals are clustered or nested within a higher-level unit (e.g., classrooms, schools, school districts), it is possible to estimate the correlation among scores within the cluster/nested structure using the intraclass correlation (ICC) (denoted ρ_I in the population). Alternatively, ρ_I can also be interpreted as a measure of the proportion of variation in the outcome variable that occurs between groups versus the total variation present. In either case, ρ_I ranges between 0 and 1, with higher values indicating a stronger within-cluster correlation and a greater proportion of variance in the dependent variable associated with cluster membership. It is calculated as

$$\rho_I = \frac{\tau^2}{\tau^2 + \sigma^2} \tag{13.1}$$

where τ^2 is the population variance between clusters and σ^2 is the population variance within clusters.

It is possible to estimate τ^2 and σ^2 using sample data and then use these values to estimate ρ_I. These estimates are similar to the sum of squared terms that we used in calculating the F statistic for ANOVA (Chapter 5). The sample estimate for variation within clusters is simply

$$\hat{\sigma}^2 = \frac{\sum_{j=1}^{C}(n_j - 1)S_j^2}{N - C} \tag{13.2}$$

where S_j^2 is the variance within cluster $j = \dfrac{\sum_{j=1}^{n_j}(y_{ij} - \bar{y}_j)}{n_j - 1}$, n_j is the sample size

for cluster j, N is the total sample size, and C is the total number of clusters.

In other words, σ^2 is simply the weighted average of within-cluster variances.

Estimation of τ^2 involves a few more steps but is not much more complex than what we have seen for σ^2. To obtain the sample estimate for variation between clusters, $\hat{\tau}^2$, we must first calculate the weighted between-cluster variance.

$$\hat{S}_B^2 = \sum_{j=1}^{C} \frac{n_j(\bar{y}_j - \bar{y})^2}{\tilde{n}(C - 1)} \tag{13.3}$$

where $\bar{y}_j$ is the mean on response variable for cluster j and $\bar{y}$ is the overall mean on response variable.

$$\tilde{n} = \frac{1}{C - 1}\left[N - \frac{\sum_{j=1}^{C} n_j^2}{N}\right]$$

We cannot use S_B^2 as a direct estimate of τ^2 because it is impacted by the random variation among subjects within the same clusters. Therefore, to remove this random fluctuation, we will estimate the population variance between clusters as

$$\hat{\tau}^2 = S_B^2 - \frac{\hat{\sigma}^2}{\tilde{n}}$$ (13.4)

Using these variance estimates, we can in turn calculate the sample estimate of ρ_I

$$\hat{\rho}_I = \frac{\hat{\tau}^2}{\hat{\tau}^2 + \hat{\sigma}^2}$$ (13.5)

Note that equation (13.5) assumes that the clusters are of equal size. Clearly, such will not always be true, in which case this equation will not hold. However, the purpose of its inclusion here is to demonstrate the principle underlying the estimation of ρ_I, the basic principles of which hold even as the equation might change. Note that ρ_I is also referred to as the ICC, which we will do throughout this chapter.

Given that $\hat{\rho}_I$ is a sample estimate, we know that it is subject to sampling variation, which can be estimated with a standard error, which is calculated as

$$S_{\rho_I} = (1 - \rho_I)(1 + (n-1)\rho_I)\sqrt{\frac{2}{n(n-1)(N-1)}}$$ (13.6)

The terms in (13.6) are as defined previously, and the assumption is that all clusters are of equal size. As noted earlier in the chapter, this latter condition is not a requirement; however, an alternative formulation exists for cases in which it does not hold. However, equation (13.6) provides sufficient insight for our purposes into the estimation of the standard error of the ICC. We can use the standard error to test hypotheses about the ICC and to create confidence intervals for it.

The ICC is an important tool in multilevel modeling, in large part because it is an indicator of the degree to which the multilevel data structure might impact the outcome variable of interest. Larger values of the ICC are indicative of a greater impact of clustering. Thus, as the ICC increases in value, we must be more cognizant of employing multilevel modeling strategies in our data analysis. In the next section, we will discuss the problems associated with ignoring this multilevel structure before we turn our attention to methods for dealing with it directly.

Multilevel Linear Models

In the following section, we will review some of the core ideas that underlie multilevel linear models (MLM). We will first focus on the difference between random and fixed effects, after which we will discuss the basics of parameter estimation, focusing on the two most commonly used methods, maximum likelihood (ML) and restricted maximum likelihood (REML), and conclude with a review of assumptions underlying MLMs and an overview of how they are most frequently used. Included in this section, we will also address the issue of centering and explain why it is an important concept in MLM. We will then consider alternative estimation techniques for cases where outliers are present and/or the data are not normally distributed and then conclude with a full example demonstrating the various approaches.

Random Intercept Model

As we transition from the single-level regression framework of Chapter 9 to the MLM context, let's first revisit the basic simple linear regression model:

$$y = \beta_0 + \beta_1 x + \varepsilon \tag{13.7}$$

Here, the dependent variable y is expressed as a function of an independent variable, x, a slope, β_1, an intercept, β_0, and random variation from subject to subject, ε. We defined the intercept as the conditional mean of y when the value of x is 0. In the context of a single-level regression model such as this, there is one intercept that is common to all individuals in the population of interest. However, when individuals are clustered (e.g., within classrooms, schools, or organizational units within a company), there will potentially be a separate intercept for each cluster; that is, there may be different means for the dependent variable for $x=0$ across the different clusters. We say *potentially* here because if there is in fact no cluster effect, then the single intercept model will suffice. In practice, assessing whether there are different means across the clusters is an empirical question, which we describe below.

Allowing for cluster-specific intercepts leads to the following model:

$$y_{ij} = \beta_{0j} + \varepsilon_{ij} \tag{13.8}$$

Here, the j subscript refers to cluster membership for individual i. The intercept term (β_{0j}) is said to be random in this case because it is allowed to vary across the J clusters. We can express β_{0j} as

$$\beta_{0j} = \gamma_{00} + U_{0j} \tag{13.9}$$

In this framework, γ_{00} represents an average or general intercept value that holds across clusters, whereas U_{0j} is a group-specific effect on the intercept. We can think of γ_{00} as a fixed effect because it remains constant across all clusters and U_{0j} as a random effect because it varies from cluster to cluster.

Therefore, for an MLM, we are interested not only in some general mean value for y when x is 0 for all individuals in the population (γ_{00}) but also in the deviation between the overall mean and the cluster-specific effects for the intercept (U_{0j}). If we go on to assume that the clusters are a random sample from the population of all possible clusters (e.g., all possible schools), then we can treat the U_{0j} as a part of the residual effect on y_{ij}, very similar to how we think of ε. In that case, U_{0j} is assumed to be drawn randomly from a population with a mean of 0 (recall U_{0j} is a deviation from the fixed effect) and a variance, τ^2. Furthermore, we assume that τ^2 and σ^2, the variance of ε, are uncorrelated. In addition to its role in calculating the ICC, τ^2 can also be viewed as the impact of the cluster on the dependent variable. Therefore, testing it for statistical significance is equivalent to testing the null hypothesis that the cluster (e.g., school) has no impact on the dependent variable. Substituting equation (13.9) in place of the intercept in the linear regression model gives us the full random intercept model

$$y = \gamma_{00} + U_{0j} + \beta_1 x + \varepsilon \tag{13.10}$$

Often in MLM, we begin our analysis of a dataset with this simple random intercept model, known as the null model, which takes the form

$$y_{ij} = \gamma_{00} + U_{0j} + \varepsilon_{ij} \tag{13.11}$$

While the null model does not provide information regarding the impact of specific independent variables on the dependent variable, it does yield important information regarding how variation in y is partitioned between variance among the individuals σ^2 and variance among the clusters τ^2. The total variance of y is simply the sum of σ^2 and τ^2. In addition, as we have already seen, these values can be used to estimate the ICC. The null model, as we will see later in the chapter, is also used as a baseline for model building and comparison.

Random Slopes Model

It is a simple matter to expand the random intercept model to accommodate one or more independent predictor variables. As an example, if we add a single predictor (x_{ij}) at the individual level (level 1) to the model, we obtain

$$y_{ij} = \gamma_{00} + \gamma_{10} x_{ij} + U_{0j} + \varepsilon_{ij} \tag{13.12}$$

This model can also be expressed in two separate levels as

$$\text{Level-1}: y_{ij} = \beta_{0j} + \beta_{1j} x + \varepsilon_{ij} \tag{13.13}$$

$$\text{Level-2}: \beta_{0j} = \gamma_{00} + U_{0j} \tag{13.14}$$

$$\beta_{1j} = \gamma_{10} \tag{13.15}$$

This model now includes the predictor and the slope relating it to the dependent variable, γ_{10}, which we acknowledge as being at level-1 by the subscript 10. We interpret γ_{10} in the same way that we did β_1 in the linear regression model; i.e., a measure of the impact on y of a 1-unit change in x. In addition, we can estimate the ICC just as we did before, though now it reflects the correlation between individuals from the same cluster after controlling for the independent variable, x. In this model, both γ_{10} and γ_{00} are fixed effects, while σ^2 and τ^2 remain random.

One implication of the model in (13.12) is that the dependent variable is impacted by variation among individuals (σ^2), variation among clusters (τ^2), an overall mean common to all clusters (γ_{00}), and the impact of the independent variable as measured by γ_{10}, which is also common to all clusters. In practice, there is no reason that the impact of x on y would need to be common for all clusters, however. In other words, it is entirely possible that rather than having a single γ_{10} common to all clusters, there is actually a unique effect for the cluster of $\gamma_{10} + U_{1j}$, where γ_{10} is the average relationship of x with y across clusters, and U_{1j} is the cluster-specific variation of the relationship between the two variables. This cluster-specific effect is assumed to have a mean of 0 and to vary randomly around γ_{10}. The random slopes model is

$$y_{ij} = \gamma_{00} + \gamma_{10}x_{ij} + U_{0j} + U_{1j}x_{ij} + \varepsilon_{ij} \tag{13.16}$$

Written in this way, we have separated the model into its fixed ($\gamma_{00} + \gamma_{10}x_{ij}$) and random ($U_{0j} + U_{1j}x_{ij} + \varepsilon_{ij}$) components. The model in equation (13.16) states that there is an interaction between the cluster and x, such that the relationship of x and y is not constant across clusters.

Heretofore we have discussed only one source of between-group variation, which we have expressed as τ^2, and which is the variation among clusters in the intercept. However, the random slopes model adds a second such source of between-group variance in the form of U_{1j}, which is cluster variation on the slope relating the independent and dependent variables. To differentiate between these two sources of between-group variance, we now denote the variance of U_{0j} as τ_0^2 and the variance of U_{1j} as τ_1^2. Furthermore, within clusters, we expect U_{1j} and U_{0j} to have a covariance of τ_{01}. However, across different clusters, these terms should be independent of one another (i.e., their correlation is 0), and in all cases, it is assumed that ε remains independent of all other model terms. In practice, if we find that τ_1^2 is not 0, we must be careful in describing the relationship between the independent and dependent variables, as it is not the same across clusters.

Given the variance components in equation (13.16), it is possible to estimate the variances associated with U_{1j} and U_{0j}, τ_1^2 and τ_0^2, respectively. Assuming that the clusters (e.g., schools) have the same number of level-1 units (e.g., students), the calculation of these variances is a straightforward matter for the slopes:

$$\frac{\sum \left(U_{1j} - \bar{U}_1\right)^2}{J-1} \tag{13.17}$$

An analogous equation can be used to estimate the variance associated with the intercept random variance.

Centering

Centering simply refers to the practice of subtracting the mean of a variable from each individual value. This resets the mean for the sample of the centered variables to 0 and implies that each individual's (centered) score represents a deviation from the mean, rather than whatever meaning its raw value might have. In the context of regression, centering is commonly used, for example, to reduce collinearity caused by including an interaction term in a regression model. If the raw scores of the independent variables are used to calculate the interaction, and then both the main effects and interaction terms are included in the subsequent analysis, it is very likely that collinearity will cause problems in the standard errors of the model parameters. Centering is a way to help avoid such problems (e.g., Iversen, 1991). Such issues are also important to consider in MLM, in which interactions are frequently employed. In addition, centering is also a useful tool for avoiding collinearity caused by highly correlated random intercepts and slopes in MLMs (Wooldridge, 2004). Finally, centering provides a potential advantage in terms of interpretation of results. Recall that the intercept is the value of the dependent variable when the independent variable is set equal to 0. In many applications, the independent variable cannot reasonably be 0 (e.g., a math test score); however, this essentially renders the intercept as a necessary value for fitting the regression line but not one that has a readily interpretable value. However, when x has been centered, the intercept takes on the value of the dependent variable when the independent is at its mean. This is a much more useful interpretation for researchers in many situations, and yet another reason why centering is an important aspect of modeling, particularly in the multilevel context.

Probably the most common approach to centering is to calculate the difference between each individual's score and the overall, or grand, mean across the entire sample. This *grand mean centering* is not the only way to center our data, however. An alternative approach, known as *group mean centering*, involves calculating the difference between each individual score and the mean of the cluster to which they belong. In our school example, grand mean centering would involve calculating the difference between each score and the overall mean across schools, while group mean centering would lead the researcher to calculate the difference between each score and the mean for their school. While there is some disagreement in the literature regarding which approach might be best at reducing the harmful effects of collinearity (Raudenbush & Bryk, 2002; Snijders & Bosker, 1999), researchers have demon strated that in most cases either will work well in this regard (Kreft, de Leeuw, & Aiken, 1995). Therefore, the choice of which approach to use must be made on substantive grounds regarding the nature of the relationship between *x* and *y*. By using grand mean centering, we are implicitly comparing individuals to one another (in the form of the overall mean) across the entire sample. However, when using group mean centering, we are placing each individual in a relative position on *x* within their cluster. Thus, in our school example, using the group mean-centered values of the test scores would reflect an individual's relative vocabulary score in their school. In contrast, the use of grand mean centering would reveal one's relative standing in the sample as a whole. This latter interpretation would be equivalent conceptually (though not mathematically) to using the raw score, while the group mean centering would not. Throughout the rest of this chapter, we will use grand mean centering, per recommendations by Hox (2002), among others. We should remember, however, that there are some applications in which interpretation of the impact of an individual's relative standing in their cluster might be more useful than their relative standing in the sample as a whole.

Basics of Parameter Estimation with MLMs

In the context of linear regression, we described the estimation of model parameters using OLS. However, as we move from these fairly simple applications to more complex models, OLS is not typically the optimal approach to use for parameter estimation. Instead, we will rely on ML estimation and REML. In the following section, we review these approaches to estimation from a conceptual basis, focusing on the generalities of how they work, what they assume about the data, and how they differ from one another. For the technical details, we refer the interested reader to Bryk and Raudenbush (2002) or de Leeuw and Meijer (2008), both of which provide excellent resources for those desiring a more in-depth coverage of these methods. Our purpose here is to provide the reader with a conceptual understanding that will aid in their understanding of the application of MLM in practice.

ML has as its primary goal the estimation of population model parameters that maximize the likelihood of our obtaining the sample that we in fact obtained. In other words, the estimated parameter values should maximize the likelihood of our particular sample. From a practical perspective, identifying such sample values takes place through the comparison of the observed data with that predicted by the model associated with the parameter values. The closer the observed and predicted values are to one another, the greater the likelihood that the observed data arose from a population with parameters close to those used to generate the predicted values. In practice, ML is an iterative methodology in which the algorithm searches for those parameter values that will maximize the likelihood of the observed data (i.e., produce predicted values that are as close as possible to the observed) and, as such, can be computationally intensive, particularly for complex models and large samples.

REML estimation, which is a variant of ML, has been shown to be more accurate with regard to the estimation of variance parameters than ML (Kreft & De Leeuw, 1998). In particular, the two methods differ with respect to how degrees of freedom are calculated in the estimation of variances. As we saw in Chapter 1, the sample variance is typically calculated by dividing the sum of squared differences between individual values and the mean by the number of observations minus 1. This is a REML estimate of variance. In contrast, the ML variance is calculated by dividing the sum of squared differences by the total sample size, leading to a smaller variance estimate than REML and, in fact, one that is biased in finite samples. In the context of multilevel modeling, REML takes into account the number of parameters being estimated in the model when determining the appropriate degrees of freedom for the estimation of the random components such as the parameter variances described above. In contrast, ML does not account for these, which can lead to an underestimation of the variances that does not occur with REML. For this reason, REML is generally the preferred method for estimating multilevel models, though for testing variance parameters (or any random effect), it is necessary to use ML (Snijders & Bosker, 1999). It should be noted that as the number of level-2 clusters increases, the difference in value for ML and REML estimates becomes very small (Snijders & Bosker, 1999).

Assumptions Underlying MLMs

As with any statistical model, the appropriate use of MLMs requires that several assumptions about the data hold true. If these assumptions are not met, the model parameter estimates may not be trustworthy, just as would be the case with standard linear regression. Indeed, while they differ somewhat from the assumptions for the single-level models, the assumptions underlying MLM are akin to those for the single-level models that have been the focus in most of this book. When these assumptions are not met, we may want to consider using one of the alternative estimation approaches described later in this chapter.

The first assumption underlying MLM is that the level-2 residuals are independent between clusters. In other words, there is an assumption that the random intercept and slope(s) at level-2 are independent of one another across clusters. Second, the level-2 intercepts and coefficients are assumed to be independent of the level-1 residuals; i.e., the errors for the cluster-level estimates are unrelated to errors at the individual level. Third, we must assume that the level-1 residuals are normally distributed and have a constant variance. This assumption is very similar to the one we make about residuals in the standard linear regression model. Fourth, the level-2 intercept and slope(s) are assumed to come from a multivariate normal distribution with a constant covariance matrix. Each of these assumptions can be directly assessed for a sample, as we shall see later in this chapter. Indeed, the methods for checking the MLM assumptions are not very different from those for checking the regression model that we used in Chapter 9.

Extending 2 Level MLMs

To this point, we have described the specific terms that make up the MLM, including the level-1 and level-2 random effects and residuals. We will close out this portion of the chapter by extending the model to include multiple independent variables at both level-1 and level-2. Previously in equation (13.16), we considered the random slopes model $y_{ij} = \gamma_{00} + \gamma_{10}x_{ij} + U_{0j} + U_{1j}x_{ij} + \varepsilon_{ij}$ in which the dependent variable, y_{ij} is a function of an independent variable x_{ij}, as well as a random error at both the examinee and school level. We can extend this model a bit further by including multiple independent variables at both level-1 (e.g., student) and level-2 (e.g., school). Thus, for example, in addition to estimating the relationship between an individual's vocabulary and reading scores, we can also determine the degree to which the average vocabulary score at the school as a whole is related to an individual's reading score. This model has two parts, one explaining the relationship between the individual-level vocabulary (x_{ij}) and reading, and the other explaining the coefficients at level-1 as a function of the level-2 predictor, average vocabulary score (z_j). The two parts of this model are expressed as

$$\text{Level-1}: y_{ij} = \beta_{0j} + \beta_{1j}x_{ij} + \varepsilon_{ij} \tag{13.18}$$

and

$$\text{Level-2}: \beta_{0j} = \gamma_{h0} + \gamma_{h1}z_j + U_{hj} \tag{13.19}$$

The additional piece of equation (13.19) is $\gamma_{h1}z_j$, which represents the slope for (γ_{h1}), and the value of the level-2 independent variable (z_j). We can combine equations (13.18) and (13.19) to obtain a single equation for the two-level MLM:

$$y_{ij} = \gamma_{00} + \gamma_{10}x_{ij} + \gamma_{01}z_j + \gamma_{1001}x_{ij}z_j + U_{0j} + U_{1j}x_{ij} + \varepsilon_{ij} \tag{13.20}$$

Most of these model terms have been defined previously in the chapter. The additional pieces of equation (13.20) are γ_{01} and γ_{11}, where γ_{01} represents the fixed intercept of the level-2 variable z on the outcome, and γ_{11} represents the fixed level-2 slope. In addition, equation (13.20) also includes a cross-level interaction, $\gamma_{1001}x_{ij}z_j$. As the name implies, the cross-level interaction is simply the interaction between the level-1 and level-2 predictors. The coefficient for this interaction term, γ_{1001}, assesses the extent to which the relationship between the level-1 independent variable and the outcome is moderated by the level-2 variable. A large statistically significant value for this coefficient would indicate the presence of such a moderating effect.

Estimating the Variance in the Dependent Variable Accounted for by the MLM

As was the case with single-level regression models, it is possible to estimate the proportion of variance in the outcome variable that is accounted for by each level of the MLM. In Chapter 9, we saw that with single-level OLS regression models, the proportion of the dependent variable's variance accounted for by the model is expressed as R^2. In the context of multilevel modeling, identifying the proportion of variance accounted for by the independent variables is complicated by the need to partition the total explained variance into within-cluster and between-cluster components. There are multiple approaches for calculating these R^2 values. One technique provides estimates for each level of the model (Snijders & Bosker, 1999). For level 1, we can calculate

$$R_1^2 = 1 - \frac{\sigma_{M1}^2 + \tau_{M1}^2}{\sigma_{M0}^2 + \tau_{M0}^2} \tag{13.21}$$

We can also calculate a level-2 R^2:

$$R_2^2 = 1 - \frac{\sigma_{M1}^2/B + \tau_{M1}^2}{\sigma_{M0}^2/B + \tau_{M0}^2} \tag{13.22}$$

where B is the average size of the level-2 units.

Raudenbush and Bryk (2002) proposed a different approach for estimating the variance accounted for by the model at both levels 1 and 2. The level-1 value is calculated as

$$RB_1 = \frac{\sigma^2_{M0e} - \sigma^2_{M1e}}{\sigma^2_{M0e}} \tag{13.23}$$

where σ^2_{M0e} is the error variance for the null model and σ^2_{M1e} is the error variance for the model including fixed and random slope effects.

The level-2 R^2 is calculated as

$$RB_2 = \frac{\sigma^2_{M0U0} - \sigma^2_{M1U0}}{\sigma^2_{M0U0}} \tag{13.24}$$

where σ^2_{M0U0} is the random intercept variance for null model and σ^2_{M1U0} is the random intercept variance for model including fixed and random slope effects.

LaHuis et al. (2014) proposed an approach for estimating the R^2 in a MLM that is based on the variance in model-predicted values of the outcome variable. The equation for this statistic takes the form

$$R^2_{MVP} = \frac{\sigma^2_{\hat{Y}}}{\sigma^2_{\hat{Y}} + \sigma^2_{M1U0} + \sigma^2_{M1e}} \tag{13.25}$$

where $\sigma^2_{\hat{Y}}$ is the variance in predicted values of the outcome variable.

LaHuis et al. conducted a simulation study comparing a number of these methods and found that they all provide relatively similar results and levels of accuracy.

An alternative framework for calculating R^2 in the context of MLMs was described by Rights and Sterba (2019). These authors note that prior work in the area of variance accounted for with MLMs had several shortcomings, chief among these being a lack of agreement on how to interpret this construct, as well as gaps in the sources of variance that can be represented in the measures. To address these problems, Rights and Sterba aimed to develop a more comprehensive and well-defined framework for MLM R^2. We will not delve into the full technical details of this approach here, but the interested reader is encouraged to read the paper by Rights and Sterba, which describes the approach in full detail.

Rights and Sterba (2019) defined the variance explained by level-1 fixed effects ($f1$), level-2 fixed effects ($f2$), random coefficient variance (v), and random intercept variance (m) as below:

$$R^{2(f1)}_{total} = \frac{\sigma^2_{f1}}{\sigma^2_{f1} + \sigma^2_{f2} + \sigma^2_{v} + \sigma^2_{m} + \sigma^2_{r}} \tag{13.26}$$

$$R^{2(f2)}_{total} = \frac{\sigma^2_{f2}}{\sigma^2_{f1} + \sigma^2_{f2} + \sigma^2_{v} + \sigma^2_{m} + \sigma^2_{r}} \tag{13.27}$$

$$R_{\text{total}}^{2(v)} = \frac{\sigma_v^2}{\sigma_{f1}^2 + \sigma_{f2}^2 + \sigma_v^2 + \sigma_m^2 + \sigma_r^2} \tag{13.28}$$

$$R_{\text{total}}^{2(m)} = \frac{\sigma_m^2}{\sigma_{f1}^2 + \sigma_{f2}^2 + \sigma_v^2 + \sigma_m^2 + \sigma_r^2} \tag{13.29}$$

The residual variance (σ_r^2) includes all of the variability in the outcome that is not accounted for by the fixed effects, as well as random intercept and slope variance.

Multilevel Model Example

We will demonstrate the use of MLMs with a dataset containing reading test scores for 334 students attending four schools in a single district. We are interested in examining relationships between the reading scores, student gender (male or female), age, and an index of socioeconomic status (higher values indicate a higher SES). The histogram of reading scores appears in Figure 13.1.

The histogram is negatively skewed, with most students having scores of 10 or more but some having scores below 8. The descriptive statistics appear in Table 13.1.

The skewness estimate is −0.65, suggesting a moderate level of negative skewness in the scores. The kurtosis value of 2.52 indicates that the data are slightly platykurtic.

The boxplot of reading test scores by gender appears in Figure 13.2. We can see from this graph that in this sample, females had a somewhat higher median score than males.

Females had a higher median reading test score in this sample, and the two groups have similar amounts of variability.

The scatterplots for reading by SES and age (standardized) appear in the two panels of Figure 13.3. For each plot, there is a simple regression line included as well.

These plots suggest that, for this sample, there exists a positive relationship between both SES and age and reading test scores. However, it should be noted that we do not know whether this relationship generalizes to the population and will need to explore this issue using the MLM.

The process of investigating multilevel data using MLMs involves fitting a sequence of models with increasing levels of complexity. The fits of the models to the data are then compared to one another. The first of these models includes no independent variables and is referred to as the null model (Model0). The second model includes the set of independent variables (Model1) and the third model includes the independent variables and any

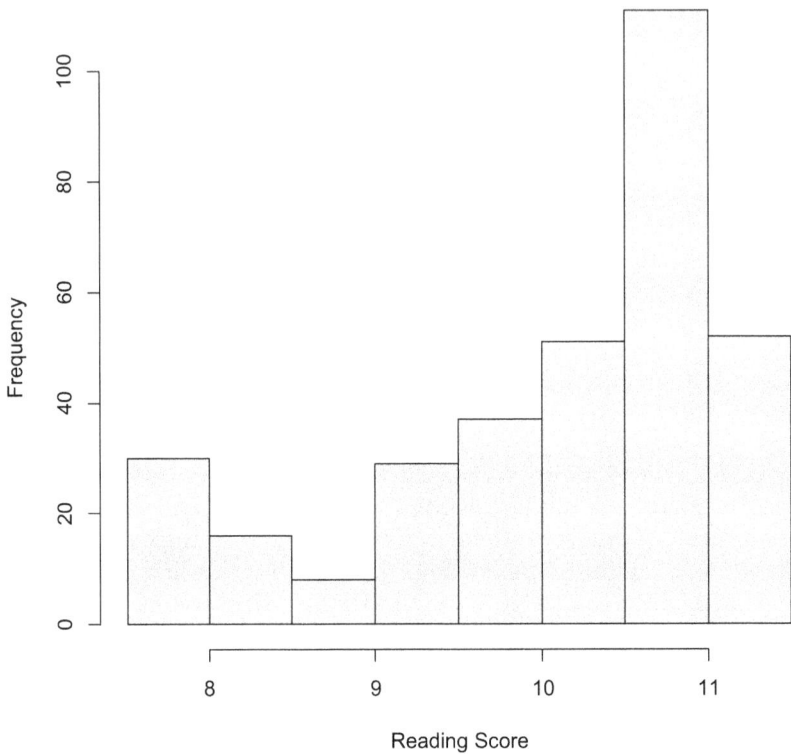

FIGURE 13.1
Histogram of reading test scores.

TABLE 13.1

Mean, Median, Standard Deviation, and MAD Values for Reading Score

Statistic	Value
Mean	10.05
Median	10.09
Standard deviation	1.01
MAD	1.24
Skewness	−0.65
Kurtosis	2.52

MLM, multilevel linear model.

random coefficient terms of interest (Model2). In this case, we will be interested in including a random coefficient for SES.

The results for Model0 can be used to obtain the ICC for reading scores associated with the school that students attend. As we discussed above, the

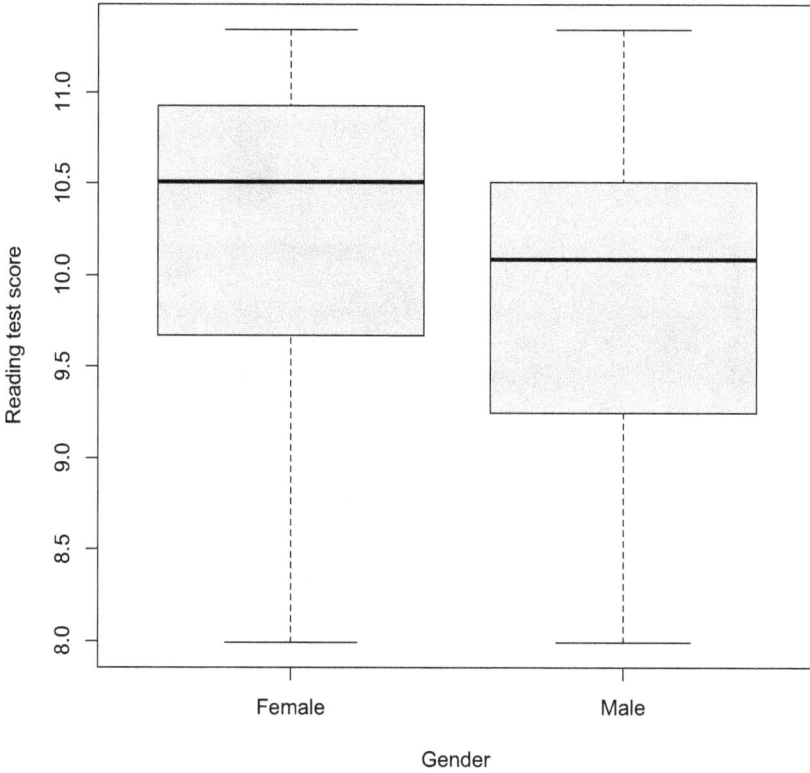

FIGURE 13.2
Boxplot of reading test scores by gender.

ICC is the ratio of the variability in scores due to the cluster variable (school for this example) to the total variability in the sample. For the reading test score, the variance due to school is 0.01, and the random error variance due to individual students is 1.01. Thus, the ICC is $\frac{0.01}{1.01} = 0.01$. In other words, school membership accounts for only 1% of the variance in the reading scores.

Now that we have obtained the ICC, we can determine which model yields the best fit to our data. The fit statistics of the three models for the reading test data appear in Table 13.2.

We learned in Chapter 10 that AIC and BIC are measures of model fit with a penalty added to account for model complexity. A smaller value for each statistic indicates a better-fitting model. The AIC for this example was minimized for Model1, whereas BIC was minimized for Model0, leaving it unclear as to which might yield the best fit to the data. Furthermore, based on the results of the chi-square tests comparing model fit, we would conclude that Model1 provides the best fit to the data. (Remember that the null hypothesis for the chi-square difference test is that the models fit the data equally well).

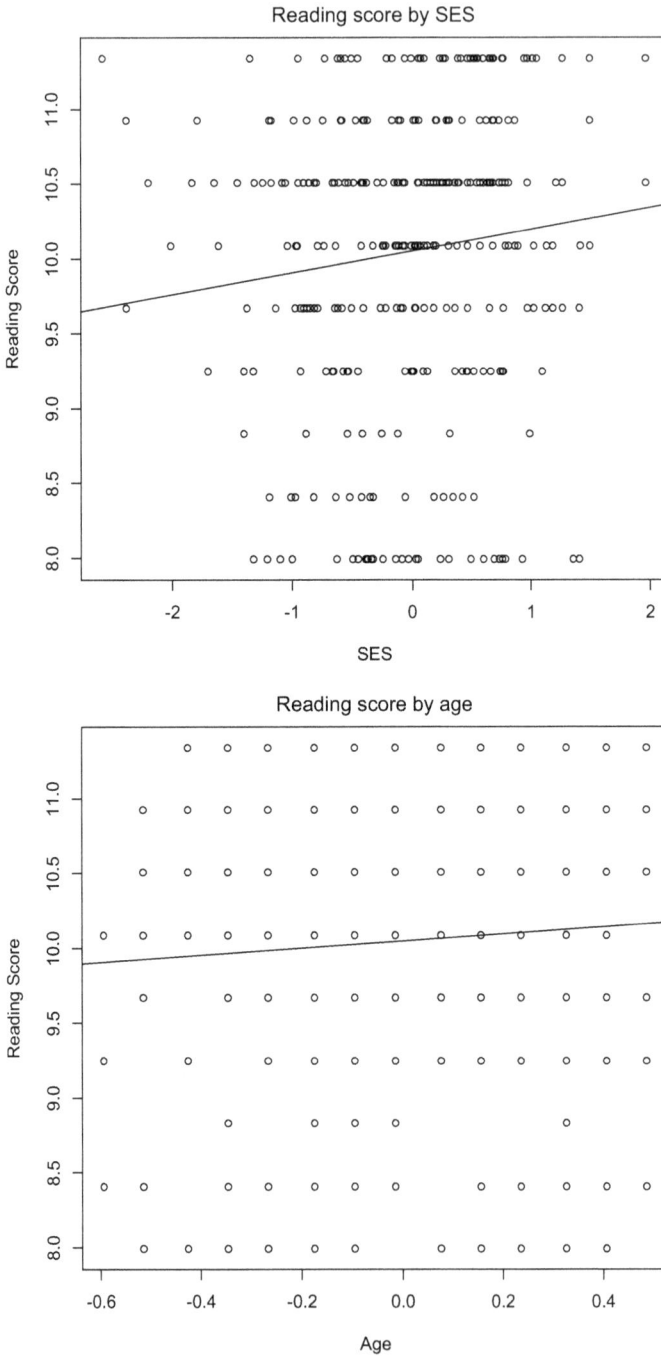

FIGURE 13.3
Scatterplots for reading by socioeconomic status (SES) and age.

TABLE 13.2

Model Fit Statistics for MLMs Applied to Reading Test
Data

Model	AIC	BIC	Chi-Square	p-Value
Model0	956.51	967.94		
Model1	948.15	971.02	14.36	0.003[a]
Model2	952.14	982.63	0.01	0.99[b]

[a] Comparison of Model1 to Model0.
[b] Comparison of Model2 to Model1.
MLM, multilevel linear model; AIC, Akaike information
criterion; BIC, Bayesian information criterion.

TABLE 13.3

Fixed and Random Effects Estimates for Model1

Effect	Coefficient	Standard Error	95% Confidence Interval
Fixed intercept	10.27	0.09	10.09, 10.44
SES	0.15	0.07	0.01, 0.29
Male	−0.34	0.11	−0.56, −0.11
Age	0.21	0.18	−0.15, 0.57
Random intercept	0.001	0.03	0.00, 0.27
Random error	0.98	0.98	0.91, 1.06

Considering these results together, we will select Model1 as providing the
best fit to the data of the models that we have considered here. This means
that we will focus on the model including the set of three independent
variables (gender, SES, and age) and the random intercept term, but not the
random coefficient for SES. Based on this result, we can conclude that the
relationship between SES and reading test score was not significantly different across schools.

The model parameter estimates, standard errors, and 95% confidence intervals for the model parameters for Model1 appear in Table 13.3.

As a reminder, the fixed intercept represents the mean reading test score
across schools when the independent variables are set to 0. The coefficients
for SES, Male, and Age reflect the relationship between each of these variables and the reading test score. The fact that the confidence interval (0.01,
0.29) does not include 0 means that students from families with higher SES
tend to have higher reading test scores. Likewise, because the slope for Males
is negative and the confidence interval (−0.56, −0.11) doesn't include 0, we
would conclude that male examinees had lower mean reading scores than
females. Finally, age was not found to be significantly related to reading
score, given that its confidence interval (−0.15, 0.57) includes 0.

With respect to the random effects, the random intercept (0.001) was not
statistically significant with a confidence interval of (0.00, 0.27). This means

TABLE 13.4

R^2 Estimates for Model1

Statistic	Value
RB_1	0.03
SB	0.04
MVP	0.04
RS fixed	0.04
RS random intercept	0.001
RS total	0.04

SB, Snijders and Bosker level-1 estimate; MVP, multi-level variance partitioning; RS, Rights and Sterba fixed effects.

that after accounting for SES, gender, and age, we do not have evidence to support mean differences in reading test scores across schools. The random error was significantly different from 0 (0.91, 1.06), indicating that there were differences among individual examinee reading scores after we accounted for school, SES, gender, and age. This last result is not particularly surprising, given that we would anticipate reading test performance to be related to factors other than just those included in this model.

The estimates for the model R^2 values appear in Table 13.4.

RB_1 corresponds to the level-1 variance explained according to the formula presented in Raudenbush and Bryk (2002). SB corresponds to the Snijders and Bosker (1999) level-1 estimate (as computed above), and the MVP is the total variance explained through multilevel variance partitioning introduced by LaHuis et al. (2014). The RS rows correspond to the Rights and Sterba fixed effects, random intercept, and total variance accounted for by the model. Considering these results together, it appears that the fixed effects in Model1 account for approximately 4% of the variance in the level-1 reading test scores. As would be expected given the non-significant results for the random intercept effect, the model accounts for only 0.1% of the variance in the scores.

Identifying Potential Outliers in Single-Level Data

The presence of outliers can have a deleterious impact on the estimation of model parameters and standard errors for multilevel models, just as they can for single-level models (Staudenmayer et al., 2009). Thus, researchers and data analysts should screen their data for the presence of outliers prior to conducting their target analyses. If outliers are detected, a decision must be made regarding how to deal with them. The general state of the art at this time is to attempt modeling of the data with the outliers included using an appropriate model, rather than removing outlying observations and fitting a standard multilevel model on the truncated dataset (Staudenmayer et al., 2009). Given this emphasis on using an appropriate model that accommodates outliers, the researcher must have some sense as to whether outliers are in fact present so that she can select the optimal analytic approach. In this and the next section of the chapter, we describe some of the more common approaches for identifying outliers, first for single-level data and then in the context of multilevel data. We will then focus on multilevel modeling strategies that are available to the researcher, given that outliers are present in the data.

Several of the more common outlier detection methods that are used with multilevel data have direct antecedents in the single-level regression literature, which we discussed in Chapter 9. One of these, Cook's distance (D), compares residuals (difference between observed and model-predicted outcome variables) for individual cases when another observation is included in the data versus when it is removed while correcting for the leverage of the data point. Recall from Chapter 9 that leverage is a measure of how far an individual's set of values on the independent variables is from the mean of these values. When there are multiple independent variables, the leverage statistic measures the distance between an individual's values on the independent variables and the set of means for those independent variables, which is commonly referred to as the centroid. Large leverage values indicate that an observation may have a great impact on the predicted values produced by the model. Remember from Chapter 9 that D is calculated as

$$D_i = \frac{\sum_{i=1}^{N}\left(e_i - e_{ij}\right)^2}{k\,\mathrm{MSr}} \tag{13.30}$$

where e_i is the residual for observation i for model containing all observations, e_{ij} is the residual for observation i for model with observation j removed, k is the number of independent variables, and MSr is the mean square of the residuals.

There are no hard and fast rules for how large D_i should be in order for us to conclude that it represents an outlying observation. Fox (2016) recommends that the data analyst flag observations that have D_i values that are unusual when compared to the rest of those in the dataset, and this is the approach that we will recommend as well.

Another influence diagnostic, which is closely related to D_i, is DFFITS. This statistic compares the predicted value for individual i when the full dataset is used ($\hat{y}_i$), against the prediction for individual i when individual j is dropped from the data ($\hat{y}_{ij}$). For individual j, DFFITS is calculated as

$$\mathrm{DFFITS}_j = \frac{\hat{y}_i - \hat{y}_{ij}}{\sqrt{\mathrm{MSE}_j\, h_j}} \tag{13.31}$$

where MSE_j is the mean squared error when observation j is dropped from the data and h_j is the leverage value for observation j.

As was the case with D_i, there are no hard and fast rules about how large $DFFITS_j$ should be to flag observation j as an outlier. Rather, we examine the set of $DFFITS_j$ and focus on those that are unusually large (in absolute value) when compared to the others. One final outlier detection tool for single-level models that we will discuss here is the COVRATIO, which measures the impact of an observation on the precision of model estimates. For individual i, this statistic is calculated as

$$\text{COVRATIO}_i = \frac{1}{(1-h_i)}\left(\frac{n-k-2+E_i^{*2}}{n-k-1}\right)^{k+1} \tag{13.32}$$

where E_i^* is the studentized residual and n is the total sample size.

Fox (2016) states that COVRATIO values greater than 1 improve the precision of the model estimates, whereas those with values less than 1 decrease the precision of the estimate. Clearly, it is preferable for observations to increase model precision rather than decrease it.

Identifying Potential Outliers in Multilevel Data

Detection of potential outliers in the context of multilevel data is, in most respects, similar to the single-level methods described above. The primary difference with multilevel data is in how observations are defined when it comes to removing them in the calculations. In the single-level case, an observation corresponding to one member of the sample is removed when calculating D_i, as an example. However, for multilevel data, the observation in question is typically associated with level-2, rather than level-1. Thus, when calculating statistics such as D_i, the data to be removed corresponds to an entire cluster, and not just to a single data point. Furthermore, for multilevel models, D_i provides information about outliers with respect to the fixed effects portion of the model only.

As was the case for single-level models, D_i for multilevel models is based on the leverage values for the observations. Demidenko and Stukel (2005) defined the fixed effects leverage value for individual i as

$$H_{i,\text{Fixed}} = X_i\left(\sum_{i=1}^{N} X_i'V_i^{-1}X_i\right)^{-1} X_i'V_i^{-1} \tag{13.33}$$

where X_i is the matrix of fixed effects for subject i, V_i is the covariance matrix of the fixed effects for subject i, and N is the number of level-2 units.

Note that in equation (13.33), the subject refers to the level-2 grouping variable. Similarly, the leverage values based on the random effects for subjects in the sample can be expressed as

$$H_{i,\text{Random}} = Z_i DZ_i' V_i^{-1} \left(I - H_{i,\text{Fixed}} \right) \qquad (13.34)$$

where Z_i is the matrix of random effects for subject i and D is the covariance matrix of the random effects for subject i.

The actual fixed and random effects leverage values correspond to the diagonal elements of $H_{i,\text{Fixed}}$ and $H_{i,\text{Random}}$, respectively. The multilevel analog of Cook's D_i can then be calculated using the following equation:

$$D_{\text{Mi}} = \frac{1}{mS_e^2} r_i' \left(I - H_{i,\text{Fixed}} \right)^{-1} V_i^{-1} X_i \left(\sum_{i=1}^{N} X_i' V_i^{-1} X_i \right)^{-1} X_i' V_i^{-1} \left(I - H_{i,\text{Fixed}} \right)^{-1} r_i$$

$$(13.35)$$

where m is the number of fixed effects parameters, S_e^2 is the estimated error variance, and I is the identity matrix $r_i = y_i - \hat{y}_i$

where y_i is the observed value of dependent variable for observation i and $\hat{y}_i$ is the model-predicted value of dependent variable for observation i.

The interpretation of D_{Mi} is similar to that for D_i, in that individual values departing from the main body of results for the sample are seen as indicative of potential outliers.

In addition to the influence statistics at each level and D_{Mi}, there is also a multilevel analog for DFFITS, known as MDFFITS. As in the single-level case, MDFFITS is a measure of the amount of change in the model-predicted values for the sample when an individual is removed versus when they are retained. This measure is considered an indicator of potential outliers with respect to the fixed effects, given the random effect structure present in the data. Likewise, the COVRATIO statistic can also be used to identify potential outliers with respect to their impact on the precision with which model parameters are estimated. The same rules of thumb for interpreting MDFFITS and the COVRATIO that were described for single-level models also apply in the context of multilevel modeling.

With respect to the random effects, the relative variance change (RVC; Dillane, 2005) statistic can be used to identify potential outliers. The RVC for one of the random variance components for a subject in the sample is calculated as

$$\text{RVC}_i = \frac{\hat{\theta}_i}{\hat{\theta}} - 1 \qquad (13.36)$$

where $\hat{\theta}$ is the variance component of interest (e.g., residual, random intercept) estimated using full sample and $\hat{\theta}_i$ is the variance component of interest estimated excluding subject i.

When RVC is close to 0, the observation does not have much influence on the variance component estimate; i.e., is not likely to be an outlier with respect to the random effects. As with the other statistics for identifying potential outliers, there is not a single agreed-upon cut-value for identifying outliers using RVC. Rather, observations that have unusual such values when compared to the full sample warrant special attention in this regard.

Identifying Potential Multilevel Outliers for the Example

We can now apply these tools for identifying outliers and unusually influential cases for the MLM of the reading test data. To gain a better understanding of the data, we can examine the histogram in Figure 13.1. We saw that the data were negatively skewed with a number of individuals having relatively low scores in comparison to those for the majority of students whose scores fell between 10 and 12.

As we noted above, outlier detection in the multilevel modeling context can be conducted at each level of the MLM. The actual statistics/approaches used are quite similar to those that we reviewed in Chapter 9, allowing us to identify unusual cases in terms of their position vis-à-vis other cases in the sample, their residual values, and their influence on the model parameters. And, as we noted earlier, the focus of outlier detection with MLMs tends to be at level-2, though we will examine the information at each level of the model.

We will begin our exploration of unusual cases by examining the Cook's D values for the individual students, which appear in Figure 13.4.

There are several students whose Cook's D values were large when compared to their classmates. Recall that there are not generally agreed-upon cutoffs for identifying a "large" value for D. However, the R software function applied in this example uses the following value to denote outliers based on D (Loy & Hofmann, 2014):

$$Q_3 + 3*\text{IQR}$$

where Q_3 is the third quartile of Cook's D values for the sample and IQR is the interquartile range of Cook's D values for the sample.

Although we cannot verify that this cutoff is applicable in all cases where D is used, it does provide us with a reasonable heuristic for interpreting the values that we obtain for our sample. Based on this cutoff and a simple examination of Figure 13.4, we see nine or ten individuals who appear to be outliers. An examination of their school membership reveals that of these ten examinees, six attended school 3, three attended school 1, and one attended school 2.

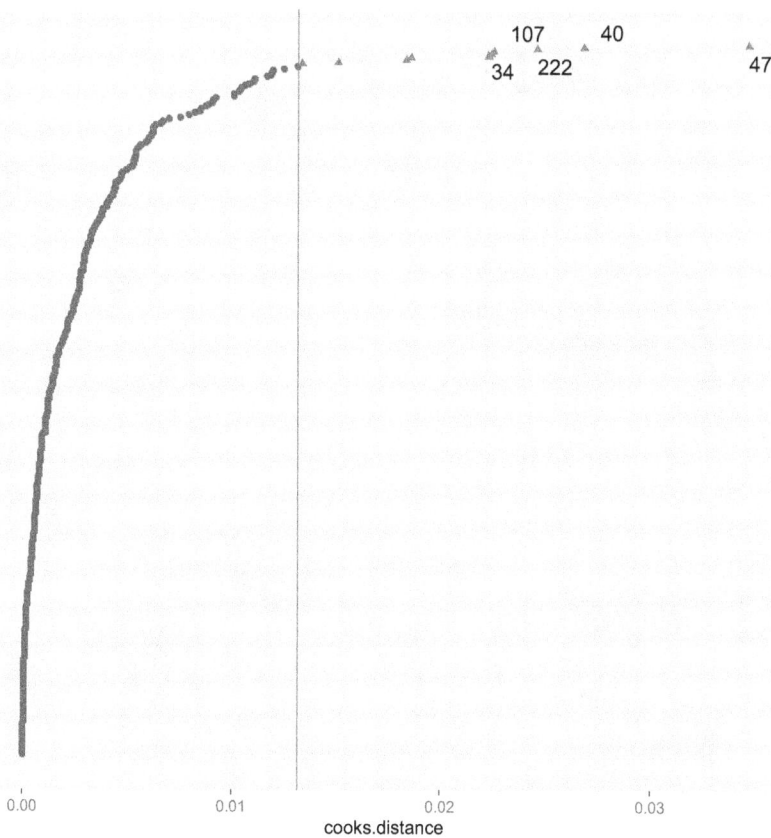

FIGURE 13.4
Cook's *D* values for level-1 (students) in the MLM. MLM, multilevel linear model.

Figures 13.5 and 13.6 display similar information for the MDFFITS and COVRATIO statistics for the student-level data.

The MDFFIT values identified the same ten individuals as outliers as did Cook's *D*. The COVRATIO results revealed that there were four individuals who had an outsized impact on the precision with which model parameters are estimated; i.e., coefficient standard errors. The four students with the largest COVRATIO values also had the largest Cook's *D* and MDFFIT values. Thus, it seems that these individuals have an outsized impact on the fixed effect estimates after accounting for the random effects.

The level-1 (student) RVC values for the random error variance estimate appear in Figure 13.7.

There are several individuals with error variance RVC values that might be outliers. The boxplot in Figure 13.7 reveals that there are also RVC outliers in terms of the random intercept variance. Indeed, there appear to be 60 students who had a large impact on this error term.

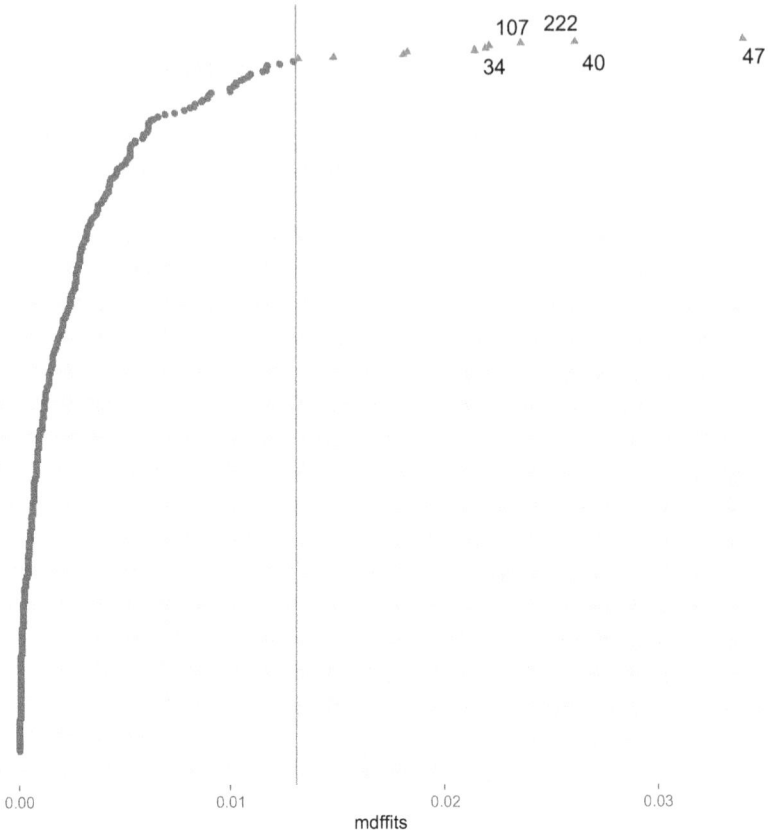

FIGURE 13.5
MDFFIT values for level-1 (students) in the MLM. MLM, multilevel linear model.

When we consider the full set of outlier investigations at level-1, it appears that there are a number of students who may be having an outsized impact on the estimates for the random effects (error and intercept variances). There were also four to ten students who were unusual with respect to their impact on the fixed effects coefficients. These results will lead us to consider using alternative approaches to estimating MLMs in subsequent sections of this chapter.

Next, we will turn our attention to the issue of outliers at level-2 (schools) of our MLM. The Cook's D, MDFFITS, and COVRATIO results for the school level appear in Figures 13.8–13.10.

In contrast to the individual-level data, there were no unusual or outlying cases among the four schools in the sample, based on these statistics.

Taken together, these results indicate that there are some potential outlying examinees in the sample. Specifically, ten individuals exhibited Cook's D and MDFFIT values, suggesting that they have an outsized impact on the fixed

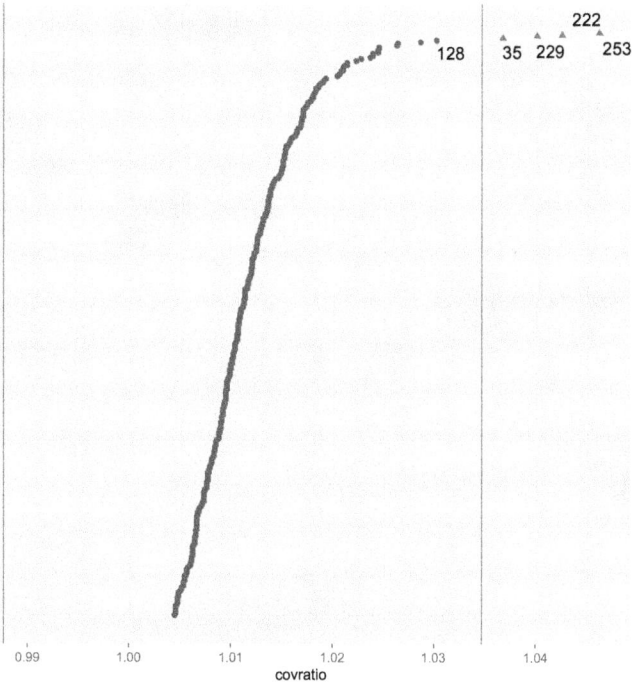

FIGURE 13.6
COVRATIO values for level-1 (students) in the MLM. MLM, multilevel linear model.

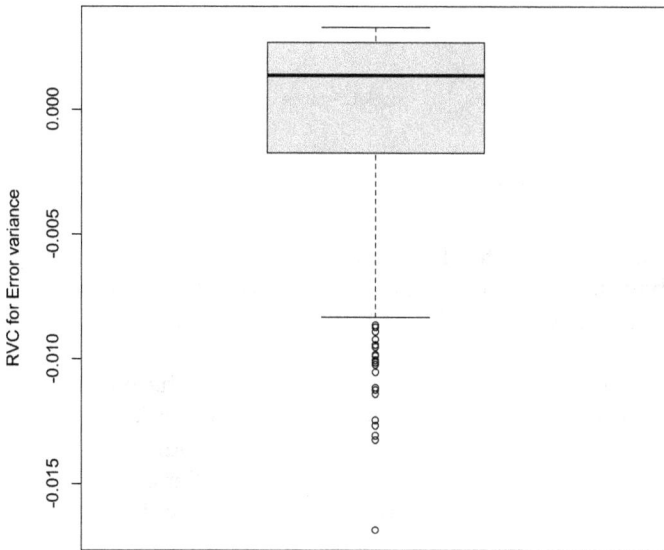

FIGURE 13.7
Boxplot of RVC values for level-1 (students) in the MLM. MLM, multilevel linear model; RVC, relative variance change.

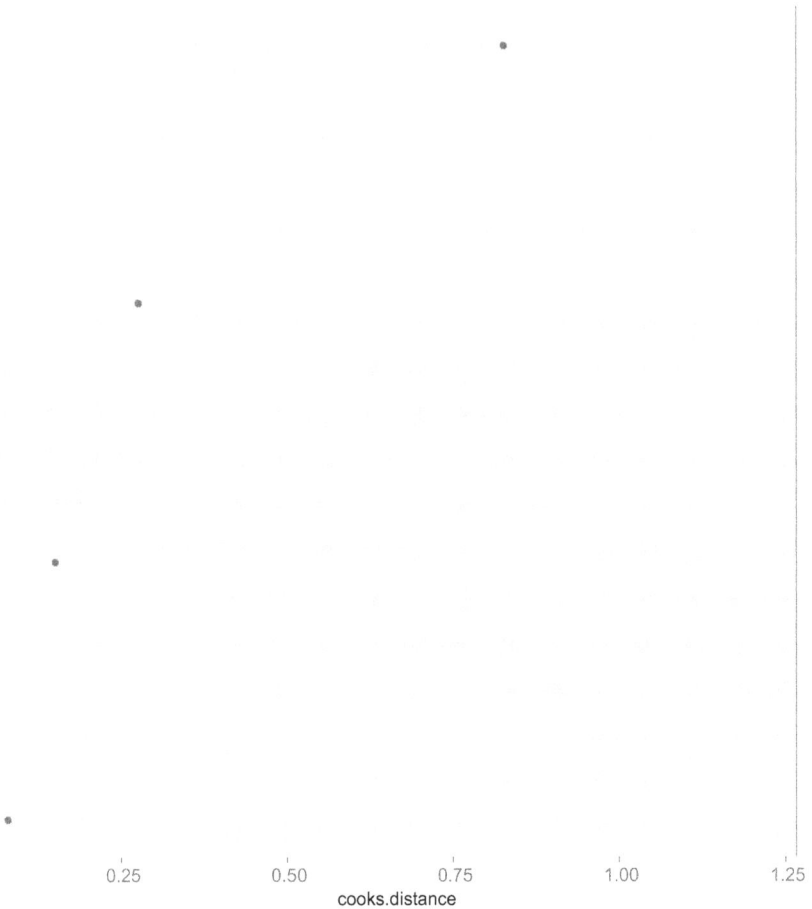

FIGURE 13.8
Cook's *D* for school-level data.

effects parameter estimates. Furthermore, four of these individuals were also found to have an unusually large impact on the standard error estimates of the fixed effects, based on the COVRATIO. For the random error effect, a large number of examinees were flagged as potential outliers. Finally, we discovered that there was quite a bit of overlap in these results, with many of the same students being identified as potential outliers in multiple ways. In contrast to the student-level data, none of the four schools yielded diagnostic statistics that would suggest them to be outliers.

  Having identified potential outliers, we must now decide how we would like to handle them. As we have discussed earlier in the chapter, it is recommended that if possible, an analysis strategy designed to appropriately model data with outliers be used, rather than simply removing them (Staudenmayer et al., 2009). Indeed, in the current example, if we were to remove the subjects

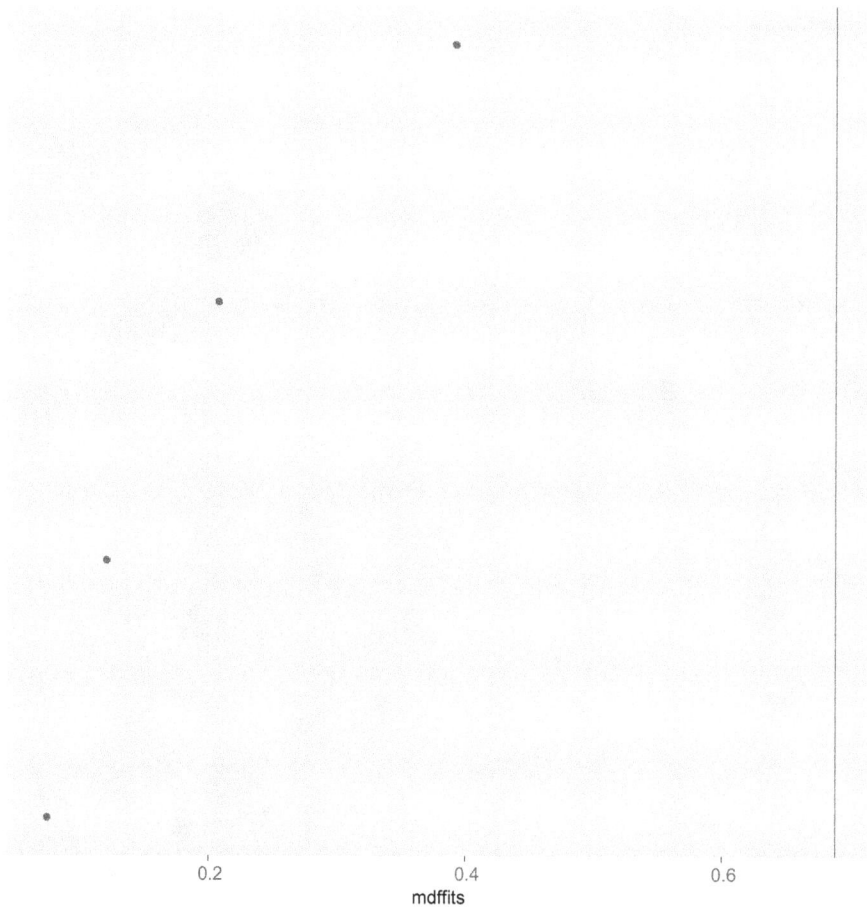

FIGURE 13.9
MDFFIT for school-level data.

who were identified as potential outliers based on the various statistics, we could delete as many as 65 individuals. Modeling the data using the full sample would seem to be the more attractive option. In the following section, we will discuss estimators that can be used for modeling multilevel data when outliers may be present. These approaches have much in common with the robust methods described earlier in the book, particularly in Chapter 9. There are fewer options in the multilevel context than were true for single-level regression modeling, but the methods that we do have access to are known to be effective for dealing with outlying and unusually influential observations at both levels 1 and 2 (Finch, 2017; Kloke et al., 2009; Pinheiro et al., 2001). We will first describe these robust approaches and then apply them to our example.

FIGURE 13.10
COVRATIO for school-level data.

Robust Estimation for Multilevel Models

Although there has been much work done in the development of models to deal with outlying observations, much of it has focused on single-level models, such as linear regression. In the context of multilevel models, Lange et al. (1989) developed an approach for modeling data with outliers that involves the adjustment of random effect distributions (e.g., ε_{ij} and U_{0j}) from the standard multivariate normal to heavy-tailed distributions, such as the multivariate t with υ degrees of freedom or the Cauchy. Models based on such heavy-tailed random effects can better accommodate extreme values (i.e., outliers).

In turn, such models should yield more accurate parameter and standard error estimates (Lange et al., 1989). Pinheiro et al. (2001) extended this work by describing a ML algorithm for estimating multilevel model parameters using the multivariate t distribution. Research has shown that these heavy-tailed random effects models yield more accurate parameter estimates and smaller standard errors than do models based on the assumption that errors are normally distributed (Tong & Zhang, 2012; Song et al., 2007; Yuan et al., 2004; Yuan & Bentler, 1998).

The standard multilevel model can be written as follows in the matrix form:

$$\begin{bmatrix} y_i, U_{0j} \end{bmatrix} \sim N \left(\begin{bmatrix} X_i \beta \\ 0 \end{bmatrix}, \begin{bmatrix} Z_j \Psi Z_j' + \Lambda_j & Z_j \Psi \\ \Psi Z_j' & \Psi \end{bmatrix} \right) \tag{13.37}$$

where β is the vector of fixed effects, Λ_j is the level-1 covariance matrix, Ψ is the level-2 covariance matrix, X_i is the design matrix for the fixed effects, and Z_j is the design matrix for the Level-2 random effects.

Pinheiro et al. (2001) showed that this model can be rewritten as:

$$\begin{bmatrix} y_i, U_{0j} \end{bmatrix} \sim t \left(\begin{bmatrix} X_i \beta \\ 0 \end{bmatrix}, \begin{bmatrix} Z_j \Psi Z_j' + \Lambda_j & Z_j \Psi \\ \Psi Z_j' & \Psi \end{bmatrix}, v \right) \tag{13.38}$$

where v is the degrees of freedom for the t distribution.

Given equation (13.38), the dependent variable, y, follows the t distribution with v degrees of freedom, as do the Level-1 and Level-2 error terms, ε_{ij} and U_{0j} (Pinheiro et al.). This stands in contrast to the model in equation (13.37), for which the dependent variable is assumed to follow the normal distribution. In turn, the variances of both ε_{ij} and U_{0j} are a function of the degrees of freedom. This last result means that there are an infinite number of possible distributions that the random effects variances can follow, all based on v degrees of freedom.

The use of a heavy-tailed distribution to model the distributions of the random effects can be extended beyond the t with v degrees of freedom to include the Cauchy (Hogg & Tanis, 1996) and the Slash (Rogers & Tukey, 1972) distributions. As with the t, these heavy-tailed distributions may be useful tools to account for the presence of outliers in the data. The Cauchy distribution has an unknown mean and variance but a defined median and mode and is symmetric with heavier tails than the normal distribution. The slash distribution is defined as the ratio of the normal (0,1) and uniform (0,1) distributions and, like the t and Cauchy distributions, has been shown to be useful for parameter estimation in the presence of outliers (e.g., Wang & Genton, 2006). For the t and Slash distributions, when v is unknown, as is most often the case, the expectation maximization algorithm that is used to estimate the model parameters includes an additional step in which the degrees of freedom are estimated as well. The logic underlying all of these approaches to parameter estimation when outliers are present is that the heavier-tailed distributions can better accommodate outliers than the normal, resulting in lower parameter estimation bias and smaller standard errors (Welsh & Richardson, 1997).

Rank-Based Methods for Fitting MLM

A second alternative to ML/REML estimation of multilevel model parameters in the presence of outliers is based on a joint rank (JR) estimator (Kloke et al., 2009). In JR, the raw scores of the dependent variable are replaced with their ranks based on a nondecreasing score function, such as the Wilcoxon (1945). Assuming a common marginal distribution of level-1 errors (ε_{ij}) across level-2 units, the estimation of the fixed effects (β_1, γ_{00}) is based on Jaekel's (1972) dispersion function:

$$\hat{\beta}_\varphi = \text{Argmin} \, Y - X\hat{\beta}_\varphi \tag{13.39}$$

where Y is the dependent variable, X is the matrix of independent variable values, and $\hat{\beta}$ is the matrix of estimates of the fixed effects for the model.

$$Y - X\hat{\beta}_\varphi = \sum_{i=1}^{N} \left[R\left(y_{ij} - \hat{y}_{ij}\right) \right]\left(y_{ij} - \hat{y}_{ij}\right)$$

where R denotes the rank, y_{ij} denotes dependent variable value for individual i in cluster j, and $\hat{y}_{ij}$ denotes the model based predicted dependent variable value for individual i in cluster j.

In short, the estimation of the fixed effects parameters is based on minimizing the ranks of the residual between the observed and predicted values of the dependent variable. This stands in contrast to standard ML (and REML), which minimizes the raw residuals and is similar in spirit to the rank-based regression estimation that we learned about in Chapter 9.

Estimation of fixed effects parameter standard errors can be done using either of two approaches (Kloke et al., 2009). The first of these was based on the assumption that the within-cluster error terms are compound symmetric; i.e., a common covariance exists between any pair of observations. The primary advantage of this compound symmetric estimator (JR_CS) is that it is computationally very efficient, requiring estimation of only one additional parameter beyond those that are a part of the standard multilevel model, the common covariance between item pairs (Kloke et al., 2009). Its main limitation is the strong assumption of exchangeability of error terms, which may frequently not hold in actual practice. Given the potential problems associated with making this exchangeability assumption, Kloke and McKean (2013) proposed a sandwich estimator (JR_SE) for the model parameter standard errors. This JR_SE estimator does not require any additional assumptions about the data beyond those of ML/REML, unlike the case for JR_CS. Kloke and McKean (2013) conducted a simulation study comparing the performance of JR_CS and JR_SE and found that JR_SE worked well for samples of 50 or more level-2 units. However, for fewer than 50 level-2 units, JR_SE yielded somewhat larger standard errors than was the case for JR_CS, thereby leading to more conservative inference with regard to the statistical significance of the parameter estimates; i.e., lower statistical power. Kloke and McKean (2013) also found that when the exchangeability assumption was violated, JR_CS standard error estimates were inflated, also reducing the power for inference regarding these parameters. Given this combination of results, Kloke and McKean recommended that JR_SE be used as the default method for estimating model parameter standard errors. However, when the level-2 sample size is small, researchers should use JR_CS, unless they know that the exchangeability assumption has been violated.

An alternative approach to obtaining parameter estimates is the iteratively reweighted generalized rank (GR). GR starts with the JR approach and then uses the rank-based prediction procedure (RPP; Bilgic, 2012). We will not get into the details in this chapter, but the interested reader is encouraged to read the aforementioned reference for an in-depth discussion of this method. GR obtains estimates using the following steps:

1. Obtain initial fixed effects estimates using the JR technique.
2. Use the fixed effects estimates to obtain residuals for each level-1 unit.
3. Use the level-1 residuals in conjunction with the RPP to obtain estimates of random effects.
4. Use the random effects from step 3 to obtain variance components for the level-2 units.
5. Compare the parameter estimates from steps 2 and 4 in the current iteration with those in the prior iteration. If the differences for both sets of estimates are less than a predetermined value (the tolerance), the estimator stops. Otherwise, it repeats steps 2–5.

Researchers have found that both JR and GR yield parameter estimation accuracy and efficiency rates similar to those for ML and REML when the outcome data are normally distributed (Abebe et al., 2016; Bilgic et al., 2013). Those authors also found that when the data included outliers, JR and GR produced estimates with less bias and smaller standard errors than ML or REML. Thus, we can conclude that rank-based approaches present a viable option for data analysts working with multilevel data, particularly when outliers are present in the data.

Bayesian Multilevel Modeling

Throughout this book, we have discussed the application of Bayesian estimation to a variety of data analysis problems. We have seen that it offers the
researcher a number of advantages, including the ability to integrate prior
knowledge and information into the data analysis process, as well as (often)
superior performance to frequentist estimation in the presence of small samples. It is not a surprise, then, that Bayesian estimation is also a viable alternative for researchers working in the multilevel modeling context. Bayesian
estimation brings many of the same advantages to fitting multilevel models as
it does for other analyses. In particular, it can prove to be particularly helpful
in situations where the number of level-2 units (e.g., schools) is small, making estimation of some model parameters difficult. Furthermore, when the
assumption of normality that underlies both ML and REML estimation has not
been met, Bayesian estimation of the multilevel model parameters is a good
alternative for estimating model parameters. Given that we described the basic
Bayesian estimation technique in Chapter 2, we will not repeat those details
here. Suffice it to say that the same ideas that we discussed earlier, including
the Markov Chain Monte Carlo (MCMC) algorithm, prior distributions, and
the particulars of fitting the model, apply in the Bayesian context as well.

Example of Fitting Robust, Rank-Based, and Bayesian
Estimators for Multilevel Data

To demonstrate the fitting of robust and rank-based multilevel models, we
will return to the school data example. Recall that based on the results of the
MLM, a number of students were flagged multiple times as potential outliers. However, none of the four schools were identified as outliers. In the
remainder of this chapter, we will fit the MLM for this data using the robust
technique based on Pinheiro et al. (2009) followed by the rank-based methods and then finish up with the Bayesian approach. We will then conclude
the chapter by comparing and contrasting the results that we obtain from the
various MLM techniques.

Robust MLM Estimation

The fixed and random effects for Model1 based on the robust MLM estimation appear in Table 13.5.

The parameter estimates and standard errors for the robust MLM are
quite similar to those from the standard estimation approach in Table 13.3.
There are some differences, however. For example, the confidence interval

TABLE 13.5

Fixed and Random Effects Estimates for Model1 Based on Robust
MLM Estimation

Effect	Coefficient	Standard Error	95% Confidence Interval
Fixed intercept	10.35	0.10	10.16, 10.55
SES	0.16	0.07	0.02, 0.30
Male	−0.34	0.12	−0.56, −0.08
Age	0.20	0.19	−0.13, 0.56
Random intercept	0.01	0.0001	0.01, 0.17
Random error	0.99	1.00	0.91, 1.07

MLM, multilevel linear model.

for the random intercept variance in the robust case does not include 0, unlike for the standard estimator, indicating that when using the robust model there does appear to be difference in reading score means across schools. However, the small value of the estimate itself (0.01) suggests that these differences are very small. This general concordance in results for the two REML and robust MLM estimators suggests that the outliers do not have a great impact on the parameter estimates. The Rights and Sterba (2019) R^2 values for the robust model were 0.04 for the fixed effects and 0.001 for the random intercept. These are identical to those obtained for the REML estimated model.

It can be very helpful to examine residual plots for the robust MLM, just as we did for the REML estimator. As for ML and REML estimation, we can assess the distributional features of the residuals and examine the weights applied to the cases to see which individuals played a relatively greater (and lesser) role in the parameter estimation process. The fitted value by residual plot for the level-1 (student) cases appears in Figure 13.11.

The fitted values by residuals plot for the robust MLM look quite similar to that for the standard REML approach and demonstrate that the variance is homogeneous across the response at level-1. Of additional interest in this graph, beyond the scatter pattern, is the color coding for the data points, which reflects the relative weights applied to the individuals. Lighter coloring indicates that an individual received a heavier weight in the estimation process from equation (13.38). Note, for example, that those with particularly large negative residuals had weights of 0.6 or below, whereas those in the middle of the residual distribution were given weights of 1.0. We can also see this weighting scheme in Figure 13.12, which is the normal QQ plot of the residuals for the robust MLM.

As we saw in the previous figure, individuals with large negative residuals were assigned lower weights than those with residuals of approximately −1 or above. Finally, given that a relatively small number of individuals were

Fitted Values vs. Residuals

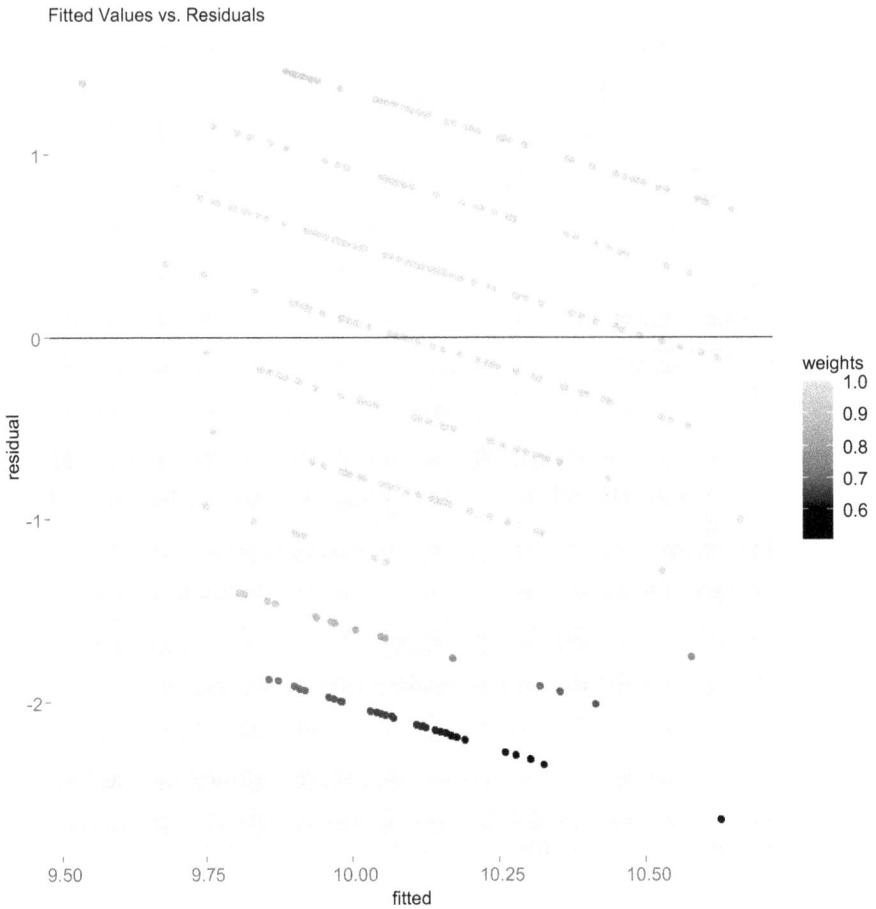

FIGURE 13.11
Fitted values by residuals plot for robust multilevel model.

thusly downweighted by the robust estimator, it is not a surprise that the results from the MLM and robust MLM are quite similar.

Rank-Based MLM Estimation

The coefficients for the JR and GR rank-based estimators are presented in Table 13.6.

The story to be told about the rank-based estimation results is quite similar to that of the robust MLM. Namely, the estimates and inference results are very similar to those for the standard MLM. Such was the case for both the JR and GR estimators. Indeed, the only marked difference between these two techniques is that the standard error for the JR fixed effect intercept estimate

Normal Q-Q vs. Residuals

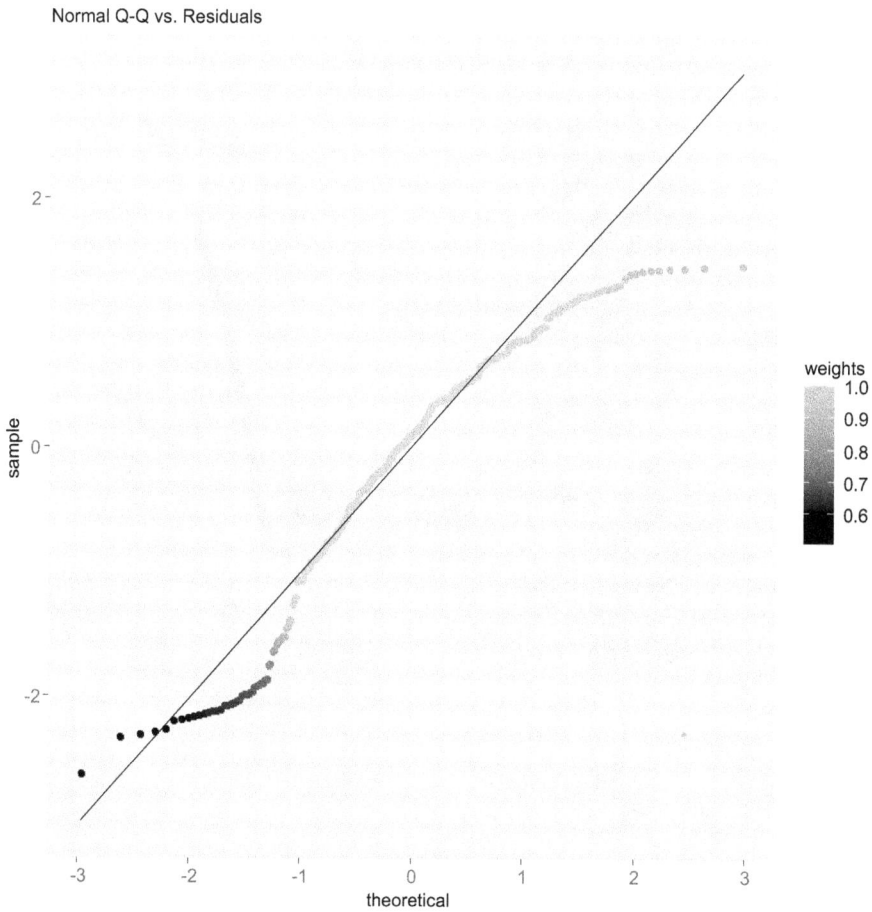

FIGURE 13.12
Normal QQ plot for residuals from robust MLM. MLM, multilevel linear model.

TABLE 13.6

Fixed and Random Effects Estimates for Model1 Based on Rank-Based MLM Estimation

Effect	Coefficient		Standard Error		*p*-Value	
	JR	GR	JR	GR	JR	GR
Fixed intercept	10.22	10.42	0.64	0.16	<0.001	<0.001
SES	0.14	0.18	0.07	0.07	0.03	0.01
Male	−0.32	−0.32	0.11	0.10	0.002	0.001
Age	0.19	0.19	0.17	0.14	0.26	0.18
Random intercept	0.02	0.02				
Random error	0.92	0.91				

MLM, multilevel linear model.

Stand. Residuals vs. Fits in GR　　　　　　Normal Q-Q Plot

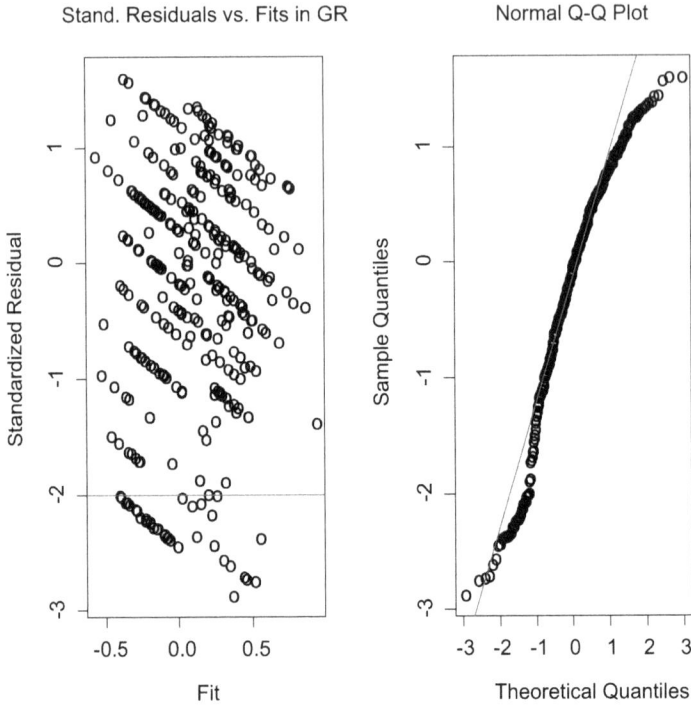

FIGURE 13.13
Fitted values by residual and QQ plots for the JR rank-based MLM. MLM, multilevel linear model; JR, joint rank.

is much larger than that for GR (and larger than the other methods featured in this chapter).

The fitted values by residual and normal QQ plots associated with the JR estimator appear in Figure 13.13. Those for the GR technique are quite similar to these and thus won't be included here.

Perhaps of particular interest in Figure 13.13 is that the QQ plot of the residuals more closely conforms to the normal line than was the case for either the standard MLM or robust MLM estimators. This result would suggest that the application of the rank-based estimator has reduced the outlying nature of some observations, particularly those at the lower end of the residual distribution. The residuals by fitted values plot for the rank-based estimator are quite similar to that of the robust and parametric methods.

Bayesian MLM Estimation

To conclude this chapter, we will turn our attention to the Bayesian estimation of MLM. The basic principles of Bayesian estimation that we have discussed in previous chapters apply in this case as well. In particular, details about parameter estimation using MCMC, including the burnin, number of iterations, thinning of the chains, the number of chains, priors, and the diagnosis of convergence, are all identical to those that we have covered earlier in the book. Therefore, we will not revisit those concepts here but rather encourage the reader to review them as needed while considering the following material. A typical default noninformative prior for the fixed effects coefficients is the normal distribution with a mean of 0 and a variance of 10,000,000,000. A common prior distribution for the random intercept (and random coefficient where appropriate) effect (together referred to as G) is expressed in two separate terms: (1) V reflecting the variation in the outcome variable across level-2 units, and (2) nu reflecting the degree of confidence in the prior value of the parameter. The default prior distribution for V is the inverse-Wishart distribution, with $V=1$ and nu$=0$. This low value for *nu* reflects the lack of information provided by the prior distribution; i.e., our lack of confidence that it is correct. The prior distribution for the residual term, R, with the defaults being precisely the same as those for G.

For the reading test data, we will use the default priors mentioned above, an MCMC chain with 100,000 iterations, a burnin period of 10,000, and thinning of 50 points. Table 13.7 includes the parameter estimates for the Bayesian estimator. As a demonstration of Bayesian estimation, we fit Model1 to the data. Recall that this model included all of the fixed effects of interest, as well as a random intercept term. However, the random slope variance was not estimated.

Overall, the results from the Bayesian estimated model are similar to those for the other MLM estimators described above. Perhaps the largest differences between the results in Table 13.7 and those for the other techniques are in the credible intervals for SES and the random intercept term. In particular, the lower bound for the SES coefficient is much closer to 0 (0.004) than was the case for the other techniques. Similarly, the credible interval for the random intercept not only includes 0 but also has an upper bound closer to 0 than was true for the other estimators. These results suggest that there is not a difference in reading score means across schools, and SES may

TABLE 13.7

Fixed and Random Effects Estimates for Model1 Based on Bayesian MLM Estimation: Noninformative Priors

Effect	Coefficient	95% Credible Interval
Fixed intercept	10.27	10.07, 10.45
SES	0.15	0.004, 0.27
Male	−0.34	−0.57, −0.12
Age	0.20	−0.15, 0.56
Random intercept	0.004	0.00, 0.01
Random error	0.98	0.84, 1.11

MLM, multilevel linear model.

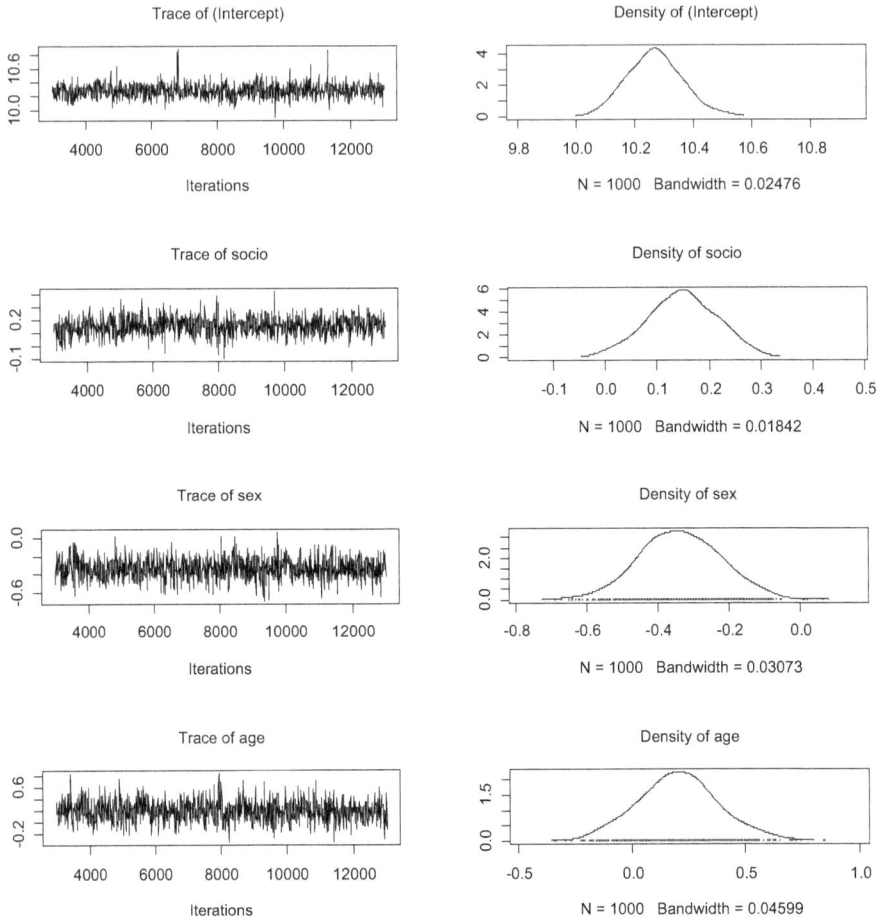

FIGURE 13.14
Trace and density plots for MCMC chains for the fixed effects with noninformative priors.
MCMC, Markov Chain Monte Carlo.

have only a very weak relationship with reading test scores. As was true for the other estimation techniques, the confidence interval for the random error term does not include 0, meaning that there are differences in test scores for students within the same school.

The trace and density plots for the Bayesian fixed effects appear in Figure 13.14.

The trace plots show that estimates for each parameter converged, and the density plots show that the estimates are symmetric around the central value. The trace and density plots for the random effects appear in Figure 13.15.

As with the fixed effects, convergence was obtained for the random effects. Altogether, these results tell us that the Bayesian estimator converged successfully on a single value for each parameter.

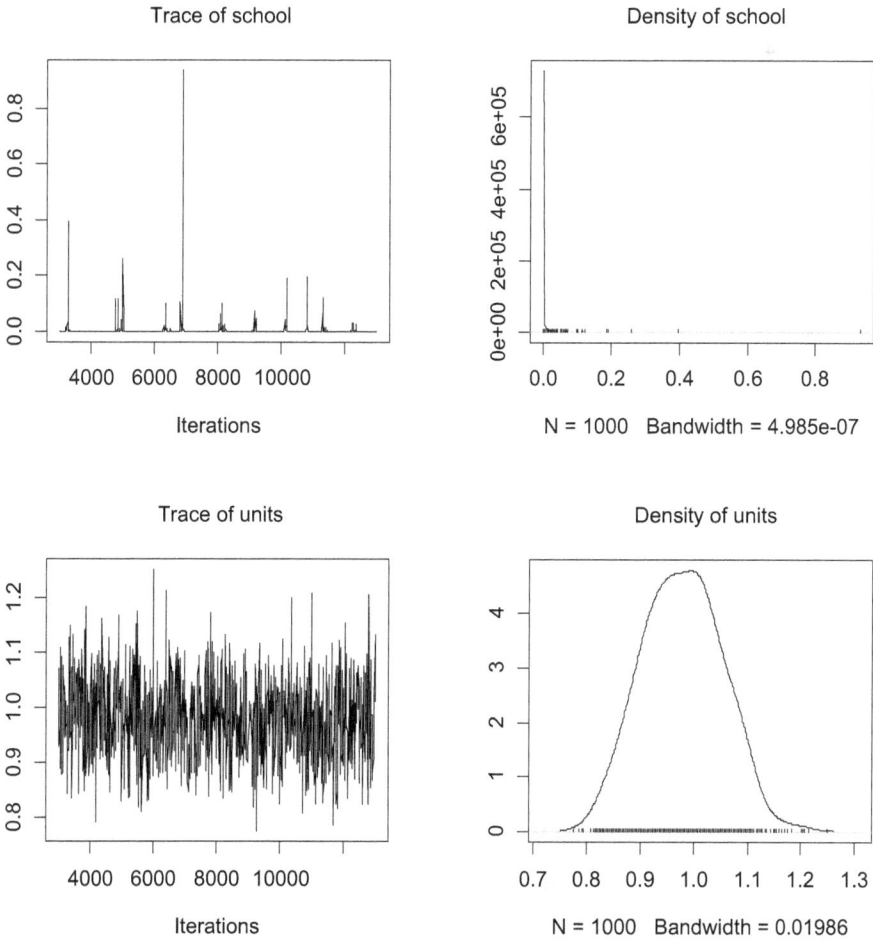

FIGURE 13.15
Trace and density plots for MCMC chains for the random effects with noninformative priors.
MCMC, Markov Chain Monte Carlo.

Next, we will fit the MLM using Bayes again, this time with informative priors for the fixed effects coefficients and random intercept variance that are based on previously published research. In particular, the informative priors for the fixed effects were drawn from the normal distribution with means of 10, 0.15, −1.0, and 0.75 and variances of 1, 0.1, 0.1, and 0.1, respectively. The informative priors for the random effects had $V=1$ and nu$=0.002$ for both the random intercept and error variance terms. All other estimation settings were the same as for the noninformative priors estimation. The diagnostic plots for the informative priors estimates appear in Figures 13.16 and 13.17.

The Bayesian estimates appear to have converged for all of the parameters.

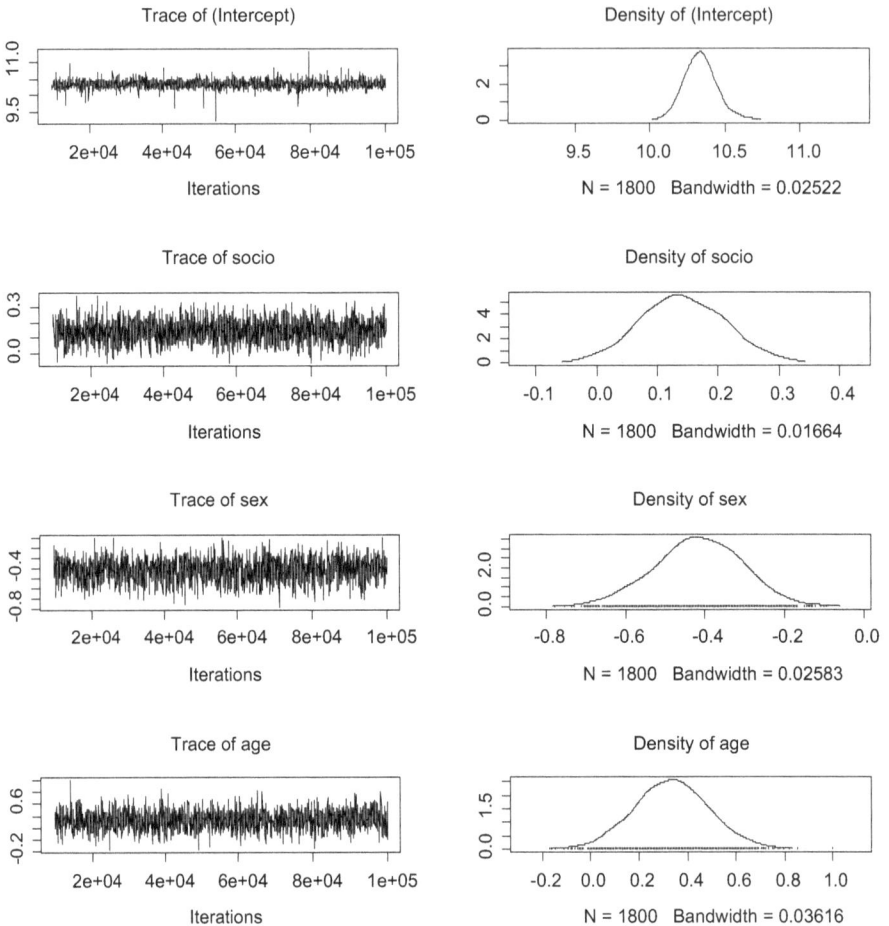

FIGURE 13.16
Trace and density plots for MCMC chains for the fixed effects with informative priors. MCMC, Markov Chain Monte Carlo.

Table 13.8 includes the parameter estimates and credible intervals for the informative priors model.

Perhaps the most notable impact of using the informative priors can be seen in the estimates and credibility intervals for males, age, and, to a lesser extent, the random intercept. When the informative prior was used for age, the resulting credible interval excluded 0, leading us to conclude that age is positively related to reading test scores. Recall that when we used noninformative priors, the credible interval for age included 0, so that we found no significant relationship between it and reading score. The coefficient estimate for males is −0.42 in the informative prior model, which suggests a stronger relationship with reading than was the case for the noninformative priors.

Trace of school

Density of school

Trace of units

Density of units

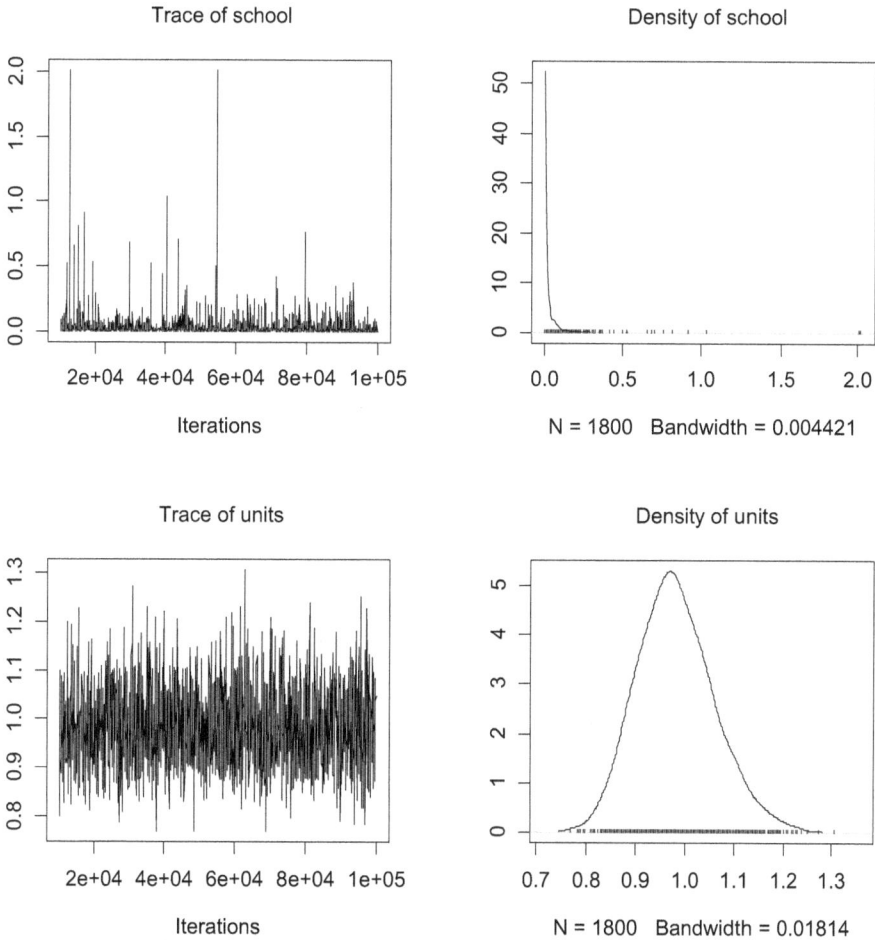

FIGURE 13.17
Trace and density plots for MCMC chains for the random effects with informative priors. MCMC, Markov Chain Monte Carlo.

TABLE 13.8

Fixed and Random Effects Estimates for Model1 Based on Bayesian MLM Estimation: Informative Priors

Effect	Coefficient	95% Credible Interval
Fixed intercept	10.34	10.11, 10.59
SES	0.14	0.01, 0.28
Male	−0.42	−0.63, −0.21
Age	0.33	0.05, 0.65
Random intercept	0.03	0.0002, 0.13
Random error	0.98	0.83, 1.13

Finally, the random intercept coefficient is slightly larger with the informative priors, and the credible interval does not include 0.

These results demonstrate the impact that using informative priors can have on the resulting parameter estimates. This sensitivity to the priors makes it very important that we employ them with careful thought. The location and scale values should be supported by prior research and should reflect what might be reasonably expected in the broader population. One useful rule of thumb is that for each informative prior value, the research should have at least one, and hopefully several, articles supporting it. If the priors are not well supported by prior research and theory, the estimates obtained using them may not be accurate or meaningful.

Summary

Given these results, the reader will correctly ask the question, which of these methods should I use? There has not been a great deal of research comparing these various methods with one another. However, one study (Finch, 2017) did compare the rank-based, heavy-tailed, and standard multilevel models with one another in the presence of outliers. Results of this simulation work showed that the rank-based approaches yielded the least biased parameter estimates and had smaller standard errors than did the heavy-tailed approaches. Certainly, those standard error results are echoed in the current example, where the standard errors for the heavy-tailed approaches were more than twice the size of those from the rank-based method. Given these simulation results, it would seem that the rank-based results may be the best to use when outliers are present, at least until further empirical work demonstrates otherwise. When the assumptions underlying ML and REML are met, they are always the optimal approach to use.

Finally, we should make a few comments about sample size and multilevel modeling. McNeish and Stapleton (2016) conducted a large-scale simulation study investigating the issue of how large the sample needs to be in order for parameter estimates in MLM to be stable. They found that for models with four or five fixed effect predictors, REML can yield unbiased and efficient estimates of the fixed effects coefficients for as few as 20 level-2 units (schools). For random intercept and slope variances, REML performs well with as few as seven to ten level-2 units. The number of individuals per level-2 unit (e.g., number of students per school) is most impactful on the estimation of the random intercept and particularly random slope effects. Bell et al. (2010) found that the power for hypothesis tests associated with fixed effects is below 0.8 for samples of fewer than 20 individuals per level-2 unit. These authors suggest that researchers strive to have, at minimum, 20–40 individuals per level-2 unit, with larger samples needed to accurately estimate random effects, particularly for the slope. Finally, we should note that there exist small sample correction procedures for cases when samples are too small for standard REML and ML estimation (e.g., Kenward & Roger, 1997).

This chapter serves as only an introduction to MLMs. There is much more to know about them, including their application to problems in which the outcome is not continuous, as was true for the models described here. However, it is hoped that this chapter does provide the reader with a worthy introduction to the topic and that it can serve as a springboard for both the fitting of such models to datasets with continuous outcomes and further reading and investigation into this fascinating and important topic.

References

Abebe, A., McKean, J.W., Kloke, J.D., & Bilgic, Y.K. (2016). Iterated reweighted rank-based estimates for GEE models. In R.Y. Liu & J.W. McKean (eds), *Robust Rank-Based and Nonparametric Methods*, 61–79. New York: Springer.

Algina, J., Keselman, H.J., & Penfield, R.D. (2005). An alternative to Cohen's standardized mean difference effect size: A robust parameter and confidence interval in the two independent case. *Psychological Methods, 10*(3), 317–328.

Azen, R., & Budescu, D.V. (2003). The dominance analysis approach for comparing predictors in multiple regression. *Psychological Methods, 8*(2), 129–148.

Azen, R., & Traxel, N. (2009). Using dominance analysis to determine predictor importance in logistic regression. *Journal of Educational and Behavioral Statistics, 34*(3), 319–347.

Baguley, T. (2009). Standardized or simple effect size: What should be reported? *British Journal of Psychology, 100*(3), 603–617

Bell, B.A., Morgan, G., Kromrey, J.D., & Ferron, J. (2010). The impact of small cluster size on multilevel models: A Monte Carlo examination of two-level models with binary and continuous predictors. *Paper Presented at the Joint Statistical Meetings*, Vancouver, British Columbia.

Belsley, D.A., Kuh, E., & Welsch, R.E. (1980). *Regression Diagnostics; Identifying Influence Data and Source of Collinearity*. New York: Wiley.

Bilgic, Y.K. (2012). Rank-Based Estimation and Prediction for Mixed Effects Models in Nested Designs. Dissertations. 40. https://scholarworks.wmich.edu/dissertations/40.

Birkes, D., & Dodge, Y. (1993). *Alternative Methods of Regression*. New York: John Wiley & sons.

Braun, M.T., Converse, P.D., & Oswald, F.L. (2019). The accuracy of dominance analysis as a metric to assess relative importance: The joint impact of sampling error variance and measurement unreliability. *Journal of Applied Psychology, 104*(4), 593–602.

Breusch, T.S., & Pagan, A.R. (1979). A simple test for heteroscedasticity and random coefficient variation. *Econometrica, 47*, 1287–1294.

Brown, D., & Rothery, P. (1993). *Models in Biology: Mathematics, Statistics and Computing*. New York: John Wiley & Sons.

Brown, M.B., & Forsythe, A.B. (1974). The small sample behavior of some statistics which test the equality of several means. *Technometrics, 16*, 129–132.

Bühlmann, P., & van de Geer, S. (2011). *Statistics for High-Dimensional Data: Methods, Theory and Applications*. Berlin: Springer.

Coakley, C.W., & Hettmansperger, T.P. (1993). A bounded influence, high breakdown, efficient regression estimator. *Journal of the American Statistical Association, 88*, 872–880.

Cohen, J. (1988). *Statistical Power Analysis for the Behavioral Sciences* (2nd ed.). Hillsdale, NJ: Lawrence Erlbaum Associates, Publishers.

Cook, R.D. (1977). Detection of influential observations in linear regression. *Technometrics, 19*, 15–18.

Cook, R.D., & Weisberg, S. (1983). Diagnostics for heteroscedasticity in regression. *Biometrika, 70*, 1–10.

Cumming, G. (2011). *Understanding the New Statistics: Effect Sizes, Confidence Intervals, and Meta-analysis.* New York: Routledge/Taylor & Francis Group.

Darlington, R.B., & Hayes, A.F. (2017). *Regression Analysis and Linear Models: Concepts, Applications, and Implementation.* New York: The Guilford Press.

de Leeuw, J., & Meijer, E. (eds.). (2008). *Handbook of Multilevel Analysis.* New York: Springer Science+Business Media.

Dillane, D. (2005). *Deletion Diagnostics for the Linear Mixed Model.* Ph.D. thesis, Trinity College, Dublin.

Demidenko, E., & Stukel, T.A. (2005). Influence analysis for linear mixed-effects models. *Statistics in Medicine, 30*(24), 893–909.

Durbin, J., & Watson, G.S. (1951). Testing for serial correlation in least squares regression. *Biometrika, 38*, 159–177.

Falk, C. F. (2018). Are robust standard errors the best approach for interval estimation with nonnormal data in structural equation modeling? *Structural Equation Modeling, 25*(2), 244–266.

Ferrari, S., & Cribari-Neto, F. (2004). Beta regression for modelling rates and proportions. *Journal of Applied Statistics, 31*, 799–815.

Finch, H. (2017). Multilevel modeling in the presence of outliers: A comparison of robust estimation methods. *Psicológica, 38*(1), 57–92.

Fligner, M.A., & Killeen, T.J. (1976). Distribution-free two-sample tests for scale. *Journal of the American Statistical Association, 71*, 210–213.

Fox, J. (2005). Nonparametric regression. In B. Everitt & D. Howell (eds.), *Encyclopedia of Statistics in Behavioral Science.* New York: John Wiley & Sons, Ltd.

Fox, J. (2016). *Applied Regression Analysis and Generalized Linear Models* (3rd ed.). United States: Sage Publications, Inc.

Fritz, C.O., Morris, P.E., & Richler, J.J. (2012). Effect size estimates: Current use, calculations, and interpretation. *Journal of Experimental Psychology: General, 141*(1), 2–18.

Gelman, A., Hill, J., & Vehtari, A. (2021). *Regression and Other Stories.* London: Cambridge University Press.

George, D., & Mallery, M. (2010). *SPSS for Windows Step by Step: A Simple Guide and Reference, 17.0 Update* (10th ed.). Boston: Pearson.

Greenhouse, S.W., & Geisser, S. (1959). On methods in the analysis of profile data. *Psychometrika, 24*, 95–112.

Gu, X. (2022). Assessing the relative importance of predictors in latent regression models. Structural Equation Modeling: A Multidisciplinary Journal, 29(4), 569-583.

Gu, X. (2021). Assessing the relative importance of predictors in latent regression models. *Structural Equation Modeling: A Multidisciplinary Journal, 29*(4), 569–583.

Harlow, L.L., Mulaik, S.A., & Steiger, J.H. (eds.). (1997). *What if There Were No Significance Tests?* New York: Lawrence Erlbaum Associates Publishers.

Hastie, T.J., & Tibshirani, R.J. (1990). *Generalized Additive Models.* New York: Chapman and Hall.

Hastie, T., Tibshirani, R., & Wainwright, M. (2015). *Statistical Learning with Sparsity: The Lasso and Generalizations.* New York: Chapman and Hall/CRC.

Hayes, A.F. (2023). *Introduction to Mediation, Moderation, and Conditional Process Analysis* (3rd ed.). New York: The Guilford Press.

Hedges, L.V. (1981). Distribution theory for Glass's estimator of effect size and related estimators. *Journal of Educational Statistics, 6(2)*, 107–128.

Hettmensperger, T.P., & McKean, J.W. (2011). *Robust Nonparametric Statistical Methods*. Boca Raton, FL: CRC Press.

Hogg, R.V. (1974). Adaptive robust procedures: A partial review and some suggestions for future applications and theory. *Journal of the American Statistical Association, 69(348)*, 909–923.

Hogg, R.V., & Tanis, E.A. (1996). *Probability and Statistical Inference*. New York: Prentice Hall.

Holm, S. (1979). A simple sequentially rejective multiple test procedure. *Scandinavian Journal of Statistics, 6(2)*, 65–70.

Hox, J.J. (2002). *Multilevel Analysis: Techniques and Applications*. Mahwah, NJ: Lawrence Erlbaum Associates Publishers.

Huynh, H., & Feldt, L.S. (1976). Estimation of the Box correction for degrees of freedom from sample data in randomized block and split-plot designs. *Journal of Educational Statistics, 1(1)*, 69–82.

Iversen, G.R. (1991). *Contextual Analysis*. Thousand Oaks, CA: Sage Publications.

Jaekel, L.A. (1972). Estimating regression coefficients by minimizing the dispersion of residuals. *Annals of Mathematical Statistics, 43*, 1449–1458.

Jennions, M.D., & Møller, A.P. (2003). A survey of the statistical power of research in behavioral ecology and animal behavior. *Behavioral Ecology, 4*, 438–445.

Johnson, N.L., & Kotz, S. (1970). *Distributions in Statistics: Continuous Univariate Distributions-2*. New York: Wiley.

Johnson, P.O., & Neyman, J. (1936). Tests of certain linear hypotheses and their application to some educational problems. *Statistical Research Memoirs, 1*, 57–93.

Kaplan, D. (2014). *Bayesian Statistics for the Social Sciences*. New York: The Guilford Press.

Kelley, T.L. (1935). *Essential Traits of Mental Life, Harvard Studies in Education* (Vol. 26). Cambridge, MA: Harvard University Press.

Kenward, M.G., & Roger, J.H. (1997). Small sample inference for fixed effects from restricted maximum likelihood. *Biometrics, 53*, 983–997.

Kloke, J.D., McKean, J.W., & Rashid, M.M. (2009). Rank-based estimation and associated inferences for linear models with cluster correlated errors. *Journal of the American Statistical Association, 104*, 384–390.

Kloke, J., & McKean, J.W. (2013). Small sample properties of JR estimators. *Paper Presented at the Annual Meeting of the American Statistical Association*, Montreal, QC, August.

Kreft, I., & de Leeuw, J. (1998). *Introducing Multilevel Modeling*. Thousand Oaks, CA: Sage Publications, Inc.

Kreft, I.G.G., de Leeuw, J., & Aiken, L.S. (1995). The effect of different forms of centering in hierarchical linear models. *Multivariate Behavioral Research, 30(1)*, 1–21.

Kroes, A.D.A., & Finley, J.R. (2023). Demystifying omega squared: Practical guidance for effect size in common analysis of variance designs. *Psychological Methods* (Advance online publication).

Kruschke, J.K. (2015). *Doing Bayesian Analysis*. Amsterdam: Elsevier.

LaHuis, D.M., Hartman, M.J., Hakoyama, S., & Clark, P.C. (2014). Explained variance measures for multilevel models. *Organizational Research Methods, 17(4)*, 433–451.

Lange, K.L., Little, R.J.A., & Taylor, J.M.G. (1989). Robust statistical modeling using the *t* distribution. *Journal of the American Statistical Association, 84*, 881–896.

Levene, H. (1960). Robust tests for equality of variances. In I. Olkin (ed), *Contributions to Probability and Statistics*, 278–292. Palo Alto: Stanford University Press.

Lovakov, A., & Agadullina, E.R. (2019). Empirically derived guidelines for effect size interpretation in social psychology. *European Journal of Social Psychology, 51*(3), 485–504.

Loy, A., & Hofmann, H. (2014). HLMdiag: A suite of diagnostics for hierarchical linear models in R. *Journal of Statistical Software, 56*(5), 1–28.

Luo, W., & Azen, R. (2013). Determining predictor importance in hierarchical linear models using dominance analysis. *Journal of Educational and Behavioral Statistics, 38*(1), 3–31.

MacKinnon, D., et al. (2002) A Comparison of Methods to Test Mediation and Other Intervening Variable Effects. Psychological Methods, 1, 83–104.

Mann, H.B., & Whitney, D.R. (1947). On a test of whether one of two random variables is stochastically larger than the other. *Annals of Mathematical Statistics, 18*, 50–60.

Maritz, J.S., & Jarrett, R.G. (1978). A note on estimating the variance of the sample median. *Journal of the American Statistical Association, 73*, 194–196.

McKean, J.W., & Schrader, R.M. (1984). A comparison of methods for studentizing the sample median. *Communications in Statistics – Simulation and Computation, 13*, 751–773.

McNeish, D.M., & Stapleton, L.M. (2016). The effect of small sample size on two-level model estimates: A review and illustration. *Educational Psychology Review, 28*(2), 295–314.

Myers, J.L., & Well, A.D. (2003). *Research Design and Statistical Analysis* (2nd ed.). New York: Lawrence Erlbaum Associates Publishers.

Pena, E.A., & Slate, E.H. (2006) Global validation of linear model assumptions. *Journal of the American Statistical Association, 10*, 341–354.

Pinheiro, J.C., Liu, C., & Wu, Y.N. (2001). Efficient algorithms for robust estimation in linear mixed-effects models using the multivariate t distribution. *Journal of Computational and Graphical Statistics, 10*(2), 249–276.

Raftery, A.E. (1995). Bayesian model selection in social research. *Sociological Methodology, 25*, 111–164.

Raudenbush, S.W., & Bryk, A.S. (2002). *Hierarchical Linear Models. Applications and Data Analysis Methods* (2nd ed.). Thousand Oaks, CA: Sage Publications.

Rigby, R.A., & Stasinopoulos, D.M. (2005). Generalized additive models for location, scale and shape. *Journal of the Royal Statistical Society Series C: Applied Statistics, 54*(3), 507–554.

Rights, J.D., & Sterba, S.K. (2019). Quantifying explained variance in multilevel models: An integrative framework for defining R-squared measures. *Psychological Methods, 24*(3), 309–338.

Rogers, W.H., & Tukey, J.W. (1972). Understanding some long-tailed symmetrical distributions. *Statistica Neerlandica, 26*(3), 211–226.

Rom, D.M. (1990). A sequentially rejective test procedure based on a modified Bonferroni inequality. *Biometrika, 77*, 663–666.

Rousseeuw, P.J., & Yohai, V. (1984). Robust regression by means of S estimators in robust and nonlinear time series analysis. In J. Franke, W. Härdle, & R.D. Martin (eds.), *Lecture Notes in Statistics, 26*, 256–274. New York: Springer-Verlag.

Schad, D.J., Nicenboim, B., Bürkner, P.-C., Betancourt, M., & Vasishth, S. (2023). Workflow techniques for the robust use of bayes factors. *Psychological Methods, 28*(6), 1404–1426.

Schelldorfer, J., Buhlmann, P., & van de Geer, S. (2011). Estimation for high-dimensional linear mixed-effects models using l_1-Penalization. *Scandinavian Journal of Statistics, 38*(2), 197–214.

Snijders, T.A.B., & Bosker, R.J. (1999). *Multilevel Analysis: An Introduction to Basic and Advanced Multilevel Modeling.* Thousand Oaks: Sage Publications.

Song, P. X-K., Zhang, P., & Qu, A. (2007). Maximum likelihood inference in robust linear mixed-effects models using multivariate t distributions. *Statistica Sinica, 17*, 929–943.

Sprent, P. (1989). *Applied Nonparametric Statistical Methods.* London: Chapman and Hall.

Staudenmayer, J., Lake, E.E., & Wand, M.P. (2009). Robustness for general design mixed models using the *t*-distribution. *Statistical Modeling, 9*(3), 235–255.

Steiger, J.H. (2004). Beyond the F test: Effect size confidence intervals and tests of close fit in the analysis of variance and contrast analysis. *Psychological Methods, 9*(2), 164–182.

Tibshirani, R. (1996). Regression shrinkage and selection via the lasso. *Journal of the Royal Statistical Society, B, 58*(1), 267–288.

Tomczak M. & Tomczak E. (2014). The need to report effect size estimates revisited. *Trends in Sport Sciences, 1*, 19–25.

Tong, X., & Zhang, Z. (2012). Diagnostics of robust growth curve modeling using student's *t* distribution. *Multivariate Behavioral Research, 47*, 493–518.

Tonidandel, S., LeBreton, J.M., & Johnson, J.W. (2009). Determining the statistical significance of relative weights. *Psychological Methods, 14*(4), 387–399.

Wang, J., & Genton, M.G. (2006). The multivariate skew-slash distribution. *Journal of Statistical Planning and Inference, 136*, 209–220.

Welch, B. (1951). On the comparison of several mean values: An alternative approach. *Biometrika, 38*, 330–336.

Welsh, A.H. & Richardson, A.M. (1997). 13 Approaches to robust estimations of mixed models. In: *Handbook of Statistics*, Vol. 15, 343–384. Amsterdam: Elsevier Science.

Wilcox, R. (2012). *Introduction Robust Estimation & Hypothesis Testing.* Amsterdam: Elsevier.

Wilcox, R.R., & Tian, T.S. (2011). Measuring effect size: A robust heteroscedastic approach for two or more groups. *Journal of Applied Statistics, 38*(7), 1359–1368.

Wilcoxon, F. (1945). Individual comparisons by ranking methods. *Biometrics Bulletin, 1*(6), 80–83.

Wood, S.N. (2006) *Generalized Additive Models: An Introduction with R.* Texts in Statistical Science. Chapman & Hall/CRC, Boca Raton.

Wooldridge, J.M. (2004). *Econometric Analysis of Cross Section and Panel Data.* Cambridge, MA: The MIT Press.

Yuan, K.-H., & Bentler, P.M. (1998). Structural equation modeling with robust covariances. *Sociological Methodology, 28*, 363–396.

Yuan, K-H., Bentler, P.M., & Chan, W. (2004). Structural equation modeling with heavy tailed distributions. *Psychometrika, 69*(3), 421–436.

Yuen, K.K. (1974). The two-sample trimmed T for unequal population variances. *Biometrika, 61*, 165–170.

Zhang, X., Astivia, O.L.O., Kroc, E., & Zumbo, B.D. (2023). How to think clearly about the central limit theorem. *Psychological Methods, 28*(6), 1427–1445.

Zhao, P., & Yu, B. (2006). On model selection consistency of Lasso. *Journal of Machine Learning Research, 7*, 2541–2563.

Zou, G.Y. (2007). Toward using confidence intervals to compare correlations. *Psychological Methods, 12*(4), 399–413.

Zou, H., & Hastie, T. (2005). Regularization and variable selection via the elastic net. *Journal of the Royal Statistical Society: Series B, 67*, 301–320.

Index

Note: **Bold** page numbers refer to tables and *italic* page numbers refer to figures.

For Product Safety Concerns and Information please contact our EU
representative GPSR@taylorandfrancis.com
Taylor & Francis Verlag GmbH, Kaufingerstraße 24, 80331 München, Germany